# MATERIALS FOR ADVANCED BATTERIES

## NATO CONFERENCE SERIES

I Ecology
II Systems Science
III Human Factors
IV Marine Sciences
V Air–Sea Interactions
VI Materials Science

### VI MATERIALS SCIENCE

*Volume 1* Molecular Metals
Edited by William E. Hatfield

*Volume 2* Materials for Advanced Batteries
Edited by D. W. Murphy, J. Broadhead, and B. C. H. Steele

# MATERIALS FOR ADVANCED BATTERIES


Edited by

D.W. Murphy and J. Broadhead

*Bell Laboratories*
*Murray Hill, New Jersey*

and

B.C.H. Steele

*Imperial College*
*London, United Kingdom*



Published in coordination with NATO Scientific Affairs Division by

PLENUM PRESS · NEW YORK AND LONDON

Library of Congress Cataloging in Publication Data

Nato Conference on Materials for Advanced Batteries, Aussois, France, 1979.
Materials for advanced batteries.
(NATO conference series: VI, Materials science; v. 2)

"Proceedings of a NATO conference on Materials for Advanced Batteries, held September 9–14, 1979 in Aussois, France."
Includes index.
1. Electric batteries—Congresses. I. Murphy, D. W. II. Broadhead, J. III. Steele, B. C. H. IV. Title. V. Series.
TK2901.N38 1979 621.31'242'028 80-19967

ISBN 978-1-4684-3853-6 ISBN 978-1-4684-3851-2 (eBook)
DOI 10.1007/978-1-4684-3851-2

Proceedings of a NATO Symposium on Materials for Advanced Batteries, sponsored by the NATO Special Conference Panel on Materials Science, and held in Aussois, France, September 9–14, 1979


Softcover reprint of the hardcover 1st edition 1980
A Division of Plenum Publishing Corporation
227 West 17th Street, New York, N.Y. 10011

# PREFACE

The idea of a NATO Science Committee Institute on "Materials for Advanced Batteries" was suggested to JB and DWM by Dr. A. G. Chynoweth. His idea was to bring together experts in the field over the entire spectrum of pure research to applied research in order to familiarize everyone with potentially interesting new systems and the problems involved in their development. Dr. M. C. B. Hotz and Professor M. N. Ozdas were instrumental in helping organize this meeting as a NATO Advanced Science Institute.

An organizing committee consisting of the three of us along with W. A. Adams, U. v Alpen, J. Casey and J. Rouxel organized the program. The program consisted of plenary talks and poster papers which are included in this volume. Nearly half the time of the conference was spent in study groups. The aim of these groups was to assess the status of several key aspects of batteries and prospects for research opportunities in each. The study groups and their chairmen were:

Current status and new systems
J. Broadhead

High temperature systems
W. A. Adams

Interface problems
B. C. H. Steele

Electrolytes
U. v Alpen

Electrode materials
J. Rouxel

These discussions are summarized in this volume.

We and all the conference participants are most grateful to Professor J. Rouxel for suggesting the Aussois conference site, and to both he and Dr. M. Armand for handling local arrangements.

We would also like to thank all of the conference participants for their hard work and valuable contributions.

D. W. Murphy
Murray Hill

J. Broadhead
Murray Hill

B. C. H. Steele
London

CONTENTS

## I. PLENARY REVIEW PAPERS

Requirements of Battery Systems . . . . . . . . . . . . 3
Elton J. Cairns

Reflections on Some Recent Studies of Materials of Importance in Aqueous Electrochemical Energy-Storage Systems . . . . . . . . . . . . 29
E.J. Casey

Lithium-Aluminum/Iron Sulphide Batteries . . . . . . . 47
Donald R. Vissers

Solid Electrolytes . . . . . . . . . . . . . . . . . . 91
R.A. Huggins

Molten Salt Electrolytes in Secondary Batteries . . . . . . . . . . . . . . . . . . . . . 111
Gleb Mamantov

The Recharging of the Lithium Electrode in Organic Electrolytes . . . . . . . . . . . . . 123
S.B. Brummer, V.R. Koch, and R.D. Rauh

Intercalation Electrodes . . . . . . . . . . . . . . . 145
M.B. Armand

Interface Phenomena in Advanced Batteries . . . . . . . 163
F.G. Will

## II. SHORT COMMUNICATIONS

Lead-Acid Battery Overview . . . . . . . . . . . . . . 199
S.M. Caulder and A.C. Simon

Design and Development of Micro-Reference Electrodes for the Lithium/Metal-Sulphide Cell System . . . . . . . . . . . . . . 205
L. Redey and D.R. Vissers

Molten Salt Electrochemical Studies and High Energy Density Cell Development . . . . . . 211
A. Bélanger, F. Morin, M. Gauthier, W.A. Adams, and A.R. Dubois

Selenium (IV) Cathode in Molten Chloroaluminates . . . . . . . . . . . . . . . . . . 223
R. Marassi, D. Calasanzio, and G. Mamantov

Behavior of Hard and Soft Ions in Solid Electrolytes . . . . . . . . . . . . . . . . . . . . 229
F.W. Poulsen

Use of the Proton Conductor Hydrogen Uranyl Phosphate Tetrahydrate as the Solid Electrolyte in Hydride-Air Batteries and Hydrogen-Oxygen Fuel Cells . . . . . . . . . . 235
P.E. Childs and A.T. Howe

Behaviour of the Passivating Film in $Li/SOCl_2$ Cells Under Various Conditions . . . . . . . . . . 241
G. Eichinger

Novel Solid Electrolytes in the System $Li_4GeO_4$ - $Li_2Zn\ GeO_4$ - $Li_3\ PO_4$ . . . . . . . . . . 247
J. Schoonman, J.G. Kamphorst, and E.E. Hellstrom

The Analysis of Surface Insulating Films on Lithium Nitride Crystals . . . . . . . . . . . . 253
A. Hooper, C.J. Sofield, J.F. Turner, and B.C. Tofield

Kinetic and Morphological Aspects of Lithium Electrodes at Ambient Temperature in Propylene Carbonate . . . . . . . . . . . . . . . . 257
I. Epelboin, M. Froment, M. Garreau, J. Thevenin, and D. Warin

Investigation of the $Li-Bi_2O_3$ Battery System Using the A.C. Impedance Technique . . . . . . . 263
U. Von Alpen, M.F. Bell, and F.J. Kruger

Enhanced Ionic Motion in Solid Electrode Materials . . . . . . . . . . . . . . . . . . . . 269
W. Weppner

Dynamic Considerations in the Design of Batteries with Intercalation Cathodes . . . . . . 275
S. Atlung, K. West, and T. Jacobsen

Strain Effects in Intercalation Systems . . . . . . . . 283
W.R. McKinnon and R.R. Haering

Amorphous Cathodes for Lithium Batteries . . . . . . . 291
M.S. Whittingham, R.R. Chianelli, and A.J. Jacobson

A Study of Electrointercalation and Diffusion of Silver in Tantalum Disulphide Using Solid State Cells . . . . . . . . . . . . . . . . . 301
F. Bonino, M. Lazzari, and C.A. Vincent

III. STUDY GROUP REPORTS

State of the Art and New Systems . . . . . . . . . . . 309

High Temperature Systems . . . . . . . . . . . . . . . 315

Solid Electrolytes . . . . . . . . . . . . . . . . . . 331

Electrode Materials . . . . . . . . . . . . . . . . . 343

Interface Problems . . . . . . . . . . . . . . . . . . 349

PARTICIPANTS . . . . . . . . . . . . . . . . . . . . . 359

INDEX . . . . . . . . . . . . . . . . . . . . . . . . 367

# SECTION I

# PLENARY REVIEW PAPERS

# REQUIREMENTS OF BATTERY SYSTEMS

Elton J. Cairns

Lawrence Berkeley Laboratory, and
University of California
Berkeley, CA 94720

## I. Introduction

Since 1973, the energy economy of the U.S. has been changing very rapidly. Our perceptions of the value of energy, and the price of petroleum are very different from what they were just a few short years ago. It is interesting to examine the total energy economy of the U.S., and to consider in which directions the shifts of supply and demand will go. As members of the scientific and electrochemical communities, we ask ourselves where our field can contribute to achieving a more nearly optimal distribution of energy supplies and energy demands, and how can electrochemical energy storage help them mesh in an effective, economically sound manner.

A diagram of the energy economy of the United States is shown in Figure 1, based upon projections made a few years ago. On the left hand side of the diagram are shown the primary energy sources, and the numbers indicate the size of the energy supply in units of millions of barrels per day in oil equivalent. The widths of the various bands on the diagram are drawn proportional to the size of the energy supply. Overall, the projections made in Figure 1 are rather accurate with a few minor exceptions, such as the fact that the U.S. is already importing more oil than it is producing, and the transportation sector is using slightly less energy than the projected 12.0 million barrels per day oil equivalent. The general features of the diagram are clear: About 45% of all of the energy consumed in the U.S. is in the form of oil. About half of this oil is used in the transportation sector. About 27% of all of the energy is consumed for the purpose of generating electrical energy. This sector of the energy economy has been growing faster than any other, historically at the rate of 7% per year except during the

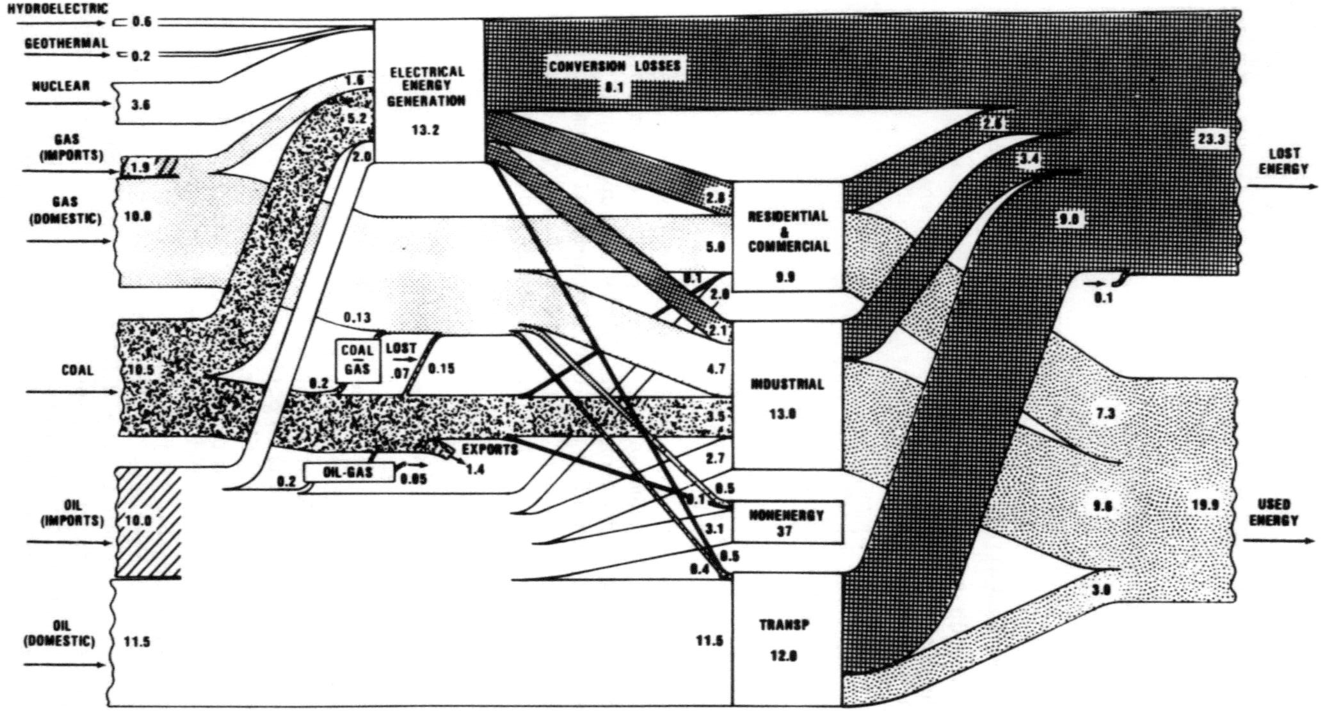


(UNITS: MILLION BBLS/DAY OIL EQUIVALENT)

Fig. 1 Diagram of the energy economy of the U.S., projected to 1980, in units of millions of barrels of oil equivalent per day, after A.L. Austin, B. Rubin, and G.C. Werth, "Energy: Uses, Sources, Issues" in Lawrence Livermore Laboratory Report, UCRL-51221, May 30, 1972.

last few years. The combination of rapid growth in the electrical generating industry and its lesser dependence upon oil make this an attractive opportunity for more effective utilization of non-petroleum energy resources.

The varying load placed upon a typical electric utility is shown as a function of the time of day in Figure 2. The four curves show the load for days selected from each of the four seasons of the year for a midwestern utility. In each curve there is a much larger demand during the daytime and evening hours than there is during the middle of the night. This large variation in load results in a large excess of electrical generating capability during the nighttime hours because the size of the equipment must exceed the peak demand by some safety margin. If it were possible to operate electrical generating equipment at the same power level for nearly all of the time, then less generating equipment would be required. In principle, the storage of energy in batteries could make an important contribution in this direction. Large blocks of batteries located at the substation level could accept energy during the nighttime hours, and could deliver that energy during the heavy load periods in the daytime and early evening. This would level the load on the central power plants and on the distribution network down to the substation level. Additional advantages to this approach are the building-block nature of batteries, allowing them to be used only in the precise numbers necessary to meet the local demand at the substation. As demands change with time, additional battery modules could be installed or removed as necessary. Other unique opportunities exist for energy storage in batteries in the centers of large metropolitan areas where addition of more electrical capacity is extremely expensive because of the expense of installing additional underground distribution lines and equipment. Battery modules could be installed, for example, in the basements of large buildings, leveling the load for that building and effectively increasing the peak capability of the distribution network. For these special situations rather high cost could be justified.

Referring again to the energy diagram of Figure 1, it is clear that there are strong advantages to be realized by shifting at least a part of the transportation energy demand to electricity instead of oil. This can of course be accomplished by the use of battery-powered electric vehicles. The batteries could be recharged during the low demand period at night, making use of excess generating capacity. The relatively high efficiency of the electric vehicles compensates for the efficiency losses in the generation of the electricity, the result being merely a shift in energy demand away from petroleum toward coal and nuclear primary energy sources. The main problem with this idea is the fact that batteries of sufficiently high energy storage capabilities to provide an attractive vehicle range, for example 100 kilometers or more, have not been available.

One additional growing area of opportunity for batteries to contribute to easing the energy problem is in the area of storage of the

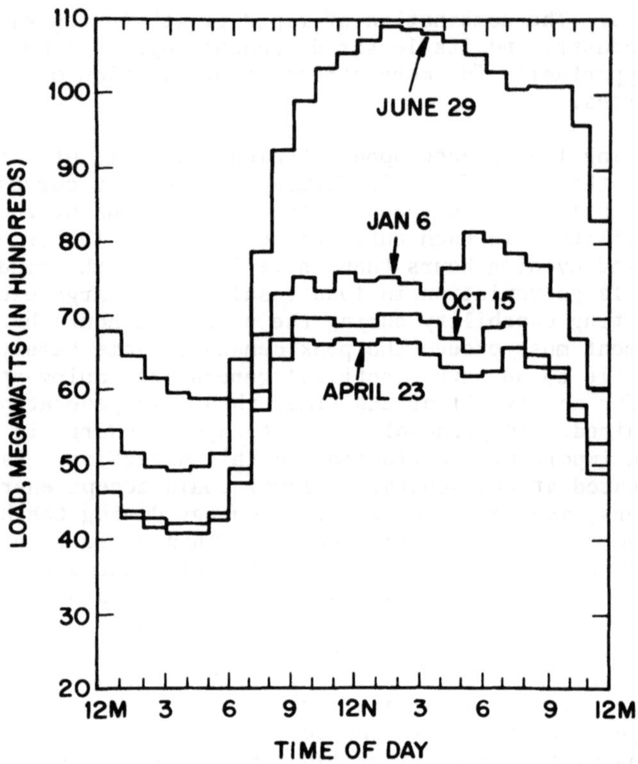


Fig. 2 Daily load profiles of the Commonwealth Edison Company for sample days in each of the four seasons of 1971, from M.L. Kyle, E.J. Cairns, and D.S. Webster, Argonne National Lab Reports, ANL-7958, March, 1973.

energy generated by solar-powered and wind-powered electrical energy generators. The attractiveness of these methods of energy generation includes the fact that they don't use conventional fuels, and the primary energy source is available even in remote locations. Storage of energy from solar- and wind-powered devices is an attractive opportunity for batteries. Whenever the energy demand is less than the energy being supplied, the excess energy being generated can be stored in the battery for later use. Depending upon the design of the system the battery could supply energy for extended dark or cloudy or windless periods.

The above three major energy storage opportunities which could make significant contributions to the U.S. energy independence will be discussed in terms of specific needs and specific battery requirements in the sections below.

## II. Energy Storage for Electric Utilities

The electric utility load curves shown in Figure 2 offer the opportunities discussed above for energy storage during the low demand period and energy delivery during the peak demand period. Examination of the curves of Figure 2 indicates that a period of 5 to 7 hours is available during the late night hours for battery recharge. This means that the battery must be capable of accepting a full charge of energy during those hours. The daytime and evening peak demand period varies in duration from season to season. It may be as short as 3 hours in the afternoon in the summertime or as long as 8 or 10 hours during spring or fall. In addition to having the charge acceptance and charge delivery capabilities just discussed, the battery must operate very efficiently in order to avoid loss of the energy which is stored. The efficiency of pumped hydroelectric storage is between 65 and 70%, so batteries should do at least this well in order to be fully competitive in the storage of large blocks of energy. For storage of energy at the substation level it may be possible to tolerate somewhat lower efficiencies depending upon the features of the local utility network.

In principle, utilities have the choice of either storing energy or generating energy as needed. The main source of electrical energy during the peak demand periods is gas turbines, which are rapidly started and very responsive to changes in load. These units are relatively inexpensive and can provide energy at a cost which is somewhat higher than that generated by the large plants. Unfortunately, gas turbines rely upon oil as the fuel making this less attractive than it was a few years ago. The combination of the disadvantage of using oil for gas turbines and the relative lack of availability of appropriate sites for pumped hydroelectric storage make the battery choice an attractive one providing that the performance, durability, and cost requirements can be met.

The economics of providing land and buildings for battery stations indicate that it is necessary for the battery to be compact enough that floor space of only one square meter for every 80 kilowatt hours of energy storage capability be allowed. This corresponds to something above 30 Wh/l, depending upon the details of the system design. The durability and cost requirements for the battery are set in such a manner that the battery is competitive with alternative means of providing power during peak demand periods, and the lifetime is compatible with reasonable projections of present technology. The longer the battery life, the higher is the tolerable initial cost. For a cycle life of 2000, an initial cost of $30/kWh is acceptable, corresponding to a storage cost of 1.5¢/kWh for the battery alone.* If the battery is cycled 200 times per year, then it should last 10 years. Longer lives, of course, would result in lower storage costs, or might allow a higher initial cost.

---

*The cost of peaking energy from a gas turbine using $3.50/10^6BTU fuel is about 6¢/kWh. [1]

If batteries were to be used for dispersed energy storage in locations having very high costs for installation of additional distribution hardware, such as in the center of a large city, where these costs may be over \$200/kW, then higher battery costs are allowable--perhaps \$100/kWh or more, depending upon the situation and the load curve.

At the small substation level, the battery may be required to provide 20-50 MW for a few hours, requiring an availability of 100-200 MWh on a daily basis. Smaller installations may be used in large office buildings, apartment complexes, and shopping centers.

A summary of the requirements for off-peak energy storage batteries is presented in Table 1, reflecting the characteristics discussed above. At present, no batteries meet all of the requirements. Many batteries can meet the charge and discharge times and the efficiency (e.g., $Pb/PbO_2$), and a few can meet the cycle life and lifetime values ($Pb/PbO_2$, Fe/NiOOH), but none can meet all of these, plus the cost goal. As costs of a given battery type are reduced, the cycle life and performance also tend to be reduced, making the simultaneous achievement of performance, durability, and cost goals a difficult task, requiring complex compromises.

An example of a system which may meet the goals of Table 1 in the future is the high-temperature battery $Li_4Si/LiCl\text{-}KCl/FeS_2$, which operates at 450°C. The active materials are sufficiently inexpensive that the \$30/kWh cost goal may be met if the materials of construction problems can be solved (see below).

A schematic cross section of a typical $Li_4Si/FeS_2$ cell is shown in Figure 3. [2] The positive electrode (center) is comprised of powdered $FeS_2$ mixed with graphite powder (as a current collector), in contact with a molybdenum mesh current collector. A zirconia cloth serves as a particle retainer around the positive electrode. Between the electrodes is a boron nitride fibrous mat separator, which contains the molten LiCl-KCl electrolyte. The negative electrodes are comprised of $Li_4Si$ powder and a fibrous mickel current collector, attached to the stainless steel cell case. The cell is hermetically sealed.

Typical discharge curves for a $Li_4Si/FeS_2$ cell are shown in Figure 4. [2] The plateaus correspond to various steps in the overall discharge reaction:

$$Li_4Si + FeS_2 \rightarrow 2Li_2S + Fe + Si \qquad (1)$$

The theoretical specific energy for this reaction is 944 Wh/kg; cells such as that of Figure 3 have achieved 180 Wh/kg, [2] or 19% of theoretical. It should be feasible to obtain up to 25% of the theoretical value, or 230-240 Wh/kg.

Cycle lives in excess of 700 cycles have been demonstrated for high specific energy $Li_4Si/FeS_2$ cells, corresponding to a lifetime of almost

TABLE 1

REQUIREMENTS FOR OFF-PEAK ENERGY STORAGE BATTERIES

| | |
|---|---|
| Discharge time | 3-8 hours |
| Charge time | 5-7 hours |
| Overall efficiency | >70% |
| Energy/floor area (6.1 m max. height) | 80 kWh/m$^2$ |
| Typical size | 100-200 MWh |
| Cycle life | 2000 |
| Lifetime | 10 years |
| Cost | \$30/kWh |

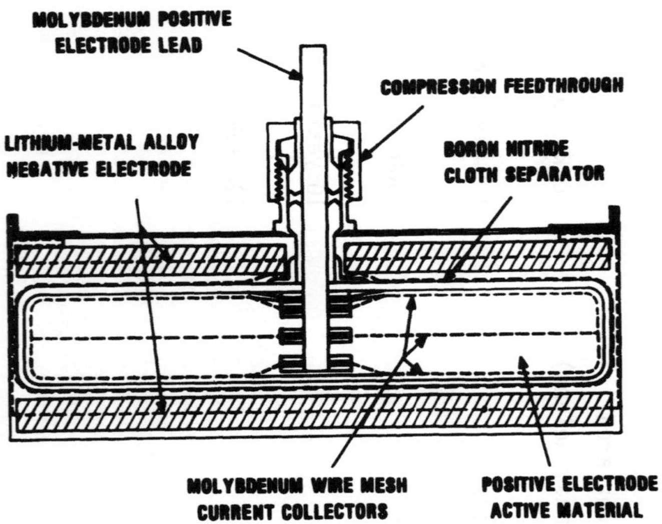


Fig. 3 Schematic cross section of a typical $Li_4/Si/$ $LiCl$-$KCl/FeS_2$ cell, from E.J. Zeitner and J.S. Dunning, "High Performance Lithium/Iron Disulfide Cells," in Proceedings of 13th IECEC, SAE, Warrendale, PA,1978, p.697

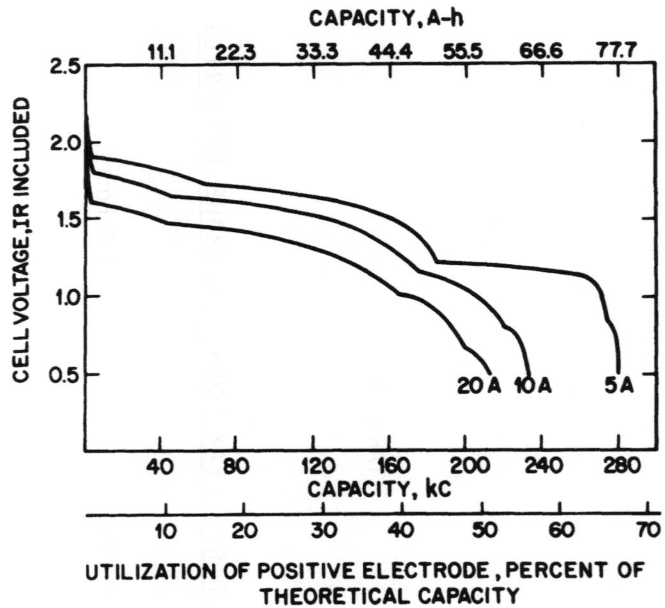


Fig. 4 Typical voltage vs. capacity curves for constant current discharges of a $Li_4Si/LiCl$-$KCl/FeS2$ cell like that of Figure 3. See reference in caption of Figure 3.

two years. Energy efficiencies of 80-90% have been achieved, exclusive of thermal losses. [2] The current status of the $Li_4Si/FeS_2$ cell is summarized in Table 2. A gradual increase of internal resistance results in a decline of cell performance, especially at high specific power. The present cost of these cells is very high because no mass production facilities exist. The active materials costs, however, are compatible with the goal of $30/kWh.

The use of molybdenum and stainless steel in significant amounts is not consistent with the cost goal; substitutes must be found. The presently-used boron nitride separator is far too expensive. Lower cost forms of boron nitride separator, or substitute materials are necessary. Recent cost projections for BN felt are encouraging, [3] indicating a cost below $\$10/m^2$ in large volume. Materials problems in general hold the key to economic viability of this system. Both corrosion-resistant electronic conductors and electronic insulators are needed. Probably the current collector and feedthrough problem for the positive electrode will prove to be the most difficult.

If the problems above can be solved, then energy storage systems such as the one in Figure 5 may be feasible.

## III. Batteries for Electric Automobiles

In the introduction to this paper, it was indicated that the use of electric vehicles could help to shift the energy demand away from petroleum, and toward such primary energy sources as coal and nuclear fuels. This is a very attractive concept, especially if the overall effectiveness of energy utilization is not reduced. This means that the amount of primary energy to accomplish a given vehicle mission should not be increased by the shift to electric vehicles. Two major components comprise the overall consideration: the efficiency of energy conversion for the overall process (energy resource in the earth to energy at the wheels of the vehicle), and the energy required by the vehicle (at the wheels) in executing its mission.

With regard to the energy efficiency issue, Figure 6 summarizes the efficiencies for each step in the process for conversion of petroleum to energy at the wheels of a vehicle, comparing the standard spark-ignition (SI) vehicle as it is now used to the electric vehicle. Note that the overall efficiencies are similar--about 13%. Of course, the objective is not to use petroleum, but to shift to other sources. Figure 7 shows the overall efficiency for the use of coal in vehicles. The overall efficiency of the SI engine vehicle suffers because of the efficiency loss in converting coal into a liquid fuel for a vehicle, yielding the efficiency advantage to the electric vehicle (EV). In Figure 8, the efficiency values for the nuclear fuel situation are shown. Again, the advantage goes to the EV, by a significant margin: 4% vs. 14%. In Figures 7 and 8, relatively high efficiencies were estimated for the preparation of liquid fuels, so the actual efficiency advantage of EV's is likely to be somewhat greater than shown.

TABLE 2

$Li_4Si/LiCl\text{-}KCl/FeS_2$

$Li_4Si + FeS_2 \rightarrow 2Li_2S + Fe + Si$

E = 1.8, 1.3 V; 944 Wh/kg Theoretical

| | |
|---|---|
| **Status** | |
| Specific Energy | 120 Wh/kg @ 30 W/kg |
| | 180 Wh/kg @ 7.5 W/kg |
| Specific Power | 100 W/kg peak |
| Cycle Life | 700 @ 100% DOD |
| Lifetime | ∿15,000 h |
| Cost | >100/kWh |
| **Recent Work** | |
| Bipolar cells | |
| Li-Si electrodes | |
| BN felt separators | |
| 70 Ah cells | |
| **Problems** | |
| Materials for $FeS_2$ current collector | |
| Leak-free feedthroughs | |
| High internal resistance | |
| Low-cost separators needed | |
| Thermal control | |

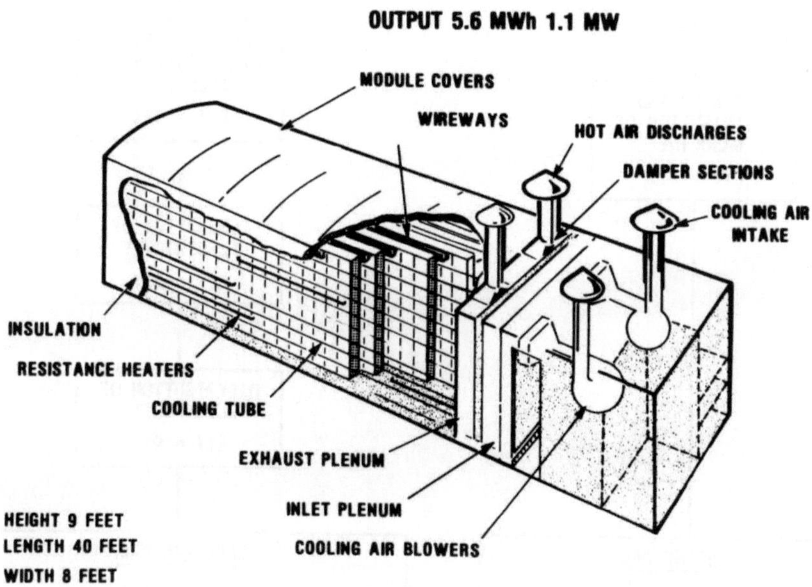


Fig. 5 Artist's concept of a truckable lithium/iron sulfide battery module for off-peak energy storage in the electric utility network. See S.M. Zivi, in Annual DOE Review of the Lithium/Metal Sulfide Battery Program, June 20, 21, 1979.

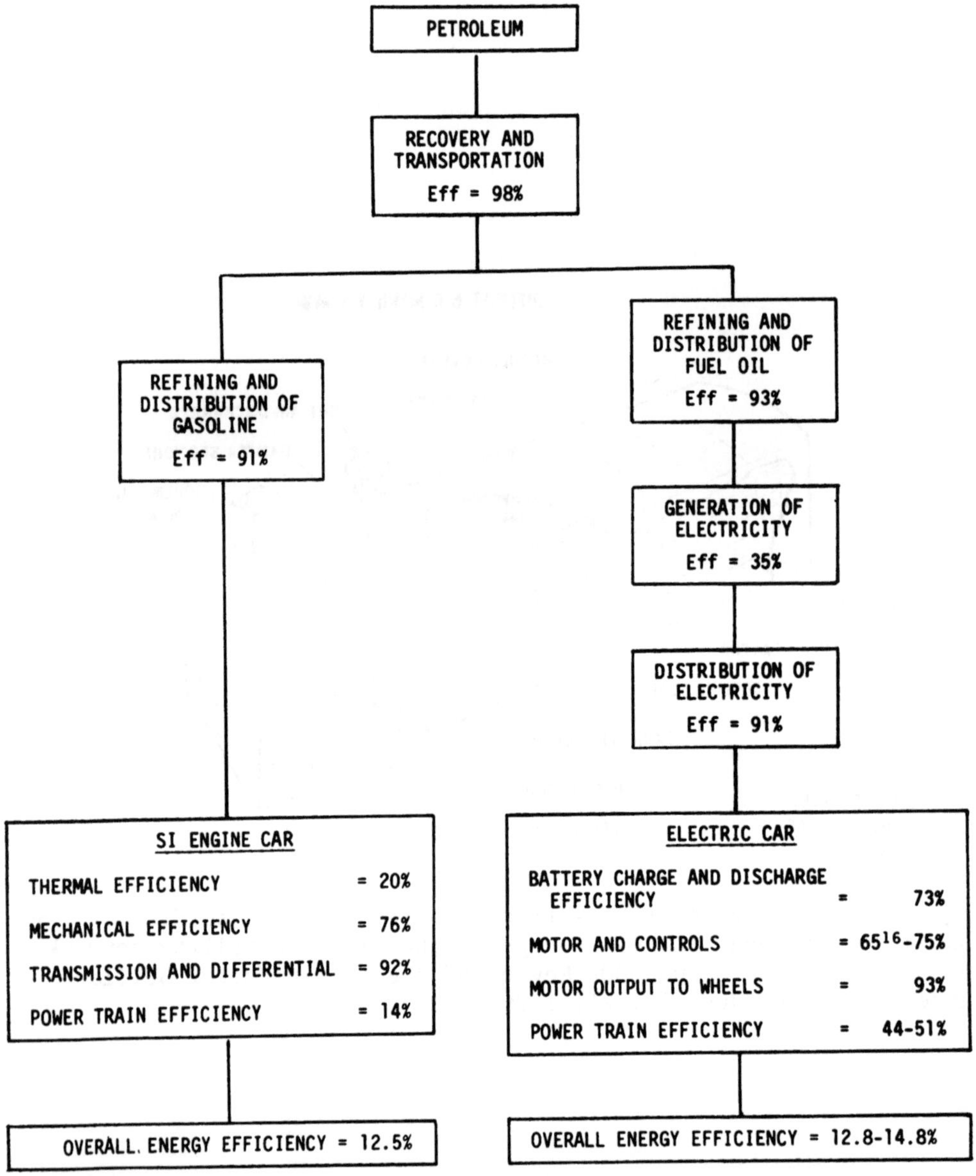


Fig. 6 Overall energy efficiency comparison for the use of petroleum to power spark-ignition engine cars and electric cars. Based on M.C. Yew and D.E. McCulloch, in Proceedings of 11th IECEC, AIChE, NY, 1976, p. 363; and E.J. Cairns and E.H. Hietbrink, in Volume VII of Comprehensive Treatise of Electrochemistry, Bockris, Conway, and Yeager, eds., Wiley & Sons, 1980.

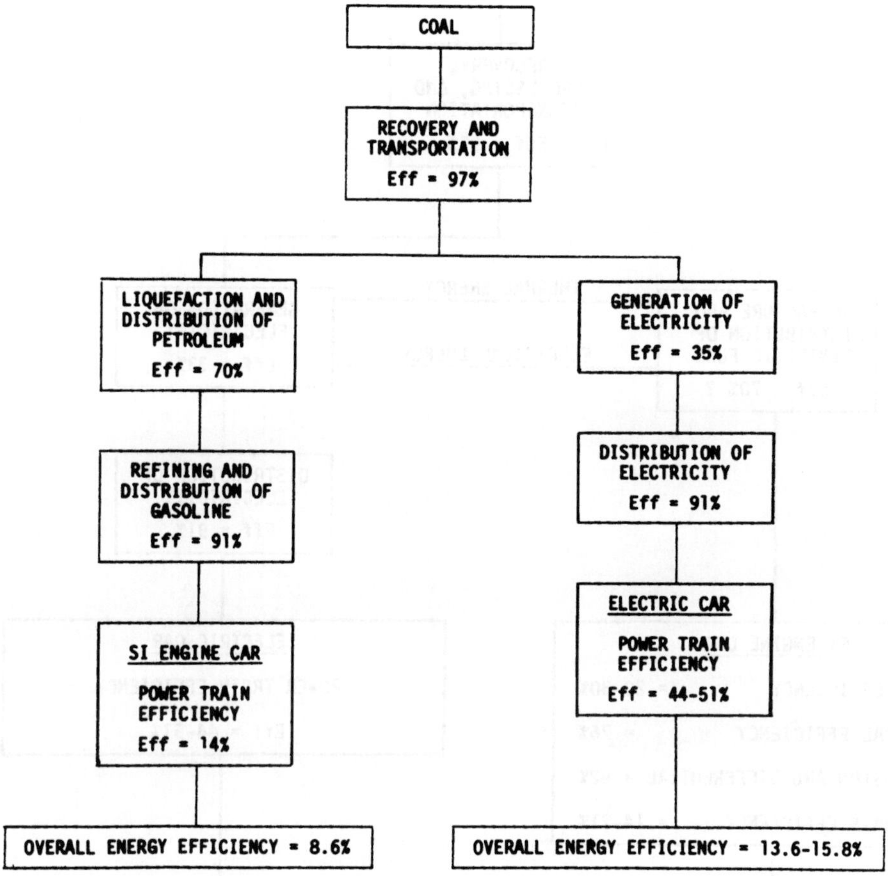


Fig. 7 Overall energy efficiency comparison for the use of coal to power spark-ignition engine cars and electric cars. See references in caption of Figure 6.

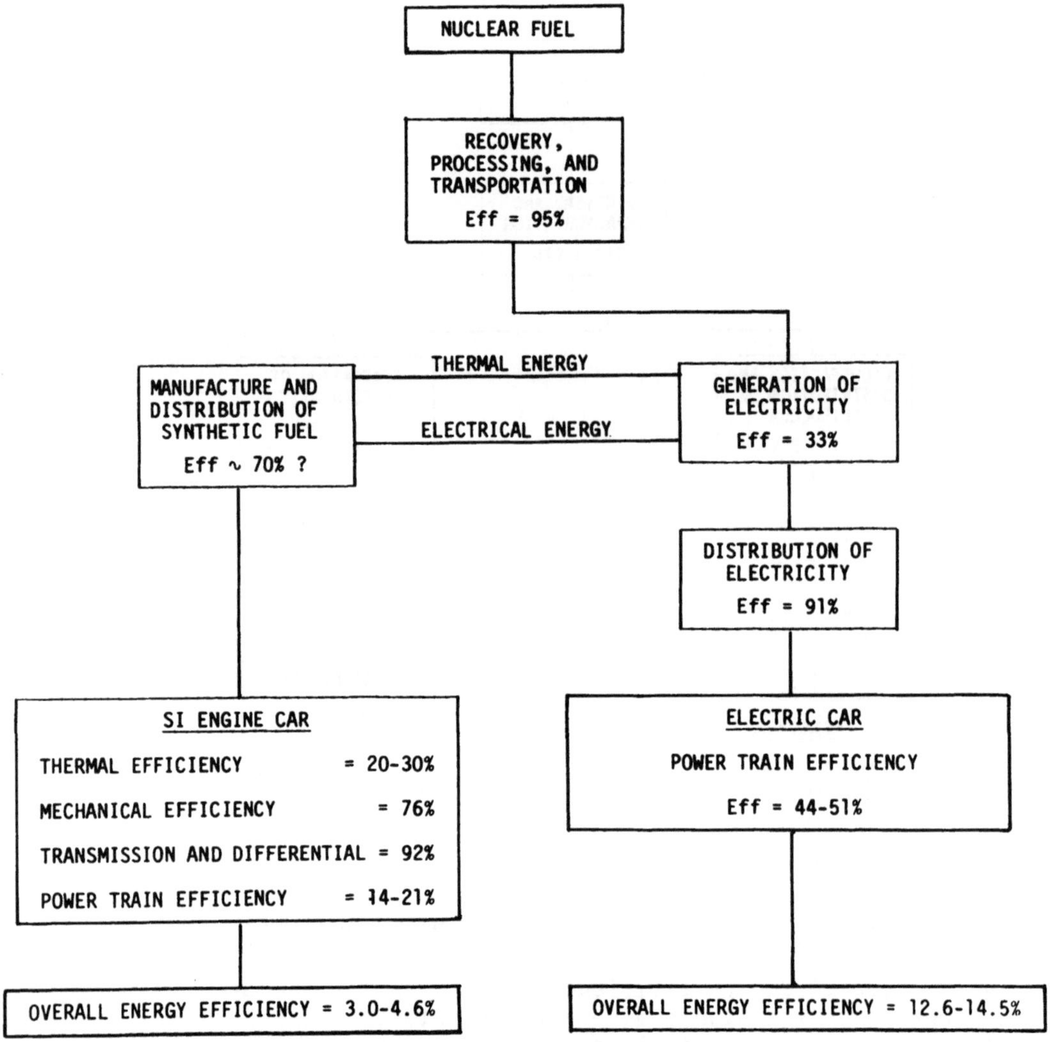


Fig. 8 Overall energy efficiency comparison for the use of nuclear fuel to power spark-ignition engine cars and electric cars. See references in caption of Figure 6.

With regard to the issue of energy consumption by the vehicle, it is possible to calculate with good accuracy the amount of energy required, knowing a few characteristics of the vehicle, and the velocity vs. time profile (driving profile). Since electric vehicles are limited to modest range and performance by the batteries, it is reasonable to perform calculations for urban and suburban driving profiles only. The applicable equations are:

$$P_b = \frac{P_r}{E_m \cdot E_e} + \frac{P_a}{E_a} \tag{2}$$

$$P_r = 9.8\ V\ (R_r + R_w + R_g + 1.1\ R_a) \tag{3}$$

$$R_r = \frac{W}{65}\ (1 + 4.68 \times 10^{-3} V + 1.3 \times 10^{-4} V^2) \tag{4}$$

$$R_w = \frac{\rho_a}{g}\ C_d\ A_f\ \frac{V^2}{2} \tag{5}$$

$$R_g = W \sin\Theta \tag{6}$$

$$R_a = \frac{W}{g}\frac{dV}{dt} \tag{7}$$

where $P_b$ = power from the battery, W

$P_r$ = power required at the wheels, W

$E_m$ = mechanical efficiency of the transmission and differential

$E_e$ = electrical efficiency of the drive train

$P_a$ = power required by the accessories

$E_a$ = electrical efficiency of the accessories, W

$V$ = velocity, m/s

$R_r$ = rolling resistance of the tires, $kg_f$

$R_w$ = wind resistance, $kg_f$

$R_g$ = gravitational resistance, $kg_f$

$R_a$ = acceleration resistance, $kg_f$

$W$ = vehicle test weight, kg

$\rho_a$ = density of the air, $kg/m^3$

$C_d$ = air drag coefficient, demensionless

$A_f$ = frontal area of the vehicle, $m^2$

$\Theta$ = angle of inclination

g = gravitational acceleration, 9.8 $m/s^2$

When the above equations are applied point-by-point (by computer) to driving profiles such as those of Figure 9, it is found that the energy and power required by the vehicle are essentially proportional to the vehicle mass, so that the requirements may be expressed simply in terms of kWh/T-km and kW/T, summarized as shown in Table 3. [4] Note from the table that the battery must provide about 0.15 kWh/T-km for urban driving, that the peak power needed is up to 35 kW/T, and the average power is 4-5 kW/T.

From the standpoint of good vehicle design practice, it is desirable not to exceed 0.25-0.30 of the vehicle mass as the fraction assignable to the battery. The urban range of an electric vehicle may be calculated from the expression:

$$R = \frac{S_pE}{0.150} \; \frac{M_b}{M_v} \tag{8}$$

where R = vehicle range, km

$S_pE$ = specific energy of battery, Wh/kg

$M_b$ = battery mass, kg

$M_v$ = vehicle test mass, kg

Equation 8 is plotted in Figure 10. Note that a range of 100 km requires a battery having a specific energy of 60 Wh/kg if the battery fraction is 0.25, and a 150 km range requires a specific energy of 75 Wh/kg at a battery fraction of 0.3. These considerations provide the battery performance requirements shown in Table 4: a specific energy of at least 70 Wh/kg, for an urban range of 140 km at a battery fraction of 0.3. In order to provide the battery with an acceptably small volume (from a vehicle design point of view) the energy density should be at least 140 Wh/l. The specific power for safe acceleration (0 to 50 km/h, ∿9 sec.) should be about 130 W/kg.

Aside from vehicle and battery performance, the efficiency, durability, and cost are all important to the acceptability of a battery for use in an electric vehicle. The efficiency should be at least 70%, keeping the heat rejection rate to an acceptable value. As before, the cycle life and cost are compromises, based on what might be achieved. A

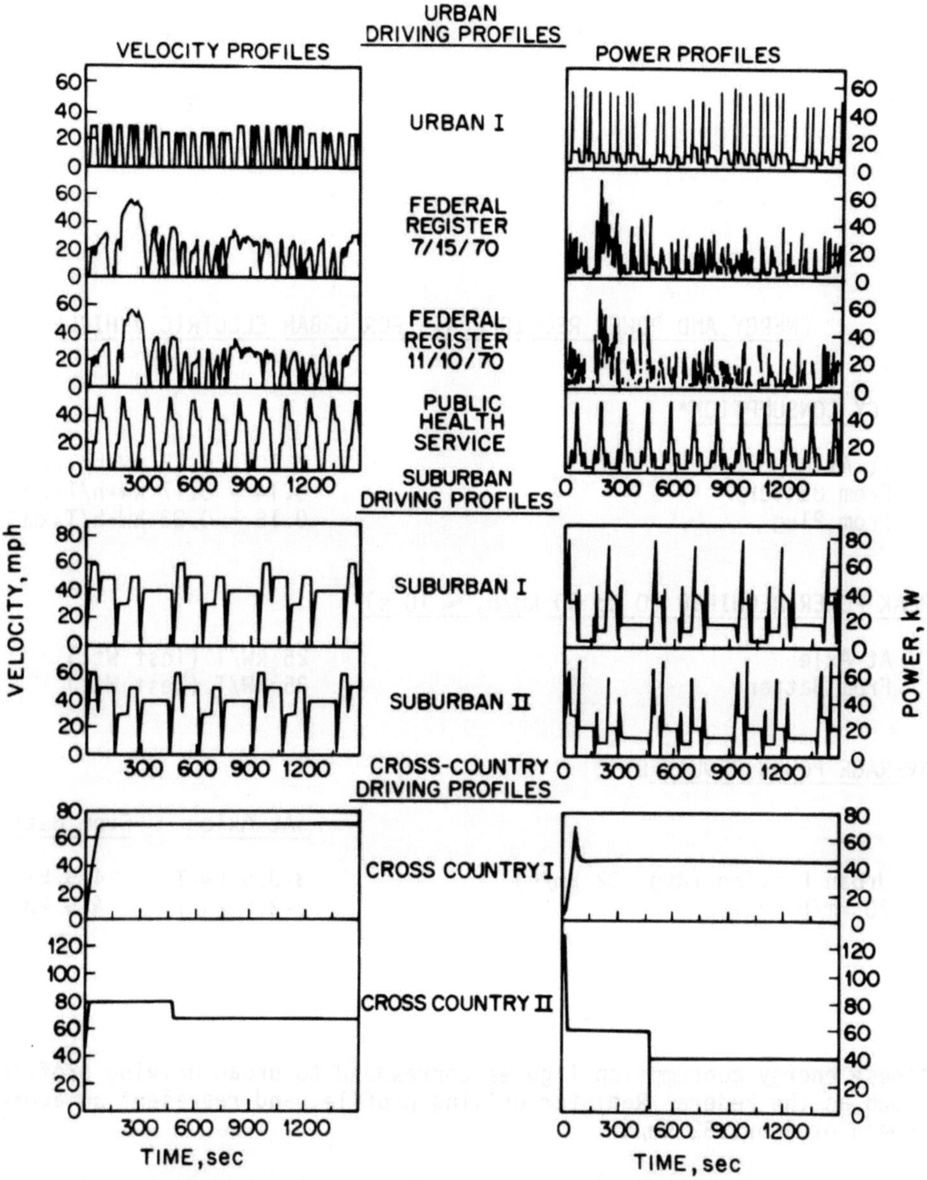


Fig 9 Driving profiles (velocity vs. time) of several types, with corresponding power profiles for a 2000 kg automobile, as described in E.J. Cairns, et al., "Development of High-Energy Batteries for Electric Vehicles," Argonne National Laboratory Report, ANL- 7888, December 1971.

TABLE 3

ENERGY AND POWER REQUIREMENTS FOR URBAN ELECTRIC VEHICLE

| | | |
|---|---|---|
| **ENERGY CONSUMPTION*** | | |
| At Axle | 0.10 - 0.12 kW•h/T•km | |
| From Battery | 0.14 - 0.17 kW•h/T•km | |
| From Plug | 0.18 - 0.23 kW•h/T•km | |
| **PEAK POWER REQUIRED (0 to 50 km/h, ≤ 10 s)** | | |
| At Axle | 25 kW/T (Test Wt.) | |
| From Battery | 35 kW/T (Test Wt.) | |
| **AVERAGE POWER REQUIRED** | | |
| | At Axle | From Battery |
| Urban Driving (Avg. 32 km/h) | 3-3.5 kW/T | 4-5 kW/T |
| 50 km/h Cruise | 3-3.5 kW/T | 4-5 kW/T |

*These energy consumption figures correspond to urban driving profiles such as the Federal Register driving profile, and represent an average speed of about 32 km/h.

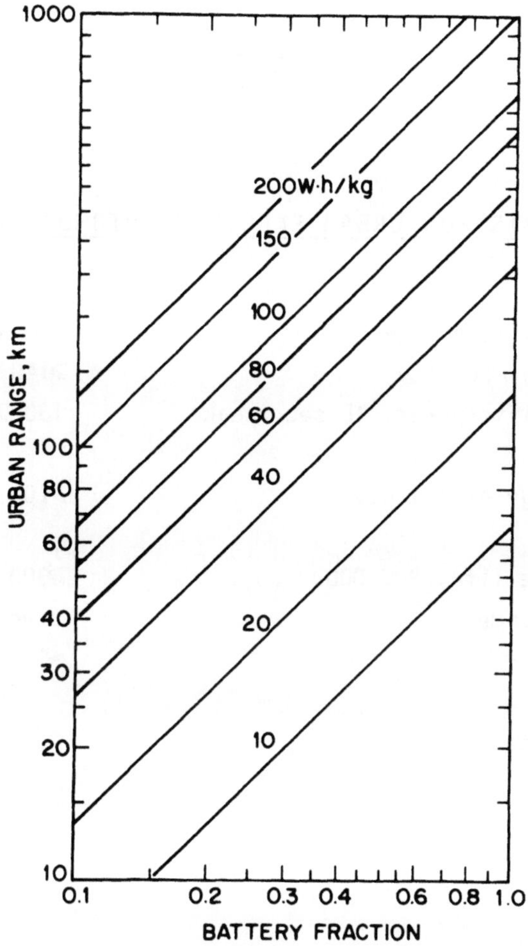


Fig. 10 Urban range for electric vehicles as a function of both battery fraction and battery specific energy, using Equation 8.

TABLE 4

REQUIREMENTS FOR URBAN ELECTRIC AUTOMOBILE BATTERIES

| | |
|---|---|
| Specific energy | ⩾70 Wh/kg* |
| Energy density | ⩾140 Wh/l |
| Specific power, 15 sec. peak | 130 W/kg |
| Energy efficiency | ⩾70% |
| Cycle life, 80% DOD | ⩾300 |
| Lifetime | 3 years |
| Cost | ⩽$70/kWh |
| Typical size | 20-40 kWh |

*Corresponds to 140 km range for a battery mass of 30% of the vehicle test mass.

minimum life of 300 deep cycles, and a cost of \$70/kWh correspond to an amortized battery cost of 3.5¢/km. This is about the maximum tolerable cost, unless new factors come into consideration.

An example of an ambient-temperature battery that might be acceptable for urban electric automobiles is the zinc/nickel oxide battery, which has a potassium hydroxide electrolyte, and operates according to the overall reaction:

$$Zn + 2NiOOH = H_2O \rightarrow ZnO + 2Ni(OH)_2 \quad (9)$$

and has a theoretical specific energy of 373 Wh/kg. At the present stage of development, these cells display 55-75 Wh/kg (15-20% of theoretical), making them capable of giving a range in excess of 100 km in an electric vehicle. Specific power values in the range of 150 W/kg have been achieved by Zn/NiOOH cells of light-weight construction.

The cycle life of the zinc electrode has been shorter than needed: 100-200 deep cycles. Failure is traceable to combinations of the following problems: [5] a) dendrite formation, b) zinc redistribution (shape change), c) densification, and d) passivation. Dendrite formation can be minimized or eliminated by proper choice of separators with extremely small pore diameters. Zinc redistribution is the gradual movement of zinc away from the edges of the electrode toward the center as cycling proceeds. It is related to the formation of soluble zinc species on discharge, and their redeposition during recharge on sites closer to the center of the electrode. No existing theory is capable of a quantitative explanation of this complex process. Its rate is significantly reduced by making the current density uniform, and by the use of $K_2TiO_3$ fibrous mats against the zinc electrode. Densification occurs with repeated cycling of the zinc electrode, and is the loss of porosity and surface area in the zinc deposit, finally resulting in passivation of the zinc because the current density exceeds the critical value ($\sim$20 mA/cm$^2$) for the formation of a passive oxide film, preventing further electrochemical reaction. These problems continue to receive attention, and gradually the cycle life of the zinc electrode is being improved. For a more detailed discussion, see reference 6.

The current cost of Zn/NiOOH cells is significantly above \$100/kWh, but projected values are near \$70/kWh, for cells with polymer-bonded electrodes. Achievement of the performance and durability goals requires better separators and zinc electrodes. These are the areas of current emphasis. The current status of the Zn/NiOOH cell is summarized in Table 5. Batteries of more than 10 kWh have been tested in electric automobiles, and have yielded the expected range and performance: more than twice the range available from the $Pb/PbO_2$ cell, and better acceleration. [7]

## IV. Energy Storage for Solar and Wind Powered Systems

In contrast to the two areas for the application of batteries

TABLE 5

Zn/KOH/NiOOH

$Zn + 2NiOOH + H_2O \rightarrow ZnO + 2Ni(OH)_2$

E = 1.74 V; 373 W•h/kg Theoretical

<u>Status</u>

| | |
|---|---|
| Specific Energy | 55-75 W•h/kg @ 30 W/kg |
| Specific Power | 80-150 W/kg @ 35 W•h/kg |
| Cycle Life | 100-200 @ 25-50 W/kg |
| | 80% DOD |
| Cost | >$100/kW•h |

<u>Recent Work</u>

Inorganic separators (e.g., $K_2TiO_3$, $ZrO_2$, others)

Sealed cells

Nonsintered electrodes

<u>Problems</u>

Sealing of cells - $O_2$ evolution and recombination

Shape change and densification of zinc electrode

Separators

discussed above, the storage of energy generated by solar- and wind-powered systems is not an established technology with clearly-defined requirements. At this point, it is not clear what the sizes, load profiles, and other characteristics of the systems will be because very few solar- and wind-powered installations exist. Therefore, it is extremely difficult to list the performance, durability, and cost requirements for the energy storage batteries that might be used.

In spite of the above difficulties, some general indications of desirable battery features can be given. There are a number of options available:

1) Short-term (minutes to hours) storage to level the supply as well as the demand, to provide better matching between the generator and load.

2) Intermediate term (several hours) storage to provide energy for a period such as the whole evening or night, when solar (or wind) input is unavailable.

3) Longer-term storage to provide energy for days, when solar or wind energy may be insufficient to meet the demand.

In the case of short-term storage, the battery would probably be designed in such a manner that is is usually on "float charge," only providing a small fraction of its capacity before being recharged. This sort of service is similar to automotive starting-lighting-ignition service, and could probably be handled well by a $Pb/PbO_2$ battery.

Intermediate-term storage, to a first approximation, is similar to the off-peak energy storage for electric utilities, and probably would be satisfied by a battery having the characteristics given in Table 1. A higher cost and/or a shorter lifetime than those shown in the table might be acceptable, especially for remote installations with no reasonable alternative for energy storage.

Longer-term storage of energy in relatively large amounts has usually been outside of the proposed area of applicability of conventional batteries, partly because of the relatively high cost implied. There are some battery systems that could prove attractive in this application, however. These are flow systems, in which the reactants and products are stored in tanks. The tanks are sized for the desired <u>capacity</u>, without affecting the electrochemical cells, which are sized for the desired <u>power</u>. This feature tends to reduce the total system cost for large capacities below what it would be for a system in which the size of the electrochemical converter is proportional to capacity. Examples of such systems are redox systems, and to some degree zinc/halogen systems (in which the halogen only is stored externally).

As more experience is gained in the use and the identification of specific applications for solar- and wind-powered systems, a clearer definition of battery requirements can be provided. In the meantime, analytical studies and experimental programs can be carried out to gain more information regarding the power vs. time profiles of both the generators and the loads.

## V. Conclusions

Based on the above discussion, the following points can be made in the way of summary and conclusions.

- The general requirements for rechargeable batteries in electric utility networks, electric vehicles, and solar/wind-electric systems have been presented and discussed.
- No presently-available batteries meet all of the performance, durability, and cost requirements for use in off-peak energy storage, electric vehicles, or solar/wind energy storage systems.
- Materials problems are among the most important in achieving the goals for widespread use of batteries in major energy storage applications. The rate of progress in this area may determine the rate at which the goals can be met.
- Overall, batteries have a number of important opportunities to contribute to energy independence by shifting part of the energy demand away from petroleum, both in the transportation and in the electrical energy generation sectors.

## References

1. J.R. Birk and N.P. Yao, "Batteries for Utility Applications: Progress and Problems," in Proceedings of Symposium on Load Leveling, The Electrochemical Society, Princeton, NJ, 1977, p. 229.

2. E.J. Zeitner and J.S. Dunning, "High Performance Lithium/Iron Disulfide Cells," in Proceedings of 13th IECEC, SAE, Warrendale, PA, 1978, p. 697.

3. R. Hamilton, "Carborundum Program," in Annual DOE Review of the Lithium/Metal Sulfide Battery Program, Argonne National Lab, Argonne, IL, June, 1979.

4. E.J. Cairns and J. McBreen, "Batteries Power Urban Autos," Ind. Res., June, 1975, p. 56.

5. E.J. Cairns, presented at the International Society of Electrochemistry Meeting, Budapest, Hungary, September, 1978; see also Extended Abstracts.

6. J. McBreen and E.J. Cairns, "The Zinc Electrode," in Advances in Electrochemistry and Electrochemical Engineering, Volume 11, H. Gerischer and C.W. Tobias, eds., John Wiley & Sons, NY, 1978.

7. E.J. Cairns and E.H. Hietbrink, in Comprehensive Treatise of Electrochemistry, Volume VII, J. O'M. Bockris, E. Yeager, and B. Conway, eds., John Wiley & Sons, in Press.

# REFLECTIONS ON SOME RECENT STUDIES OF MATERIALS OF IMPORTANCE IN AQUEOUS ELECTROCHEMICAL ENERGY-STORAGE SYSTEMS

E.J. Casey

Director, Energy Conversion Division
Defence Research Establishment Ottawa

## I. INTRODUCTION

### Delineation

In this paper an attempt is made to provoke thought on materials which are used or are formed in rechargeable batteries based on watery electrolytes. It was decided not to attempt a compendium, but rather to focus in on selected topics which in our judgment seem important and on which new information has a good probability of practical payoff in improved performance.

By performance here are meant the usual measurables:- For rechargeable cells: nominal energy-density by some selected criteria; turn-around efficiency and realizable energy-density as measured under a selected cycling regime; energy-retention during a cycling regime chosen to simulate some operational mode; incidence and types of failure; reconditioning-ability and repairability. For fuel cells: efficiency of conversion at a power density acceptable for the application; mean-time between failures; reconditioning ability and repairability.

To the user, especially the user who has an important tactical mission to achieve, the ability to recondition or repair --

---

*Prepared for presentation at NATO Advanced Study Institute on Materials for Advanced Batteries, Aussois, France, 9-13 Sept 79. Issued as DREO Report No. 811.

ideally of course the elimination of the necessity to do so -- is often more important than other considerations. Reliability and longevity are assigned high places amongst performance criteria. Reliability and longevity should not be confused, although they are often interrelated manifestations of performance, equally dependent upon materials-selection and upon quality-control during manufacture.

## Philosophical Focus

Three major questions are considered:

1. What use are we really making of the information resulting from techniques which have become so popular in recent years?

2. What information are we lacking which would enable us to make major improvements in performance?

3. What basic assumptions about systems are we making which should be challenged?

In this paper these questions are not treated systematically, or separately, one from the other, but they underlie all of the discussion to follow. "What problems demand our attention?" "If we worked on a particular problem, using certain tools, are we likely to be able to use the results in some practical way?".

This paper contains no discussion on many practical problems having fundamental overtones and concerning the systems now in use or being readied for introduction into the marketplace. We make little comment on topics such as: (a) Lighter supporting structures for the active masses in lead-acid battery plates, or the selection and characterization of suitable oxides; (b) Separator materials which will be chemically stable and will defeat the run-away (burnout) problem in the nickel-cadmium and silver-zinc systems during overcharge; (c) Long-lived and efficient air electrodes for air batteries and fuel cells; (d) Shape changes at the non-oscillating zinc plate in nickel-zinc batteries; (e) Adaptation of lead chloride for more extensive use in a variety of sea-water batteries. These topics are indeed of great practical importance even to the non-military user; and although our Laboratory has been working in most of these areas, it is not our intention here to focus on them. Let our only example be to mention the replacement of the scarce silver salt by cheap lead chloride in the seawater batteries which are now used by the thousands in most passive sonobuoys. Coleman and King did the pioneering work (1a), and others have followed it up (1b).

Further, the organizers of this Institute have invited us, indeed challenged us to stick with aqueous systems, and to try to say interesting and useful information about them. This proscription should not be interpreted as indicating that our Laboratory does not have interest in non-aqueous, molten salt and solid-polymer electrolyte systems. Indeed there is strong interest, because many new applications of the new systems can be envisaged. Much of our research effort currently is directed to these other battery systems.

Neither do we include a comparison of the state of the development of the fifty or so electrochemical couples which are receiving attention in the many laboratories engaged in research and development, world-wide. Each credible consultant or specialist in this field has his own good sources and resources, and no attempt is made here to produce yet another primer for the battery engineer. Many comparative reviews exist, but computerized expression of what system might be best for any given application is not highly developed yet.

We decided to offer information from our own research directed towards real problems facing our customers in the Canadian military community. Our collective goal surely, with rising costs of research, must be to improve our chances of coming up with practical information which will enable ourselves and our colleagues in industry to bring new and improved systems more rapidly to the marketplace for application.

<u>Inclusions</u>

Following this Introduction, five general topics selected for discussion in this paper are treated in the following order:

Deteriorating Water Quality -- (A Cause that Defreshes)
Photoactivity -- (New Light on Surface Phenomena)
Transition Region Electrolytes -- (Almost Out of Bounds)
Electrochemical Spectroscopy -- (The New Optic)
Techniques and Materials -- (The Chronic Problems)

These topics reflect some of the current interests of colleagues at DREO. The references to their work and to that of the author are intended to promote contacts with others who are interested in these same subjects.

## II. DETERIORATING WATER QUALITY

New materials are entering the surface water world-wide, some carried by rainfall, some as waste-water from domestic or industrial use and some from modern plumbing materials. Salts, organic materials and unnatural (perhaps redundant) synthetics are decreasing the purity of the water which is used as the source of electrolytes for all aqueous electrolytic cells. Some of these additions are hard to detect, let alone identify, and may persist through the various purification processes which the water undergoes before it finally enters the electrochemical cell. In Canada, polychlorinated biphenyls are found in the rain and rivers which feed the Great Lakes. Polyethylene has been found in the mid-Altantic and in lab plumbing. Traces of chloroform have been reported in the water supplies of many Canadian and other cities. These products and motes of all sorts (finely disbursed inorganics, spores, etc.) can persist through most of the purification processes which the water undergoes before it finally enters the electrochemical cell.

Years went by before the purity of the water which is used as the source of electrolytes for all aqueous electrolytic cells, batteries and fuel cells was suspected. Many of the organic compounds as added or as altered pass through ordinary distillation and even from alkaline permanganate. Criddle (2) early recognized this problem and devised a preparative procedure for "pure water" which is low in surface-active organics, products of biodegradation and stray inorganics and motes unwittingly introduced by the removal procedures and equipment themselves.

Pyrodistillation is the key new process: it includes oxidative pyrolysis of water vapour, to destroy the organic impurities of water vapour, and careful redistillation. With pyrodistilled water researchers (2) have been able to make up experimental cells which show much less evidence of poisons adsorbing gradually on the surfaces under examination. The process of Reverse Osmosis, using membranes of carefully selected and controlled chemical composition and physical properties, can produce purified water of outstanding quality. Deionizers help--or do they? How many of us really do know the quality of the water used in preparative procedures in our own laboratories? Just imagine how important the procedures quality-control should be in processing industries in which battery materials are fabricated and assembled. Without pure water, even researchers are still plagued with irreproducibility....With such variability in the surface waters of the earth, no wonder battery cells from different countries can show such variation.

Sometimes the persistence of poisons can be correlated with the infrared spectra of the evaporated residue -- there should be no residue and no spectrum, of course, but preparative procedures very often fail. One asks: Is it the care by the researcher, or is it the difficulty of getting the water to separate from the impurity, which is failing? In our experience, it is both. In our Laboratory we have found it extremely difficult to remove impurities and to keep them out. In crude experiments with complete alkali batteries the purity does not seem to be crucial. However, when exhaustive performance is intended -- at low temperatures and high power levels for example -- water quality must be assured, otherwise "oils" which have inadvertently been allowed to enter into the cells, may even cause stable foams which can eject electrolyte from gassing cells.

The effects of insufficient attention to air-quality can often be overcome by sufficient flushing, or electrochemical working, of the cell after closure, but such procedures are none too sure either. Microporous filters remove some particulates but not all. Pyrolysis can remove the organic-spore type, but other solids get through and eventually can show up on the electrode in the double layer and participate in the process under observation, usually to degrade the performance.

This problem of purity of watery solutions should not offer the non-aqueous researcher any consolation, for "oils" can be introduced into non-aqueous cells during the preparative techniques used for electrodes, chemicals, separators and seals, and the assembly of non-aqueous cells, just as easily, perhaps even more easily, than in the aqueous electrolytes under discussion here.

One can surmise that the careful researcher will be rewarded with new observations on the behaviour of the tried-but-true conventional systems when he realizes that the quality of his own water is deteriorating as a starting material, as well as that of others who process his components. More painstaking care than ever is now necessary. One must remember that Criddle did make a series of 1.6-volt oxygen-platinum electrodes fifteen years ago (Electrochim. Acta, 9, 853 (1964)), by working in wintry air that was free of enzyme producing plants and by classic distillation of deep well water from which surface waters had been excluded by ground frost.

Since DREO introduced plastic plumbing materials, the really careful researchers have had to cope with new impurities. Let us now all be more aware of "oils" in the distilled water. In our view carefully pyrolyzed water should be used for every critical and baseline experiment.

## III. PHOTOELECTROCHEMICAL PROCESSES ON BATTERY MATERIALS?

Photoassisted electrolysis of water was first reported in 1974. The prospect of practical photoassisted electrolysis has excited researchers world-wide. Even in Canada no fewer than nine research teams are active on some aspect of the phenomenon this year.

Photo-effects on electrode reactions had been observed a good many times before and, of course, up-to-date books are available (Eg. Gurevich et al., Photoelectrochemistry, Plenum Publ. Corp., NY, 1979). Some effects have practical interest. Thus color changes accompanying charge acceptance by vanadates and tungstate bronzes have elicited some interest for possible display devices. Electrolytic generation of beautifully colored polymeric intermediates, and electrochemiluminescent phenomena both caused a stir for awhile and some practical devices have resulted. The photoactivation of organic molecules absorbed on electrodes and undergoing electrooxidation or reduction receives continuing attention, as new and useful synthetic routes are sought. Photostimulated electron-ejection into electrolytes has been avidly studied by some researchers, who have been concurrently examining the surface composition by ellipsometry.

On certain surface sites of semiconductors with the right band-gap, photons can generate electrons and positive holes and separate them so that they can take different reaction paths, the electrons to reduce protons, or the holes to oxidize hydroxyl-ions in the electrolyte. The first observations of photoassisted electrolysis were made on $TiO_2$ electrodes.

For some years corrosion and erosion of the surface of the $TiO_2$ semi-conductor, under conditions of photoactivation of either the oxidation process (to $O_2$) or reduction (to $H_2$) process of water, seemed inevitable. Instability and degradation of performance resulted from continuous operation. However, it has been found that additions, particularly of soluble sulfides, to the electrolyte offer an alternative to the corrosion process. Secondly, dopants to the semiconductor have been found which move the wavelength of maximum absorption into the visible, and nearer to the solar maximum wavelength. Thirdly, in at least one case, GaAs, it has been found that the rapid surface-recombination of photogenerated electrons and holes can be inhibited by the addition of ruthenium sulfate to the electrolyte -- and the conversion efficiency greatly improved. Along with colleagues (3a), Snelling et al. (3b) at DREO have attempted to apply some of these findings to the CdSe and to CdS systems, for someone has to learn how to solve these problems and others without using the scarce gallium or the poisonous arsenic or the scarce ruthenium, if practical hydrogen production or photoelectrochemical conversion and storage of solar energy at the practical level is to be achieved. Although other problems must be overcome before photoelectrochemical conversion can be realized -- such as: (1) identification of a cheap and reliable photoelectrode production technique, (2) long-term stability (or simple regeneration) of the photoelectrodes, and (3) identification of more suitable stabilizing electrolytes than polysulfides (See Heller, Electrochim. Acta, in press) -- an excellent foundation now exists for the application of these principles in other areas.

Application of these techniques and ideas to electrocatalysts and to the study of aqueous battery electrode materials has scarcely been started. Our own interest quickened when Brossard et al. (4) confirmed Pavlov's observation that the rate (current) of anodic formation of, and cathodic reduction of, thin films of $PbO_2$ on Pb in aqueous $H_2SO_4$ is markedly photosensitive under some conditions.

It is not the photosensitivity itself which is important in this case. Rather the photosensitivity, if it is not due to thermal effects, can be diagnostic. It could be telling us that we have been searching in the wrong direction in our attempts to modify the positive plate structure and composition so as to improve the charge acceptance -- rate as well as quantity accepted -- at low temperatures. Battery plates themselves, of course, cannot be photosensitive at all, unless mass-transfer along the tortuous porous structure is not rate-determining, or unless the pores and the lead-sulfate crystals are translucent.

At this stage one can only imagine what new information might be obtained if we began to use pulse-electrochemical AND pulse-photo techniques, simultaneously, on working electrodes. From the work of Langhus and Wilson (Anal. Chem. 51, 1134 (1979)) one can see that it should be possible to sort out the change in number and kind of charge carriers induced in the $M/MO_x$ electrode during oxidation or reduction, when monochromatic coherent light, which has the same wavelength as the band-gap, is pulsed on and off. The most interesting results would be those which indicate whether and/or to what extent the $MO_x$ is changing in this respect during charging/discharging. Given some ideas about what charge carriers are important, it might be possible to inject dopants which could promote more complete oxidation/reduction of the active material, and thereby improve the charge storage per gram, as well as the turn-around efficiency at higher cycling rates.

One concludes that it is not altogether trite to view photoelectrochemical techniques as "the new light" on electrochemical interface phenomena.

## IV. TRANSITION REGION ELECTROLYTES

This is the most speculative part of this paper, in that, although the indirect evidence that this might be the way to go if new secondary battery systems of high energy density are to be developed, there is no experimental evidence on even one total battery system that could be considered the breakthrough which convinces.

The Transition Region has been defined and described by Klochko et al. (7) as the region of composition in which the solvent is small (present generally to less than 5 mole percent) and the molecular structure and properties are principally those of the solute. Limited, in this paper, to aqueous systems, transition-region electrolytes are simply moist salts, acids or bases which are in the liquid state, or perhaps as glassy solids.

The liquid state may begin at a surprisingly low temperature. The conductivity of the liquid generally is high. The chemical activity of the solvent -- in this case water -- is quite low. Chemical affinity of the light alkali and alkaline earth metals for the electrolyte is relatively low, and as a result these active metals can be "domesticated". Uncommon metals or plastics should not be required for the container or metallic contacts.

Light alkali metals are tame in aqueous solutions in which the concentration ratio of ions to water molecules approaches or exceeds unity. Both Li and Na have been shown to be workable as tame anodes in such very highly concentrated aqueous electrolytes. However, the electrodeposition of $Li^+$ or $Na^+$ from moist salts, acids and bases has not been investigated to any useful extent; few researchers have thought it likely enough to be rewarding even to try.

Conductivities, viscosities and densities are known for many low-melting hydrates, binaries, ternaries and other combinations of complex ions. Two examples will suffice:

$Ca(NO_3)_2 \cdot 4\,H_2O$ : fp 43°C,

$LiNO_3$ (25.8 mole %); $NH_4NO_3$(66.7 mole %); $NH_4Cl$ (7.5 mole %): fp 86°C dry

In the first, there is some evidence that calcium electrodeposition is possible, and the nitrate should be reducible. In the second, reduction of the nitrate is possible, as is reformation by reoxidation, and reduction of the lithium ion may be possible without destruction of the ammonium ion. The oxidation of lithium probably would proceed without incident. These systems need to be investigated.

The moist-electrolyte system based on equimolar KOH-NaOH gives transition-region electrolytes which are liquids from 25 to 200$^o$ depending upon the amount of water added. The interesting ones are the low-moisture electrolytes, from which $Na^+$ might be electrodeposited and in which an oxide clathrate like $MoO_3$ might be able to be cycled.

Anhydrous sulfuric acid also has phenomenally high electrolytic conductivity due to self-ionization and hydrogen-bridge formation. Moisture affects the conductivity but very little.

Klochko (5) has pointed out other possibilities based on moist bases or on moist acids, in which the chemical activities of the water are low enough that the essential processes of charge and discharge should be able to proceed without interference from side reactions. M. Abraham and his students at U. de Montréal are actively engaged in measurements of conductivity, density and viscosity of electrolytes in the transition region, and A.N. Campbell, U of Manitoba, retains his interest and productivity in logging baseline data in this area.

The impossibility of the applicability of dilute-electrolyte theories to the transition-region electrolytes and the lack of a good theory based on molten-salt complex theory, has probably discouraged many teachers from directing students into investigations of the transition-region. Because suitably trained investigators have been lacking, the battery community as a whole has not searched amongst transition-region electrolytes for interesting rechargeable systems. Excellent wide-open opportunities probably exist.

Some slight evidence is accumulating that some moisture is actually essential for the electrodeposition of lithium in non-aqueous systems. If this finding is confirmed, the search of the transition region becomes more cogent.

Further, the rechargeability of sodium and lithium in certain solid-polymer-electrolyte cells, may or may not have been observed in total absence of moisture. This is another reason to examine transition-region systems more thoroughly than heretofore.

Amongst possible solid systems one could suggest that the proton conductor based on hydrated uranyl and phosphorus oxides might be considered as a transition-region electrolyte. Preliminary work on the rechargeable cell

$$ZrH_x/HUO_2P_2O_4.4H_2O/PdH_2$$

has been reported. Perhaps lighter hydrides will be found to be tame in this low-moisture electrolyte.

## V. ELECTROCHEMICAL SPECTROSCOPY

New approaches are necessary if better understanding of the processes, better quality control during manufacture, and better operational control of the power source, are to be achieved. Three examples have been chosen to illustrate kinds of new information on aqueous systems which we think to be important.

(1) The first has to do with reactions which take place at the surfaces of anodically formed (hydr)oxides, during overcharge, in alkali batteries. (2) The second bears on the effects of organic degradation-products on passivation. (3) The third is concerned with the refractory materials which accumulate on acidic fuel-cell catalysts after extended use. The techniques being used are electrochemical and spectroscopic. Elaboration on each of these topics follows a remark on limitations of present methods.

Electrochemical techniques (charging curves, multi-pulse potentio-dynamic traces, cyclic voltammetry, etc.) although useful and provocative, are not sufficiently definitive to permit unique interpretation, or even description, of what is going on at the molecular level. Simultaneous identification of the reacting species <u>in situ</u> is necessary. Other than by fast spectroscopic means, such identification is next to impossible to achieve (See Fleichmann, ECS Extended Abstracts 76-1, 1976). Although the information so obtained may be only fragmentary, and is seldom unequivocal on identification, it does help provide a fresh view of the molecular mechanisms which can help us get better performance out of the system, be it higher energy density, greater longevity, etc. Although these techniques can be usefully applied to non-aqueous molten salt and solid electrolyte systems, their application to aqueous systems offers many surprises too. Almost every proposed mechanism of anodic oxidation or cathodic reduction includes free radicals as intermediates, and although the inferential evidence from rate theory and V/I curves, from cyclic voltammetry, from MPPD (multi-pulse potentiodynamic techniques), from photoelectron current transients, and other techniques, is quietly convincing, the skeptic longs for direct evidence, obtained directly from the operating cell itself. Some notes on three techniques in use at DREO are now offered.

Absorbed Species on Reactive Surfaces

The electrical performance of the Ni-Cd battery at normal and low temperature is affected in perplexing ways by what is on the reacting surfaces, both positive and negative. Hydroxycarbonates, polycarbonates, partially decomposed organic polymers and wetting agent, oxides and hydrated oxides, hydroxides and metallic species: all have been implicated as potential players in the overall processes which occur during charging and discharging of the battery. Verville is using fast-fourier-transform (FFT) analysis of repetitive infrared (IR) scans to separate out weak absorption signals by surface species from the background absorption noise of the battery plate -- with some success. The technique is elegant, but the specimen to be examined has to be dried well enough that OH bands from moisture do not bury the signal.

Studies on positive and negative plates of new and used Ni-Cds, and on specially prepared surfaces on nickel foils, have shown (6) that: (a) The surface material does indeed vary in composition over the surfaces; (b) Some effects previously ascribed to carbonates could come from hydr(oxides) only; (c) Carbonates do appear, and are indeed hard to remove; (d) Partially decomposed polymers from the separator or wetting agents can affect the behaviour of the nickel and the cadmium electrodes during the cycling operation of the battery; (e) Very careful specification and selection of the separator material and wetting agent can lead to cells of superior performance, whether in aircraft batteries (6b) or satellite batteries (6c).

In recent work, the electrophoretic deposition of active material into sintered nickel plaques has been shown to be markedly sensitive to the initial wettability of the plaque and dependent upon the organic materials which are adsorbed upon them during the deposition (6b). The identification of the controlling species is as yet incomplete.

The FFTIR reflectance technique may yet prove insufficiently discriminative, but at this stage it does appear that some real further progress is being made in identifying the materials responsible for the variable cranking performance and rechargeability of the Ni-Cd battery at low temperatures. Should it not be applicable to Fe, $LaNi_5H_x$, Fe-Ti-H, and Zn electrodes also?

It is suggested that this FFT IR reflectance technique could be applied with enthusiasm to the other negative electrodes used in alkali batteries. We might be surprised at what the organic decomposition products are affecting the limiting performance parameters which were mentioned at the beginning.

Free-Radicals in Battery Electrolyte

If a free-radical exists in a solution even in minute quantities, it can sometimes be detected by its unique electron spin resonance (ESR) spectrum. Information on its presence in the electrochemical cell, and on its movement and fate, would be unique information: at least we would know for certain what one species is doing. Many studies of the traps, and of trapped species, in and on dry solids have appeared. Many radicals have been formed electrochemically, and have been detected and followed by ESR, most of them being organic free-radicals with lifetimes of minutes or more, in aqueous and/or in non-aqueous solutions. In a few cases the role of radicals trapped in micelle polymers has been surmised from ESR results. The simple inorganic free radicals have escaped detection. Or have they?

One such discovery was reported from our Laboratory in 1971. The anodic formation of the free radical ions, ozonide ($O_3^-$) and its cousin superoxyl ($O_2^-$), in aqueous KOH, at low temperatures, was unpredicted in published oxygen evolution mechanisms. The details of the formation, of $O_3^-$ and its stability, and on the likely nature of the surfaces suitable for its formation and release, and on other aspects, have been reported by Gardner et al. (7a). One of the most interesting findings was that $O_3^-$ forms copiously from an overcharging silver oxide electrode but not at all from an overcharging nickel hydroxide electrode. Another curious finding was that either $O_2^-$ or $O_3^-$ can be ejected from the anodizing cadmium electrode by changing the KOH concentration and the temperature. The question arises: How are these surfaces so different one from the other? Or, why can (or must) one oxygen-bathed (hydr)oxide surface eject radicals while another cannot? Consider air-electrodes. One could imagine that adsorbed superoxyl radicals may well escape into the electrolyte from a silver-oxide catalyzed air electrode, or a manganese-catalyzed one. But do they?

Armstrong's (8) success in developing practical air electrodes which operate well at very low temperatures has, in our view, been a spectacular achievement. He uses "manganese oxide" catalysts and carefully controlled pore sizes. These, and Tseung's super-catalysts (JECS, 125, 1660 (1978)) based on doped perovskites, which seem to permit the direct 4-electron reduction of $O_2$ without the formation of adsorbed $HO_2$ ($O_2^-$ in alkali), may or may not eject radicals into solution. Since $O_2^-$ or $O_3^-$ could end up biting holes in the separator, these are not idle questions.

It should be mentioned that most of the current experimental work in our laboratory using ESR is concerned with the cathodic reduction of $SO_2$ in NON-aqueous cells. $SO_2^-$ and $S_2O_4^=$ have been distinguished and are being followed. In general, radicals are stable longer in non-aqueous electrolytes, and the researcher is often richly and more quickly rewarded, without suffering the experimental hurdles of trying to detect and follow short-lived species in aqueous solution. Yet, the practical rewards may be greater from work on aqueous systems.

The main message is clear: there may well be other, highly reactive, free-radicals escaping, one way or another, into battery and fuel cell electrolytes, and it would be nice to know what they are, where they go and what they do. Perhaps they can be suppressed or controlled better, or even utilized in some ingenious way to improve the performance.

## New Molecular Species

In acids things are different. Or are they? Direct and unequivocal detection of adsorbed species on surfaces is still not possible in situ, although equivocal information can be obtained by ellipsometry, by reflectance photometry, photo-stimulation and such techniques. Laser-Raman techniques for examining surface species show some real promise even though the process giving rise to the spectra is not yet well understood (Van Duyne, J. Phys., Colloq. 1977). Thus although the concentration of absorbed species is very low for conventional Raman, the inadequate theoretical treatment for adsorbed species further militates against the quality of the spectrum from the positive identification of the "scattering species". But could adsorbed species not be used as a probe to study battery-active surfaces? Via laser Raman?

Laser-Raman spectra of species in solution seem to hold more promise for battery and fuel-cell researchers at this stage of development of the technique, and program of study of complexes is being pursued at DREO. Adams et al. (9a) have recently reported on a study of the condensation reactions and the hydrolysis of phosphoric acid as a function of concentration, temperature and pressure. The main result of direct interest to this Study Institute probably is the evidence for the existence of the dimer of phosphoric acid, in hot concentrated solution. It was proposed at an ECS meeting that some such species must sit on the supported platinum electrocatalyst in the phosphoric acid fuel cell, and do two things: (a) Block active sites on the Pt, and (b) Aid migration of the Pt, through the formation of some unknown solution-complex. Hence the hydrocarbon/air fuel-cell specialists probably are on the right track in their attempts to inhibit the formation of the dimer and defeat its effects. A spectroscopic (L-R) investigation directed towards the same goal, and the desirable improvement in performance, might be very rewarding.

A L-R study of the complexes of $Br_2$ in aqueous electrolytes of interest in the $Zn/Br_2$ storage battery system is also in progress (9b), but although the information being obtained on complexing is thought-provoking, it is premature yet to discuss its practicality.

One concludes that the L-R technique can give practical information on complexing in aqueous solutions, information which can be unequivocal and which cannot be obtained by IR because of masking by water bands, or by any other technique. It is one new tool which battery chemists could use more fruitfully.

## VI. REMARKS ON TECHNIQUES & EXOTIC MATERIALS

### Other Techniques

There are two main components to electrochemistry, the electrical part and the chemical part. To do research on one without correlating with the other seems fruitless. It makes for lots of papers but not much true understanding. Techniques for continuous measurement physical and chemical properties of the active materials at the solid/liquid interphase in situ are really still quite crude and uninformative. Three possibilities which seem worthy of further development:

(a) Continuous snapshot X-ray diffraction probably could be developed to be a much more useful tool. Painstaking work has already shown the power (and the limitations) of X-ray diffraction of working surfaces (Burbank, JECS, 117, 299 (1970), 118, 525 (1971)). The sensitivity of the detector is still too low for rapid scan, however, and therefore it is difficult to follow changes in the diffracting structure of the working active mass. One can take out and dry the sample, but as soon as the current is cut it probably changes. Snapshot capability for X-ray diffraction equipment needs to be developed.

(b) Quick-freezing of surfaces under working conditions, freeze-drying and subsequent analysis--whether by micro-chemistry, Auger or other methods which yield composition and ratios--should lead to a better correlation of compositional changes with electrical performance, and hence to better control procedures. We need to rekindle the patience and persistence of earlier researchers in the matter of painstaking preparation of samples for analysis.

(c) Differential scanning calorimetry (DSC) not only of active-mass materials but also of support materials in electrochemical cells could be used more extensively than it is, now that the experimental tools and the theoretical background of thermal decomposition of solids are so much more readily available than they were a decade ago. Using the DSC, Nagy et al. (10) early recognized three states of AgO, and recommended the DSC fingerprint as a criterion for selection of AgO powders for reserve-primary torpedo batteries. Hydrides are another important class of materials of which the thermodynamic and kinetics of charge and discharge of hydrogen stored could be so fruitfully and quickly scanned by DSC and the metallurgical preparative procedures of the useful ternary and quarternary alloys usefully optimized.

## Exploratory Materials

In the sense that so little definitive and unequivocal information is known about the composition and behavior of the active masses of electrochemical cells during the operation, it could be argued that all active masses are "exotic" materials. In this paper we have concentrated on ways of learning more about these active components of aqueous cells. The theme is that, if we know how they work and what can go wrong, we should be able to modify their composition or structure and obtain improved performance.

Little has been said about side reactions which can undermine the supporting structure of the active material, either chemically or physically, or about corrosion reactions of the container or other components of the cell. By contrast with what happens in non-aqueous and molten-salt cells, in aqueous cells usually some well-known material can be selected which will not undergo destructive side reactions and it is the performance of the active masses which limits the performance of the cell. Hence, major improvements in the aqueous cell systems can be expected only from major improvements in the performance of the active masses: whether by doping, by changing the support structure or using different supporting materials, or by decreasing the amount of electrolyte or non-active components.

However, occasionally a new material is developed which meets the stiff criteria for use in electrochemical cells, and permits an advanced battery to be conceived. The term "exotic", literally "foreign", carries the further inference, "expensive". Advanced batteries can sometimes be made, at higher expense, using exotic materials, which employ conventional aqueous couples. There are two classes.

The first class contains those materials which have been developed to improve the conventional systems. Examples? (i) The radiation-grafted copolymers prepared as battery separators for the Zn-Ag system is an example. With it, this high-energy-density, high-power-density system is rechargeable for a hundred cycles, without it, maybe five. The material was developed to meet a need: if it could be done, then the system would be cyclable. (ii) Polypropylene fibre mattes (first) and then porous sheets (later) were developed to improve the longevity and the high-temperature stability of Ni-Cd satellite and aircraft batteries. (iii) The lead-calcium (-plus) alloys used as grid material in the lead-acid system, and the spirally-wound internal designs, were developed to make possible the low-maintenance and sealed lead-acid batteries which are now common. (iv) Lead chloride (cheap) was developed into suitable plate material to replace the strategic and expensive silver chloride in sonobuoys. (v) The lanthanum-nickel cobaltate electrocatalysts have been developed to permit the four-electron (more highly energetic) reduction of oxygen in the air electrode of air batteries and fuel-cells, and for energetically more efficient electrolysers. (vi) The advent of the hydrides for storage of the fuel, and in thermal management during operation. These are but some examples of the development of materials which are incorporated into (yesterday's?) conventional batteries for the purpose of offering us advanced systems.

The second class contains materials which have been developed for other purposes but which may enable us to conceive and build an advanced battery from standard aqueous systems. Amongst these support materials we list generally the following: (a) AlN and SiC and BN; (b) glasses and ceramics based on silicates and oxides; (c) polyphenylene sulfide, or oxide, and polyethersulfone; (d) vitreous carbon-fibre felt and non-conducting carbon fibre paper. It is a sad but necessary comment that it is typical that the impurity content and even the stoichiometry of these materials is usually very poorly known to the user, and that, as a result, he finds that the performance in his particular environment will vary from batch to batch. I have been impressed with attention to stoichiometry and impurity data recently published on SiC. We need more such chemical and physical characterization. But these are support materials.

From a big list of new active materials one is intrigued by new solid-ion conductors. Consider only two: (a) Nazirpsio, a solid ionic conductor based on the mixed silicates and phosphates of sodium and zirconium, and (b) NaI-doped polyethylene oxide. Both are sodium-ion conductors. Both have potential as candidate materials for separators (completely anhydrous?) or as solid electrolytes (transition region type?). Some of our views on these (5a), as well as on the use of some of the new alloys of titanium (11) in batteries have been published elsewhere. It is too bad that so little information on measurements of corrosion resistance, of elasticity and related mechanical properties, and of fracture mechanics can be found in the literature on materials such as BN, SiC and glassy carbon, since the performance in cells must be intimately related to the microcrystalline structure and to the exact composition of the material. Finally, one can also become intrigued with the nitrides of sulfur. One could surmise that the metallic ringed compound $S_4N_4$ is an electrocatalyst, and one day will show properties as interesting to fuel-cell chemists as the dichalcogendes are showing to battery chemists.

## EPILOGUE

"The scientific barriers to be overcome in advancing defense, energy, communication and production material technologies are all materials related. Similarly the engineering barriers to be overcome in obtaining increased efficiency, simplicity and reliability in systems technologies are all design related.

To attain both the cost and the confidence levels required for timely utilization -- insight not research, leadership not management, innovations not ideas, and accomplishments not theories must be stressed.

The risk is competitive obsolescence, whether measured as corporate profits, national efficiency or technology leadership. The needs....for the foreseeable future will be materials limited".

W.L. Lachman (1979)

## REFERENCES

1. (a) J.R. Coleman, Batteries (D.H. Collins, Ed.), Pergamon Press, Amsterdam, 1972; Can. Pat. No. 973 607, 26 Aug 75.
   (b) T. Gray et al., U.S. Pat. 4 074 733 Feb 21, 1978; G.J. Donaldson, et al, DREO TN 78/3.
2. (a) E.E. Criddle, in Proc. Symp. on Oxide-Electrolyte and Interfaces (R.S. Alwitt, Ed.), The Electrochem. Soc., Princeton, USA, 1973.
   (b) B.E. Conway, H. Angerstein-Kozlowska, W.B.A. Sharp and E.E. Criddle, Anal. Chem. 45, 1331 (1973).
3. (a) M. Laban and B.L. Funt, in Chem. for Energy (M. Tomlinson, Ed.), American Chemical Society Symp. Series 90, 1979.
   (b) J.S. Lewis, R.A. Sawchuk and D.R. Snelling, DREO. Unpublished work, 1979.
4. (a) G.L. Brossard and L.D. Gallop, confirmation of D. Pavlov et al, J. Electrochem. Soc. 124, 1522 (1977).
   (b) E.L.R. Valeriote and L.D. Gallop, J. Electrochem. Soc. 124, 380 (1977); Power Sources 6, 68 (1977).

5. (a) E.J. Casey and M.A. Klochko, Chem. for Energy, (M. Tomlinson, Ed), ACS Symp. Series 90, 1979).
(b) Review "Theoretical and Practical Possibilities of the Transition Region in Electrolytes for Storage Batteries and Fuel Cells", DREO Report in Preparation.
6. (a) G. Verville, R.J. Charest and R.M. Hayashi, Power Sources 7 (J. Thompson, ED.), Acad. Press, London, 1979.
(b) G. Verville and K. Feldman, DREO Tech. Note 78/6, Proc. Symp. Battery Design and Optimization (S. Gross, Ed), Electrochemical Soc. 79-1 (1979).
(c) J.L. Lackner, T.E. King and R.L. Haines, DREO Report 66, Proc. Symp. on Battery Separators, Electrochemical Society, 90 (1970).
7. (a) C.L. Gardner and E.J. Casey, Can. J. Chem. 49, 1782 (1971); 52, 930 (1974).
(b) E.J. Casey and C.L. Gardner, J. Electrochem. Soc. 122, 851 (1975).
8. W.A. Armstrong, Power Sources 5 (D.H. Collins, Ed.), Acad. Press, London, 1975, Can. Pat. No. 990 784, 8 June 76.
9. (a) W.A. Adams, C.M. Preston and H.A.M. Chew, J. Chem. Phys. 70, 2074 (1979).
(b) W.A. Adams and T. Field, unpublished work (1979).
10. (a) E.J. Casey and G.D. Nagy, Zn/AgO Batteries. Edited by A. Fleischer and J.J. Lander, J. Wiley & Sons Incorp. (1971), DREO Report 591.
(b) Proc. 20th Ann. Power Sources Conf., PSC Publ., 1968.
11. (a) T.A. Wheat, Clay and Ceramics, Vol. 52-1, January (1979); E.J. Casey, Chem. in Canada 31, 28 (1979).
(b) M.A. Klochko and E.J. Casey, J. Power Sources 2, 201 (1978).

# LITHIUM-ALUMINUM/IRON SULFIDE BATTERIES

Donald R. Vissers

Argonne National Laboratory
Argonne, Illinois 60439

## ABSTRACT

High-temperature rechargeable batteries are under development for electric-vehicle propulsion and for stationary energy-storage applications. These cells utilize a molten salt electrolyte such as LiCl-KCl eutectic (mp, 352°C), negative electrodes of either Li-Si or Li-Al alloy, and positive electrodes of either FeS or $FeS_2$. Over the past seven years, several hundred engineering cells with different designs have been tested, some of which have achieved specific energies of >100 W-hr/kg and specific powers of >80 W/kg at 50% depth of discharge, other cells have operated for >1000 cycles. During 1979, Eagle-Picher Industries completed fabrication of the first full-scale (40 kW-hr) Li/MS battery, which consisted of two 20 kW-hr modules (60 cells each) housed in a thermally insulated case. Testing of this battery has been indefinitely delayed due to the unexpected short-circuit of one of the modules during heat up. The fabrication of other full-scale batteries is planned in the coming years.

## I. INTRODUCTION

Rechargeable, high-temperature Li/MS batteries are being developed by Argonne National Laboratory (ANL) and its subcontractors.[1,2] These batteries are being developed for electric-vehicle propulsion and stationary energy-storage devices. The widespread use of electric vehicles would decrease the consumption of petroleum fuels,[3] since the electrical energy for charging the batteries could be generated by coal, hydroelectric, nuclear, or other energy sources. The use of stationary energy-storage batteries for load leveling could save petroleum by reducing the need for gas turbines to meet peak power demands.[4,5] The stationary energy-storage batteries may also find application in systems involving solar, wind, or other cyclic or intermittent energy sources.

Although the same electrochemical systems are being developed for these applications and the two types of batteries share much of the same basic technology, the performance and cost goals differ significantly. The performance goals we have set for 1985 for these two battery applications are given in part in Table 1.[6] The batteries for electric vehicles require high specific power to permit passing and hill climbing and high specific energy to provide an adequate driving range (a specific energy of about 150 W-hr/kg corresponds to about 250 km for a passenger car). Stationary energy-storage batteries have lower specific-energy and specific-power requirements, but the cost and lifetime requirements for such devices are much more stringent.

In the current ANL program on the development of an electric-vehicle battery, the fabrication of a series of full-scale lithium/metal sulfide batteries is planned. The performance goals for this series of batteries are listed in Table 2.[7] The first battery in this series, planned for testing in 1979 and designated Mark IA, will have a capacity of 40 kW-hr; the main purpose of this battery is to evaluate the technical feasibility of the system for use in electric vehicles, and to resolve interfacing problems among the battery, the vehicle, and the charging system. The Mark II battery will have higher performance (50 kW-hr capacity) than Mark IA and will be tested in 1981. The major objective of the Mark II battery is to develop designs and materials that have a low cost in mass production.

Table 1. The 1985 Projected Performance Goals for Lithium/Metal Sulfide Batteries

| Goals | Electric-Vehicle Propulsion | Stationary Energy Storage |
|---|---|---|
| Power | | |
| Peak | 60 kW | 25 MW |
| Sustained Discharge | 25 kW | 10 MW |
| Energy | 45 kW-hr | 100 MW-hr |
| Specific Energy, W-hr/kg | 150 | 100 |
| Discharge Time, hr | 4 | 5-10 |
| Charge Time, hr | 4-8 | 10 |
| Cycle Life | 1000 | 3000 |
| Lifetime, years | 3 | 5-10 |

Table 2. Program Goals for Lithium/Metal Sulfide Electric-Vehicle Batteries

| | Mark IA | Mark II | Mark III | Long Range |
|---|---|---|---|---|
| Specific Energy,[a] W-hr/kg | | | | |
| Cell (average) | 80 | 125 | 160 | 200 |
| Battery | 60 | 100 | 130 | 155 |
| Energy Density,[a] W-hr/L | | | | |
| Cell (average) | 240 | 400 | 525 | 650 |
| Battery | 100 | 200 | 300 | 375 |
| Peak Power, W/kg | | | | |
| Cell (average) | 80 | 125 | 200 | 250 |
| Battery | 60 | 100 | 160 | 200 |
| Heat Loss Through Jacket, W | 400 | 150 | 125 | 75-125 |
| Lifetime | | | | |
| Deep Discharges[b] | 200 | 500 | 1000 | 1000 |
| Equivalent Distance, km | 30,000 | 100,000 | 200,000 | 300,000 |
| Target Dates | | | | |
| Battery Test | 1979 | 1981 | 1983 | - |
| Pilot Manufacture | - | 1983 | 1985 | 1990 |

[a]Calculated at the 4-hr discharge rate.

[b]Utilization of more than 50% of the theoretical capacity every 10 cycles.

It is anticipated that these batteries may have commercial potential for certain limited applications. The Mark III battery will have a higher performance (60 kW-hr) than Mark II. This battery will be suitable for evaluation and demonstration in passenger vehicles by 1983 and pilot-plant manufacture is expected in about 1986. For stationary energy-storage application, a test of a 5 MW-hr battery module in the Battery Energy Storage Test (BEST) Facility is planned for 1982.[7] The Best Facility, which is supported jointly by the U.S. Department of Energy and the Electric Power Research Institute, is being constructed to provide for the testing of various types of batteries for load-leveling applications.

The cells currently being developed consist of solid Li-Al[8,9] or Li-Si[10,11] negative electrodes, solid FeS or $FeS_2$ positive electrodes,[12] and a molten LiCl-KCl electrolyte (mp, 352°C). The operating temperature of this cell is 400 to 500°C. A key cell component is the electrode separator, a porous material that provides electrical isolation of the electrodes but permits the migration of lithium ions between the electrodes. Separator materials currently being used include boron nitride (BN) fabric, BN felt, and MgO powder.[1,13,14] In most cell designs, it is necessary to use screens or other structures to retain particulate material within the electrodes. To enhance the electronic conductivity of the electrodes, metallic current collectors are used to provide a low-resistance current path between the active material and the electrode terminal.[2]

The overall electrochemical reaction for the Li-Al/FeS cell can be written as follows:

$$2LiAl + FeS \underset{}{\overset{2e^-}{\rightleftarrows}} Li_2S + Fe + 2Al \qquad (1)$$

The theoretical specific energy for this reaction is about 460 W-hr/kg, and the voltage *vs* capacity curve has a voltage plateau at about 1.3 V. The reaction is actually more complex than shown; for example, an intermediate compound, $LiK_6Fe_{24}S_{26}Cl$ (J phase),[15] is formed through an interaction with the KCl in the electrolyte. The overall reaction for the Li-Al/$FeS_2$ cell can be written as:

$$4LiAl + FeS_2 \overset{4e^-}{\rightleftarrows} 2Li_2S + Fe + 4Al \qquad (2)$$

The theoretical specific energy for reaction (2) is about 650 W-hr/kg. The voltage *vs* capacity curve has two voltage plateaus at approximately 1.6 and 1.3 V, respectively. Reaction (2) also involves several intermediate compounds (generally ternary compounds of lithium, iron, and sulfur). Although cells having $FeS_2$ electrodes offer a higher specific energy and voltage than those with FeS electrodes, the higher sulfur activity of $FeS_2$ leads to high corrosion rates and long-term instability of the electrode. Consequently, for the near-term, development efforts have concentrated on the FeS electrode.

A major objective of the ANL advanced battery program is to transfer the technology to interested commercial organizations as it is developed, with the ultimate goal of a competitive, self-sustaining industry for the production of lithium/metal sulfide batteries. To this end, most of the cell and battery development, design, and fabrication is contracted to industrial firms, which include Eagle-Picher Industries, and Gould Inc. The Carborundum Co. is involved in the preparation of BN separator materials and the development of production processes for these materials. Other contractors currently participating in the program include the Energy Systems Group of Rockwell International, the General Motors Research Laboratories, ESB, Inc., Illinois Institute of Technology, the University of Florida, the Institute of Gas Technology, and United Technologies Corp.

## II. CHEMISTRY OF Li/MS

While the chemistry of the lithium-alloy/metal sulfide cell systems is very complex, and much is still not understood, great advances have been made in understanding the chemistry of these systems. The chemistry of the Li-Al alloy and the FeS electrodes, for example, has been investigated quite extensively and is now quite well understood, whereas much is still to be learned about the chemistry of the $FeS_2$ electrode systems, which have also been extensively investigated.

### A. Electrolyte

The molten LiCl-KCl eutectic composition is 58.2 ± 0.3 mol % LiCl and 41.8 ± 0.3 mol % KCl. With a density of 1.65 $g/cm^3$ at 450°C, the eutectic is 17.13 molar in LiCl and 12.33 molar in KCl, and has a specific conductance[16] of 1.574 $\Omega^{-1}cm^{-1}$. The electrolyte has a decomposition potential[17] of 3.61 V; the viscosity and vapor pressure are both sufficiently low so that they present no problem to the cell. The solubilities of both FeS and $FeS_2$ in the electrolyte are quite low, 2 and 40 ppm, respectively.

### B. Negative Electrode

Lithium is attractive as a negative electrode material[18,19] because of its low weight per unit of electricity delivered, low affinity for electrons, and high electrochemical reactivity. However, containment of liquid lithium in the negative electrode over a large number of cycles is difficult because, in time, uncontained lithium bridges to the positive electrode and causes an eletrical short circuit. Another problem with liquid lithium has been the high corrosion rates of ceramic separators and electrical feed-throughs. These problems can be avoided or alleviated by the use of solid lithium-metal alloy as the negative electrode material.

The two alloys presently being used as negative electrode materials are lithium-aluminum[8,9] and lithium-silicon.[10,11]

1. Li-Al Electrode. The Li-Al negative electrode has shown excellent performance in Li/MS cells.[1,2,9] The lithium-aluminum phase diagram is shown in Fig. 1. This figure indicates that, at the cell operating temperature (425°C), 9 at. % Li dissolves in aluminum, and β-LiAl has a composition width of 9 at. % Li (47-56 at. % Li). In the two phase region between 9 and 47 at. % Li, where the Li-Al

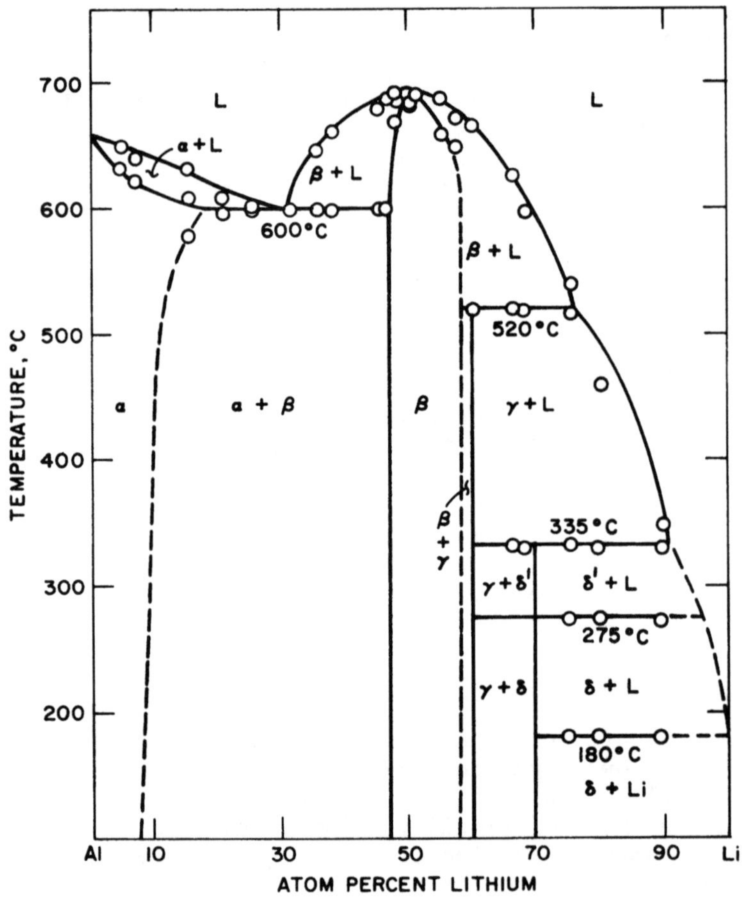


Fig. 1. Phase Diagram of the Lithium-Aluminum Alloy System ($\alpha$ = Al, $\beta$ = LiAl, $\gamma$ = $Li_3Al_2$, $\delta$ = $\delta'$ = $Li_9Al_4$).

electrode is normally cycled, the emf of Li-Al *vs* Li is a constant 0.297 V.[8] Studies of Selman *et al.*[20] have shown that, in going from the two-phase region into β phase, the potential decreases very rapidly as the lithium content increases; the use of Li-Al electrodes containing more than 50 at. % Li is not practical because of the high lithium activity at these concentrations. The theoretical capacity of Li-Al containing 47 at. % Li, which has a density of 1.75 g/cm$^3$ at 425°C, is approximately 0.74 A-hr/g or 1.3 A-hr/cm$^3$.

2. Li-Si Electrode. Negative electrodes utilizing the Li-Si alloy as the active material have also shown promise in Li/MS cells.[10,11] The phase diagram[21,22] of the Li-Si system is not as well defined as that of the Li-Al system. Sharma and Seefurth[21] have reported the compounds present in the phase diagram to be $Li_2Si$, $Li_{21}Si_8$, $Li_{15}Si_4$ and $Li_{22}Si_5$, whereas D. R. Vissers *et al.*[6] and San-Cheng Lai[23] have reported the phases to be $Li_2Si$, $Li_{2.8}Si$, and $Li_5Si$. The techniques used by these three investigators to determine the phases present in the Li-Si electrode differed significantly: Sharma and Seefurth examined silicon after a single charge; Lai examined Li-Si alloys after a single discharge; and D. R. Vissers *et al.* examined Li-Si/Li cells after cycling (with correction for irreversible losses through corrosion of the electrode's nickel and SS current collectors). The plateau emf's for the four Li-Si phases--333, 280, 155 and 46 mV *vs* Li at 425°C--showed good agreement among the three groups of investigators.

The Li-Si system has a very high theoretical capacity, for example, the $Li_{22}Si_5$ alloy has a theoretical capacity of 2.01 A-hr/g, which is more than twice that of the Li-Al alloy. However, these high theoretical capacities[10] have not been attained in Li-Si/$FeS_x$ engineering cells due to two factors: silicon has escaped from the electrode owing to corrosion of the current collector, and a high polarization of the Li-Si electrode developed during the charging.[24]

Of the four Li-Si compounds formed during charging, two have potentials within 155 mV of that for lithium alone; consequently, if even small polarization develops in the Li-Si electrode, these two Li-rich compounds are difficult to form electrochemically without producing free lithium. The Li-Al electrode does not have this problem because the Li-Al potential is nearly 300 mV more positive than the Li potential.

Further studies with the Li-Si electrode systems are needed to (1) develop a suitable corrosion-resistant current collector, (2) investigate techniques to reduce the electrode polarization during charging and (3) identify the compounds formed in the electrode.

## C. Positive Electrode

Early attempts to utilize elemental sulfur electrodes were beset with a number of practical difficulties,[18,19,25] the main one being the containment of the sulfur in the electrode. Displacement of the sulfur from the electrode support material and the formation on discharge of polysulfides[26,27] that are soluble in the electrolyte were the principal mechanisms for sulfur escape from the electrode. Investigations at ANL[12] and at Rockwell International[28] have demonstrated that transition metal sulfides can be used rather than elemental sulfur as the active material in the positive electrode. The metal sulfide electrodes prevent, or at least greatly reduce, the formation of soluble polysulfides, and also eliminate the problems of the sulfur displacement and the rather high vapor pressure of sulfur at cell operating temperature (77 kPa at 425°C).[29]

Numerous compounds were tested as positive electrode materials, including sulfides of nickel, copper, cobalt, and iron. Of these, FeS and $FeS_2$ were selected on the basis of their performance, cost, and availability. The metal sulfide electrodes, like the lithium-aluminum electrode, consist of a porous bed of the material saturated with LiCl-KCl electrolyte.

The evolution of the lithium/sulfur cell to the lithium-aluminum/iron sulfide cell has made possible the development of long-lived, practical cells; however, this was accomplished at the expense of cell performance. The performance characteristics as illustrated by the theoretical specific energy values and cell voltages are shown for the respective systems in Table 3.[30,31] The theoretical specific energies of the lithium-aluminum/metal sulfide cells, while being significantly lower than that of the lithium/sulfur cell, appear to be adequate to meet the performance goals mentioned earlier.[7]

1. $FeS_2$ Electrodes. Although the discharge of the $FeS_2$ electrode can be considered, for practical purposes, as a two step process ($FeS_2 \rightarrow FeS \rightarrow Fe$), the actual reaction sequence is much more complex. Typical charge and discharge curves (cell voltage as a function of capacity) for a Li-Al/$FeS_2$ cell are shown in Fig. 2. The discharge of the cell takes place along two voltage plateaus at average voltages of about 1.55 and 1.20 V. Cells of this type usually utilize IR-included charge and discharge cutoff voltages of 2.1 and 1.0 V, respectively.

The electrochemical reactions that occur during the charging and discharging of $FeS_2$ electrodes at cell operating temperature have been investigated at ANL. The findings of these studies have been incorporated into the 450°C phase diagram[32,33] shown in Fig. 3.

Table 3. Theoretical Performance Characteristics of Cell Systems.[30,31]

| Cell | Theoretical Specific Energy,[a] W-hr/kg | Theoretical Potential, V |
|---|---|---|
| Li/LiCl-KCl/S | 2600 | 2.2 |
| Li/LiCl-KCl/$FeS_2$ | 1321 | 1.8 |
| Li/LiCl-KCl/FeS | 869 | 1.6 |
| Li-Al/LiCl-KCl/$FeS_2$ | 650 | 1.5 |
| Li-Al/LiCl-KCl/FeS | 458 | 1.3 |

[a]Weight of electrolyte not included in calculation.

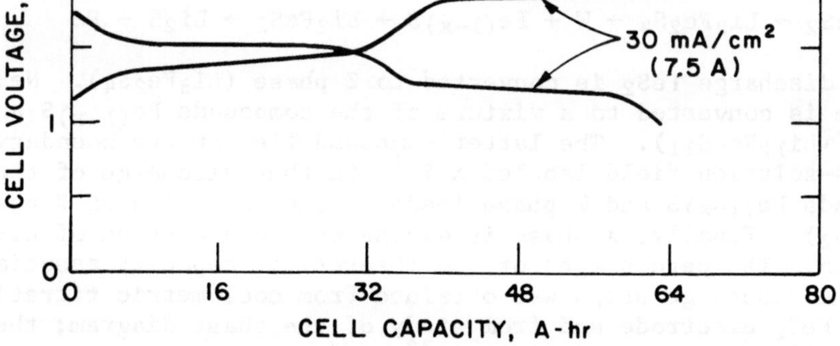


Fig. 2. Voltage *vs* Capacity Curve of Li-Al/$FeS_2$ Cell.

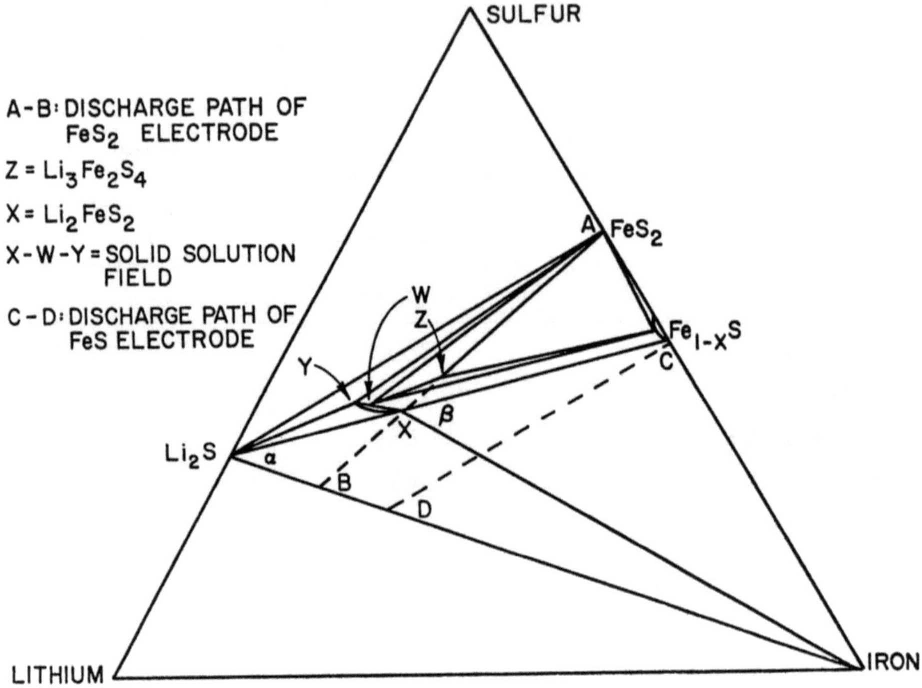


Fig. 3. Isothermal Section of Li-Fe-S Phase Diagram at 450°C

As can be seen in this figure, the $FeS_2$ electrode discharges in four steps:[32]

$$FeS_2 \rightarrow Li_3Fe_2S_4 \rightarrow W + Fe_{(1-x)}S \rightarrow Li_2FeS_2 \rightarrow Li_2S + Fe \qquad (3)$$

During discharge $FeS_2$ is converted to Z phase ($Li_3Fe_2S_4$). Next, Z phase is converted to a mixture of the compounds $Fe_{(1-x)}S$ and W phase ($\sim Li_{12}Fe_4S_{11}$). The latter compound lies on the boundary of a solid-solution field labeled X-Y.* Further discharge of the compounds $Fe_{(1-x)}S$ and W phase leads to the formation of X phase ($Li_2FeS_2$). Finally, X phase is discharged to a mixture of $Li_2S$ and iron. The percentages of the theoretical capacity associated with the discharge steps was obtained from coulometric titrations of the $FeS_2$ electrode and from study of the phase diagram; these results are presented in Table 4.[32]

It should be noted that, in the positive electrode, solid phases other than those shown on the Li-Fe-S ternary diagram can form owing to the presence of LiCl-KCl electrolyte. One such phase is a djerfisherite-like material, $LiK_6Fe_{24}S_{26}Cl$[15,34] (designated J phase), which is a minor intermediate phase in $FeS_2$ electrodes[32]

*The composition of Y-phase is estimated to be $Li_7Fe_2S_6$.

Table 4. Reactions Obtained from Coulometric Titrations and Phase Diagram of $FeS_2$ Electrode

| Reaction[a] | Emf, V *vs* LiAl | Percent Theor. Cap. | |
|---|---|---|---|
| | | Coulometric Titrations | Phase Diagram[a] |
| $FeS_2 \rightarrow Li_3Fe_2S_4$ | 1.74 | 37.2 | 37.50 |
| $Li_3Fe_2S_4 \rightarrow$ W phase + $Fe_{1-x}S$ | ∿1.60 | 4.5 | 4.25 |
| W phase + $Fe_{1-x}S \rightarrow Li_2FeS_2$ | 1.60 to 1.34 V | 7.7 | 8.25 |
| $Li_2FeS_2 \rightarrow Fe + Li_2S$ | 1.34 | 50.6 | 50.00 |

[a] Determined from phase diagram (Fig. 3).

and a major intermediate phase in FeS electrodes.[35] Another is $KFeS_2$, which forms in trace quantities in the $FeS_2$ electrodes.[32]

The phases formed in the $FeS_2$ electrode during charge are more complex than those formed during discharge. The phases present in the $FeS_2$ electrode of six small-scale cells charged to different voltages (from 1.53 to 1.85 V) are presented in Table 5.[33] As shown in this table, the major phases formed during charge are $X \rightarrow Fe_{1-x}S \rightarrow Z \rightarrow FeS_2$; however, there are also minor phases formed with each of the major phases. The $FeS_2$ forms at a potential of 1.85 V, which is about 0.11 V above the reversible potential predicted for the reaction and is formed on $Fe_{1-x}S$ (minor phase). This overpotential is a problem because it is close to the potential at which sulfur forms from $Li_2S$ (1.87 V), as well as to the potential at which FeS oxidizes to a mixture of $FeS_2$ and $FeCl_2$ (1.86 V). Either or both of these reactions may play a key role in the capacity loss with cycling typically observed in Li-Al/$FeS_2$ cells (discussed later)

2. <u>FeS Electrodes</u>. The chemistry and electrochemistry of the FeS electrode have also been investigated at ANL.[33] As indicated in Fig. 4, the voltage *vs* capacity curves for the Li-Al/LiCl-KCl/FeS cell consists of a single voltage plateau at about 1.3 V. The cell reactions deduced from the $Li_2S$-FeS-Fe triangle of the Li-Fe-S phase diagram[32] (Fig. 3) are as follows:

$$2FeS + 2LiAl \rightleftarrows Li_2FeS_2 + 2Al + Fe \tag{4}$$

$$Li_2FeS_2 + 2LiAl \rightleftarrows 2Li_2S + 2Al + Fe \tag{5}$$

These two reactions would appear as one since the Gibbs free energy of formation[36] ($\Delta G_f$) of $Li_2FeS_2$ from $Li_2S$ and FeS is less than 1 kcal/mol. However, metallographic examinations of positive electrodes taken from Li-Al/FeS cells after cycling (see Table 6[37]) have indicated that the major voltage plateau corresponds to the formation of J phase. The J phase is believed to be formed chemically by an interaction of the electrolyte and the active material present in the positive electrode at different states of charge or discharge as follows:

$$23Li_2FeS_2 + Fe + 6KCl \rightleftarrows J + 5LiCl + 20Li_2S \tag{6}$$

$$3Li_2FeS_2 + 20FeS + Fe + 6KCl \rightleftarrows J + 5LiCl \tag{7}$$

The positive-electrode reactants in Eqs. (6) and (7) are represented in triangle α and β, respectively, of the Li-Fe-S phase diagram (Fig. 3). To further complicate matters, metallographic examination of FeS electrodes showed that the surface of an FeS particle followed a different discharge path than its interior.[38] The sequence of reactions is given below:

Table 5. Phases Formed in the $FeS_2$ Electrode on Charge

| Cell Designation | Charge Potential *vs* LiAl, V | X-Ray Findings: Major Phase | X-Ray Findings: Minor Phase | Metallographic Findings |
|---|---|---|---|---|
| Z-1 | 1.53 | X | J | X + trace of J |
| Z-9 | 1.64 | $Fe_{1-x}S$ | Y and X, Z possible | X + $Fe_{1-x}S$ some Y and Z |
| Z-2 | 1.72 | Z | - | Z |
| Z-7 | 1.786 | Z | - | Z + 5% $Fe_{1-x}S$ |
| Z-3 | 1.82 | Z | $FeS_2$ + $Fe_{1-x}S$ | |
| Z-8 | 1.85 | $FeS_2$ | $Fe_{1-x}S$ | $FeS_2$ + trace $Fe_{1-x}S$ and Z |

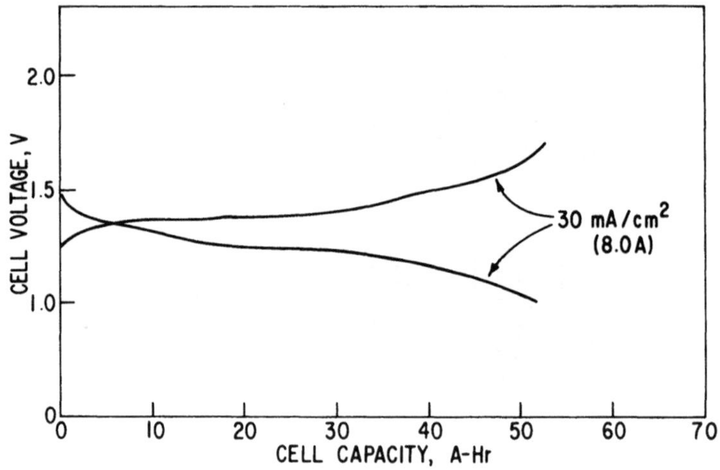


Fig. 4. Voltage *vs* Capacity Curves of Li-Al/FeS Cell.

Table 6. Phases Present in FeS Electrodes (temperature 450°C)[37]

| Condition | Voltage *vs* LiAl, V | Phases Present |
|---|---|---|
| Discharged | 1.0 | $Fe + Li_2S$ |
| Partially Charged | 1.35-1.50 | $Fe+Li_2FeS_2+LiK_6Fe_{24}S_{26}Cl$ |
| Nominally Charged | 1.50-1.60 | $Fe+LiK_6Fe_{24}S_{26}Cl$ |
| Fully Charged | 1.63-1.70 | FeS |
| Over Charged | >1.90 | $K_xFeCl_{x+2}+S+FeS_2$ |

(surface) $6Li + 26FeS + 6KCl \rightarrow LiK_6Fe_{24}S_{26}Cl + 2Fe + 5LiCl$ (8)

(interior) $2Li + 2FeS \rightarrow Li_2FeS_2 + Fe$ (9)

(interior) $2Li + Li_2FeS_2 \rightarrow 2Li_2S + Fe$ (10)

(surface) $46Li + LiK_6Fe_{24}S_{26}Cl + 5LiCl \rightarrow 26Li_2S + 24Fe + 6KCl$ (11)

The overall reaction, however, is the simple two-electron process,

$$2Li + FeS \rightarrow Fe + Li_2S \quad (12)$$

Since the formation of J phase was thought to hinder the kinetics of the FeS electrode, investigators at ANL sought methods to inhibit J phase formation. For the past several years, $Cu_2S$ has been added to the positive electrode of Li-Al/FeS cells to retard the formation of J phase; however, post-test examinations of cells of this type at ANL have indicated that cell failure is sometimes caused by deposition of metallic copper in the separator.[1,2] Small-scale cell tests[38] and thermodynamic studies[36] at ANL showed that J phase stability can be decreased in Li-Al/FeS cells by increasing the LiCl content above that of the eutectic electrolyte or operating the cell at high temperatures (>450°C). Therefore, Saboungi and Martin[39] made a careful metallographic examination of the products of reactions (6) and (7) in eutectic electrolyte and LiCl saturated electrolyte to determine the temperatures above which J phase does not form. These temperatures for reaction (6) are 455 ± 4°C for the eutectic and 419 ± 2°C for the electrolyte saturated with LiCl. For reaction (7) the temperatures are 623 ± 7°C for the eutectic and 481 ± 5°C for the electrolyte saturated with LiCl. These results clearly indicated that the LiCl content of the electrolyte and the temperature play a key role in limiting the formation of J phase in a LiAl/FeS cell. In tests of small-scale FeS cells,[40] the active material utilization of FeS electrodes was found to improve markedly as the LiCl content of the electrolyte was increased above that of the electrolyte[40] (see Table 7). It should be noted that shifting the electrolyte composition toward the LiCl-rich region, while reducing the amount of J phase formed, will also tend to reduce electrolyte polarization and partial electrolyte solidification[41,42] because a larger fraction of the current will be carried by $Li^+$ ions, thereby reducing the build-up of $K^+$ at the negative electrode during charge and at the positive electrode during discharge.

Overcharging of the FeS electrodes results in the formation of $FeCl_2(K_xFeCl_{x+2})$, elemental sulfur, and $FeS_2$.[43] A maximum charge

Table 7. Utilization of Positive Electrode in LiAl/LiCl–KCl/FeS Cells[40]

| LiCl Concentration, mol % | Temp., °C | Current Density mA/$cm^2$ | | FeS Electrode Utilization, % |
|---|---|---|---|---|
| | | Charge | Discharge | |
| 53 | 450 | 50 | 50 | 25 |
| | 500 | 50 | 50 | 44 |
| | 450 | 100 | 100 | 14 |
| | 500 | 100 | 100 | 40 |
| 58[a] | 450 | 50 | 50 | 52 |
| | 500 | 50 | 50 | 70 |
| | 450 | 100 | 100 | 44 |
| | 500 | 100 | 100 | 55 |
| 63 | 450 | 50 | 50 | 74 |
| | 500 | 50 | 50 | 77 |
| | 450 | 100 | 100 | 68 |
| | 500 | 100 | 100 | 71 |
| 67 | 450 | 50 | 50 | 91 |
| | 500 | 50 | 50 | 90 |
| | 450 | 100 | 100 | 85 |
| | 500 | 100 | 100 | 86 |

[a]Eutectic composition.

cutoff of 1.65 V is recommended for the Li-Al/LiCl-KCl/FeS cells to prevent electrochemical oxidation of the iron current collector presently used in the FeS electrode. This cutoff voltage is much below the potential at which FeS would be overcharged.

## III. MATERIALS AND COMPONENTS

The highly reactive materials (lithium metal alloys, iron sulfides, molten salt electrolyte) present in the lithium/metal sulfide cells and the high temperatures at which the cell operates place severe restrictions on the materials that can be used for cell components (*i.e.*, housings, electrical feedthroughs, insulators, separators, particle retainers, current collectors). The need to use low-cost materials in the cells places further restrictions on the available materials. Consequently, a major effort is being carried out by ANL to select, develop, and evaluate materials for low-cost cell components.

### A. Cell Housings and Current Collectors

In the present Li/MS cell design, the Li-Al electrodes are in direct contact with the cell housing, which must therefore be compatible with both the electrolyte and Li-Al alloys of varying composition. Both low-carbon and stainless steels have been found in corrosion studies to be suitable materials for cell housings.[44] Most of the recent Li/MS cells utilize the AISI-1008 or -1010 steel. When cell operating temperature approaches 500°C, however, these steels sometimes undergo slow attack by the aluminum in the negative electrode to form iron-aluminum and nickel-aluminum intermetallic compounds.[44] When cells are operated at 450°C, however, this reaction is not significant. The cell housings must also be corrosion resistant to the air environment. Stainless steel has been shown to be more corrosion resistant to the air environment than low-carbon steel; nevertheless, at ANL, cells with low-carbon steel housings have been operated in the air environment for thousands of hours without deleterious effects to the housing. Corrosion-resistant paints have also been developed for coating the low-carbon steel.

A wide variety of current collector designs has been used in Li/MS cells, including flat sheets, honeycomb structures, foamed structures, and pressed or woven configurations. Nickel and stainless steels and low-carbon steels are potential current collector materials for the negative electrode. Owing to their good electrical conductivity, low cost, and ease of fabrication, the low-carbon steels appear to be the most promising current-collector materials for the negative electrode.

The FeS and $FeS_2$ electrodes require more effective current collectors than those of the Li-Al electrodes because they do not form highly conductive structures. Unfortunately, the high sulfur activities and molten LiCl-KCl present in the porous matrices of the positive electrode create a very corrosive environment for the metallic current collectors. Therefore, static corrosion tests were conducted at ANL on a number of metallic materials to measure their corrosiveness in equal-volume mixtures of FeS and $FeS_2$ in LiCl-KCl electrolyte at 400, 450 and 500°C.[44]

The test results for the metallic materials in the FeS electrode environment are summarized in Table 8. With the exception of Armco electromagnetic iron and AISI-1008 steel, all of the test materials have acceptably low reaction rates. In the Armco electromagnet iron, this intergranular penetration, often five grains deep on the intact portion of the samples, resulted in significant grain fallout, thereby producing the high corrosion rates. This severe attack appeared relatively insensitive to temperature, being ∿450 μm/yr in the 400 to 500°C temperature range. The AISI-1008 steel (0.10% C max, 0.25-0.50% Mn, 0.04% P max, 0.05% S max) also showed the same level of susceptibility to this form of attack. Over the past several years, the AISI-1008 steel has been commonly used in FeS cells, and postoperative examinations were conducted on a number of such cells operated at or near 450°C for 17 days to 1.5 years. These examinations indicated a corrosion rate of approximately 90 μm/yr, a factor of four lower than that observed in the static tests. The reason for the significant differences in corrosion rates between the static and in-cell corrosion tests is not completely understood; one hypothesis is that FeS itself is present in the positive electrode for only a fraction of the time during electrochemical cycling.

The static test results obtained for the $FeS_2$ electrode environment are summarized in Table 9;[44] only molybdenum has an acceptably low corrosion rate. Over the past several years, molybdenum has been used as the current collector material in Li/MS cells with excellent results, but it is expensive and difficult to fabricate into the current collector design.

Carbon has been added to the positive electrode to provide added current collection by "carbon-bonding"; this structure is formed by heat treating a mixture of FeS or $FeS_2$, carbon powder, carbon cement, and a volatile material such as $(NH_4)_2CO_3$ that can be vaporized to form pores in the electrode structure.[1,2]

Corrosion-resistant coatings[2] on metallic current collectors for the iron sulfide electrodes are also receiving some attention at ANL. For the $FeS_2$ electrodes, molybdenum plating and the

Table 8. FeS/LiCl-KCl Compatibility Test Results

| | Corrosion Rate, μm/yr | | | |
|---|---|---|---|---|
| Material | 400°C | 450°C | 500°C | Remarks |
| Molybdenum | 1.8 | 3.1 | 10 | Minor surface attack |
| Nickel | +13 | +10 | +49 | Fe-Ni reaction layer |
| Niobium | +11 | +26 | +24 | Fe-Nb reaction layer |
| Armco iron | 440 | 460 | 410 | Severe intergranular attack |
| AISI 1008 | 620 | - | - | Severe intergranular attack |
| Type 304 SS | 3.8 | 3.6 | 9.0 | Minor surface attack |
| Hastelloy B | 1.8 | 2.5 | +11 | Complex reaction layer at 500°C |
| Hastelloy C | - | 0.9 | +18 | Complex reaction layer at 500°C |
| Inconel 617 | 0.2 | 2.3 | +18 | Minor surface attack |
| Inconel 625 | 0.9 | 2.7 | 4.9 | Minor surface attack |
| Inconel 706 | - | 4.6 | - | Minor intergranular penetration |
| Inconel 718 | - | 5.0 | - | Minor intergranular penetration |
| Incoloy 825 | - | 2.1 | - | Minor surface attack |

Table 9. $FeS_2$/LiCl-KCl Compatibility Test Results[44]

| Material | Corrosion Rate, μm/yr 400°C | 450°C | 500°C | Remarks |
|---|---|---|---|---|
| Molybdenum | 1.4 | +1.0 | +16 | Weakly adherent $MoS_2$ layer |
| Nickel | 2700 | 6600 | 6600 | Very severe attack; $Ni_3S_2$, NiS |
| Niobium | 720 | 2900 | 4500 | Severe attack; intergranular penetration |
| Type 304 SS | 3700 | 6000 | 6000 | Very severe attack; FeS, $FeCr_2S_4$ |
| Hastelloy B | 83 | 490 | 4000 | Intergranular penetration; NiS, FeS |
| Hastelloy C | 160 | 680 | 3200 | Minor intergranular penetration; NiS-FeS |
| Inconel 617 | 210 | 1000 | 1700 | NiS, $NiCo_2S_4$, $Cr_2S_3$ |
| Inconel 625 | 380 | 1000 | 1300 | Probable NiS, $Cr_2S_3$ |
| Inconel 706 | - | 3000 | - | Severe attack; probable NiS, $Cr_5S_6$, $FeCr_2S_4$ |
| Inconel 718 | - | 2400 | - | Severe attack; $Cr_5S_6$, $Ni_3S_2$ |
| Incoloy 825 | - | 3100 | - | Severe attack; NiS-FeS, $Ni_3S_2$ $FeCr_2S_4$ |

formation of borides and carbides coating on steel substrates are under development. For the FeS electrode, nickel coatings on low-carbon steels are under consideration. Testing of coated substrates at ANL has shown that pinholes or microcracks in the coatings can destroy the corrosion resistance of the current collector; efforts are in progress to overcome this difficulty.

In the selection of the current-collector material for the positive electrodes, one must also consider that the electrodes operate at positive potentials (up to ∿2.1 V *vs* Li-Al for $FeS_2$ and ∿1.6 V for FeS); this could result in the anodic dissolution of the metallic current collector as follows:

$$M + xCl^- \rightarrow MCl_x + xe^- \tag{13}$$

The potential (*vs* LiAl) for the anodic oxidation of several metals in LiCl-KCl electrolyte at 450°C are shown in Table 10.[17,45] As can be seen in this table, several metals (Ni, Fe, Cu, Co) have acceptable oxidation potentials for use in the FeS electrode, but only molybdenum has an oxidation potential compatible with the $FeS_2$ electrode. Other materials, not mentioned in the table but which are compatible at the $FeS_2$ electrode potential, are tungsten and carbon, but neither is practical as the primary current collector material in $FeS_2$ electrodes.

Table 10. Oxidation Potentials of Various Metals in Molten LiCl-KCl[17,45] (temperature, 450°C; reference electrode, Li-Al)

| Material | Potential, V | Probable Reaction |
|---|---|---|
| Mo | 2.40 | $Mo^\circ \rightarrow Mo^{3+}$ |
| Ni | 2.21 | $Ni^\circ \rightarrow Ni^{2+}$ |
| Cu | 2.05 | $Cu^\circ \rightarrow Cu^{+}$ |
| Co | 2.01 | $Co^\circ \rightarrow Co^{2+}$ |
| Fe | 1.83 | $Fe^\circ \rightarrow Fe^{2+}$ |
| Cr | 1.58 | $Cr^\circ \rightarrow Cr^{2+}$ |
| Al | 1.34 | $Al^\circ \rightarrow Al^{3+}$ |

B. Separator and Feedthrough Components

Ceramics are used as the electrode separator and the insulator components of the feedthrough. The high chemical activity of lithium in the Li-Al alloy precludes the use of many oxide ceramics. The free energy of formation ($\Delta G_f$) of several oxide ceramics and that of $Li_2O$ is presented in Table 11,[46,47] which indicates that only $Y_2O_3$, CaO, BeO and MgO are thermodynamically stable relative to $Li_2O$. Table 12 shows the $\Delta G_f$ for several nitride ceramics and lithium nitride;[46-48] all of the ceramic nitrides are thermodynamically stable relative to $Li_3N$.

Static corrosion tests have been conducted at ANL on various ceramic materials in molten lithium and in equal volume mixtures of Li-Al and LiCl-KCl eutectic.[44] The materials tested generally were hot-pressed to densities greater than 90 percent of the theoretical value. There was no qualitative difference between the corrosiveness of the ceramic materials tested in solid Li-Al alloy plus electrolyte and those tested in molten lithium, although the extent of corrosion was reduced by the use of the Li-Al plus electrolyte. Therefore, preliminary screening tests were often conducted in molten lithium only. The corrosion rate results for some of the ceramic materials are summarized in Table 13. Analyses of the test results indicated that failure resulted from one of the following causes: (1) thermodynamic instability, (2) impurities, and (3) formation of conductive surface layers. The corrosion resistance of thermodynamically stable ceramics is extremely sensitive to the presence of impurities. For example, impure BeO failed due to lithium attack of $SiO_2$ in the grain boundaries, whereas high purity BeO showed excellent compatibility with lithium. In addition, lithium attack on commercial BN, which normally contains about 4 percent $B_2O_3$, resulted in the formation of a black layer. However, BN that had been heat-treated at 1700°C in flowing nitrogen to remove the $B_2O_3$ showed good compatibility with lithium. Some materials, such as $CaZrO_2$ and $ZrO_2$, formed an electrically conductive surface layer that makes them unsuitable for use as insulator material. The results of these tests indicate that only BeO, $Y_2O_3$, MgO, BN, and AlN are suitable ceramic materials for the negative electrode; of these, BN and $Y_2O_3$ are the only ceramics presently available as fibers.

Many of the large engineering cells built in 1978 and 1979 had woven BN cloth separators;[1,2,49] some of these cells were successfully operated for periods as long as 8500 hr. However, the projected cost for mass production and thickness (∿2 mm) of this material precludes its use in commercial cells. In addition, the interstices of BN fabric are too large for it to serve as an effective particle retainer for the two electrodes. Cells with BN cloth separators usually have particle retainers of $Y_2O_3$ felt for LiAl and

Table 11. Free Energies of Formation of Selected Oxides at 427°C (700 K)[46,47]

| Compound | $-\Delta G_f^\circ$, kJ/g-at. O | Compound | $-\Delta G_f^\circ$ kJ/g-at. O |
|---|---|---|---|
| $Y_2O_3$ | 564 | $Li_2O$ | 509 |
| CaO | 564 | $ZrO_2$ | 482 |
| BeO | 530 | $Al_2O_3$ | 481 |
| MgO | 526 | $SiO_2$ | 392 |

Table 12. Free Energies of Formation of Selected Nitrides at 427°C (700 K)[46-48]

| Compound | $\Delta G_f^\circ$, kJ/g-at. N | Compound | $\Delta G_f^\circ$, kJ/g-at. N |
|---|---|---|---|
| ZrN | 298 | BN | 189 |
| TiN | 271 | $Si_3N_4$ | 128 |
| AlN | 243 | $Li_3N$ | 66 |

Table 13. Compatibility of Ceramic Materials With Lithium at 400°C[44]

| Material | Results |
|---|---|
| $CaZrO_3$, sintered or hot-pressed | Weight gain (<2 pct); discolored; conductive surface layer; intergranular reaction; no mechanical deterioration (200 to 4200 h) |
| $Y_2O_3$, sintered or hot-pressed | Slight discoloration and weight loss in 200 to 2550 h; nonconductive |
| $ZrO_2$ (CaO stabilized), sintered | Severe discoloration (black); conductive; mechanical failure |
| $Al_2O_3$, sintered | Severe discoloration (black); conductive; disintegraged |
| MgO, single crystal | Slight weight loss; no apparent reaction; nonconductive |
| MgO, hot-pressed | Severe weight loss (>10 pct); slight discoloration; nonconductive |
| $MgAl_2O_4$, hot-pressed | Complete disintegration to a powder |
| SiAION ($Si_3N_4$ + $Al_2O_3$) | Sample severely cracked |
| $Si_3N_4$, hot-pressed | Sample severely cracked |
| BN, untreated | Severe discoloration; conductive surface; some disintegration in <100 h |
| BN, hot-pressed, pretreatment at 1700°C in $N_2$ | Slight discoloration; nonconductive (1300 h) |
| BeO, high purity | No apparent reaction |
| SiC, hot-pressed | Disintegrated |
| SiC• AlN, hot-pressed | Circumferential surface fracture; nonconductive |
| SiC•2 AlN, hot-pressed | Sample fragmented; nonconductive |

FeS electrodes, and $ZrO_2$ felt for $FeS_2$ electrodes ($Y_2O_3$ felt cannot be used on $FeS_2$ electrodes because it tends to react with the active material to form $Y_2O_2S$[1,2]). Many recent cells are being built with separators of BN-bonded BN felt[50,51] or MgO powder,[1,2] which are amenable to low-cost mass production. Yttria felt has also been used successfully in some cells,[1,2] but its projected cost is too high for use in commercial cells. The small interstices of the felt and powder separators permit these materials to also serve as particle retainers. Boron nitride felt has been developed over the past two years by Carborundum Company;[1,2] this felt has a basic weight of 10-13 $mg/cm^2$ per millimeter of thickness, a thickness of 0.7 to 3 mm, 90 to 93% porosity, and a burst strength of 5-6.5kPa/mm. Projected cost estimates for the BN felt range from \$13.45 to 18.84/$m^2$, depending on thickness and density, at production rates of about 5000 kg/yr. Additional efforts by Carborundum are being devoted to further improve the BN felt. The development of powder separators,[1,2,11,13] such as MgO, AlN and CaO, is being pursued because separators of this type are amenable to low-cost mass production and permit the use of materials which are not available in fibrous form. The major disadvantage of powder separators is that they have a lower porosity than that of felt separators (∿40 *vs* 92%), which tends to restrict ionic transport through the separators at high current densities. The low projected cost of a MgO powder separator is very attractive--less than \$5.38/$m^2$. Efforts are continuing on the development, characterization, and optimization of powder separators.[1,2]

The present feedthrough being used in the Li/MS cell is a mechanical compression type,[52] which is similar in principle to a Conax thermocycle feedthrough. The feedthrough utilizes a compacted BN powder sealant, a lower solid insulator of BeO, and an upper solid insulator of $Al_2O_3$.[1,2] The expected cost of this feedthrough is less than \$1.25/unit in quantities of about 100,000 (representing scale mass production), and should meet the cost goals for commercial cells.

## IV. CELL DEVELOPMENT AND TESTING

The first sealed engineering Li/MS cell was built at ANL in 1974 and had a cylindrical design,[53,54] illustrated in Fig. 5. The cells were 13 cm in diameter and 3 to 5 cm thick, weighed from 1.0 to 1.7 kg, and had capacities of 112-150 A-hr. This design consisted of a horizontal central positive electrode of $FeS_2$ and facing negative electrodes of Li-Al; this electrode arrangement will hereafter be referred to as the "bicell design." In this cell design, the cell case is used as a current collector for the negative electrodes (thus only a single feedthrough is required); the lithium-aluminum electrode has stainless steel screen particle retainers; and the

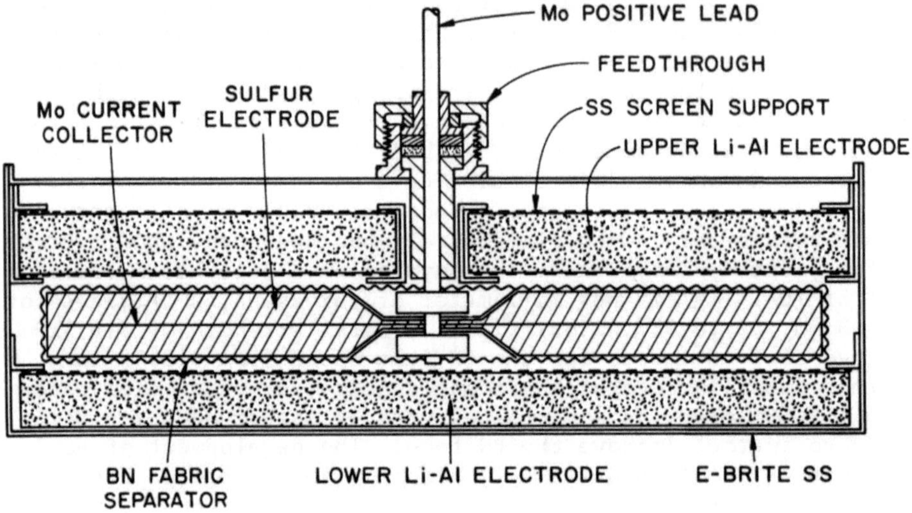


Fig. 5. Cylindrical Bicell Design

$FeS_2$ material in the positive electrode is enclosed by a disk-shaped basket of molybdenum-expanded metal mesh to give the electrode a rigid structure and to serve as a current collector. A central layer of molybdenum mesh was also used for current collection; these mesh layers were attached to a molybdenum rod that extended through an electrical feedthrough to serve as the positive terminal. The $FeS_2$ electrode was surrounded by a BN cloth separator with an inner layer of $ZrO_2$ cloth to serve as a particle retainer. Both the housing and cover were E-Brite 26-1 stainless steel. The cell used a compression-type feedthrough, with $Y_2O_3$ powder as the sealing agent.

A series of cells of this design were constructed and tested. The lithium-aluminum electrodes were prepared either by a single-step electrochemical formation procedure or by hot pressing lithium-aluminum powder. In most of the cells the capacities of the positive and negative electrodes were matched at 150 A-hr. The performances of these $LiAl/FeS_2$ cells are summarized in Table 14.[53,54] The maximum specific energy, 155 W-hr/kg, was achieved at ~10-hr rate by cell W-6, specific energies of 120-140 W-hr/kg being typical for this cell series. In general, the energy efficiency was between 70 and 80 percent and the coulombic efficiencies were as high as 100 percent. However, the cycle life and capacity retention of these cells were quite poor.

Table 14. Performance Data for Cylindrical $Li-Al/FeS_2$ Cells[53,54]

| Cell No. | No. of Cycles | Time, hr | Specific Energy, W-hr/kg Maximum | Specific Energy, W-hr/kg Typical | Energy Efficiency, % |
|---|---|---|---|---|---|
| W-2 | 41 | 1390 | 121 | 65-100 | 70 |
| W-4 | 24 | 600 | 128 | 100-125 | 78 |
| W-5 | 111 | 2137 | 155 | 110-135 | 81 |
| W-6 | 248 | 3300 | 150 | 120-140 | 76 |
| W-9 | 76 | 1511 | 98 | 80-90 | 81 |
| W-10 | 113 | 2688 | 128 | 110-120 | 78 |
| W-11 | 43 | 1128 | 122 | 100-110 | 82 |
| W-12 | 20 | 150 | 50 | 30-40 | 74 |
| W-14 | 107 | 1300 | 85 | 75-85 | 78 |

General Motors Research Laboratories is presently developing $Li-Si/FeS_2$ cells[10] with a similar design to that of the ANL W-series cells. Their cells have achieved a specific energy of 182.7 W-hr/kg at an average specific power of 8.35 W/kg. However, at higher specific power levels, relatively high internal resistance and decreased utilization of the active material in the positive electrode reduced the specific energy (*e.g.*, 75.3 W-hr/kg at 96.6 W/kg). These cells have operated for more than 650 deep cycles (7000 hr). The coulombic efficiencies of these cells are approximately 99 percent and their energy efficiencies are between 60-89 percent.

Several cylindrical Li-Al/FeS bicells[53,54] were constructed at ANL. These cells operated up to 121 cycles (3700 hr) and attained a maximum specific energy at the 10 hr rate of 82 W-hr/kg, and a typical value of 40-60 W-hr/kg. The energy efficiency was again in the range of 70-80%.

Several serious problems were observed during the cycling of the cylindrical bicells, including (1) electrode swelling and (2) the molybdenum welds on the current collector in the $FeS_2$ electrode sometimes broke. The distortion of the positive electrodes tended to cause the cell to short circuit due to extrusion of active material; the breaking of the current-collector welds caused the cell resistance to increase several fold. The swelling problem could probably be resolved by mechanical restraint; however, the current-collector problem would require a redesign of the collector to reduce the amount of stress in the weld area.

Although the cylindrical bicell has demonstrated relatively high performance, development efforts in recent years have focussed primarily on the development of vertically oriented, prismatic bicells because they can be stacked more compactly in a battery case, thereby improving the volumetric energy density. The prismatic bicells typically are either 13 x 13 cm or 13 x 18 cm and from 1 to 3 cm thick. A schematic design of a prismatic bicell is shown in Fig. 6.[50] The electrode arrangement is similar to that of the cylindrical bicell, with a central positive electrode and two facing negative electrodes. The components used in the prismatic design,[50] as indicated in Fig. 6, are very similar to those used in the cylindrical bicells.

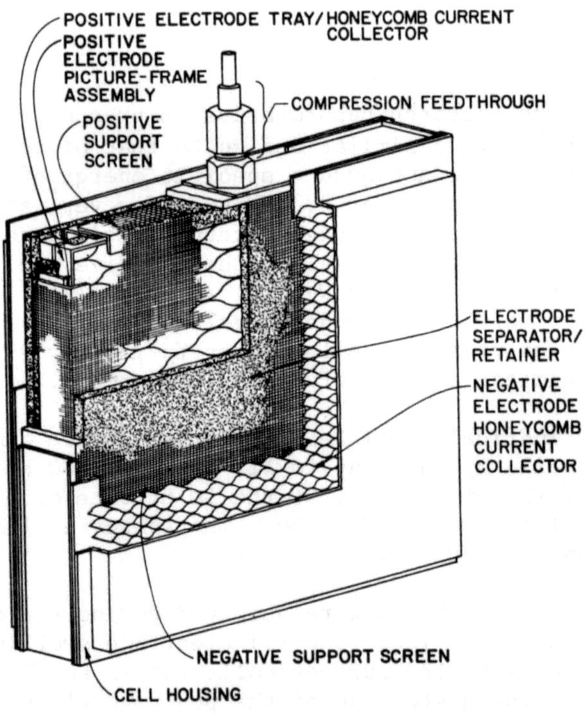


Fig. 6. Prismatic Bicell Design

Prismatic Li-Al/FeS and Li-Al/$FeS_2$ bicells with theoretical capacities of 120-150 A-hr have been fabricated by ANL[49,50] and the following contractors: Eagle-Picher Industries Inc.,[2] Rockwell International,[1,2,11] and Gould Inc.[1,2] These cells have been evaluated for performance and lifetime at ANL and the laboratories of the cell manufacturers. To date, several hundred prismatic bicells of varying designs have been built and tested. The design parameters under investigation included the following: electrode thickness, current collector materials and configurations, electrolyte composition, positive-to-negative electrode capacity ratio, loading densities of active material (A-hr/$cm^3$), separator materials, feedthrough construction, composition of active materials, particle retainers, and cell assembly techniques.

The lifetime and performance characteristics of typical Li-Al/FeS bicells fabricated by ANL, Gould, and Eagle-Picher are shown in Table 15. Many of the Li-Al/FeS bicells operated for 500 to 1200 cycles; however, the specific power of these cells was very low. As shown in Table 16, the Li-Al/$FeS_2$ bicells achieved very high performances--peak specific powers in excess of 170 W/kg at full charge (100 W/kg at 50% charged) and specific energies above 100 W-hr/kg--but the cycle life of these bicells was usually 200 cycles or less. Post-test examination of these test cells indicated that the principal cause of cell failure is extrusion of active material from the positive electrode and deposition of iron in the separator (Li-Al/$FeS_2$ bicells only).

The following observations have been made from the test results: (1) BN felt, $Y_2O_3$ felt, and MgO powder are suitable separator materials, (2) the performance of Li-Al/FeS bicell is markedly improved by using the LiCl-rich electrolyte (67 mol % LiCl), (3) the peak power of the Li-Al/FeS bicell is improved by use of carbon-bonded FeS electrodes, (4) FeS electrode loadings of 1.4 A-hr/$cm^3$ appear to be optimal, and (5) nickel is a suitable current collector for FeS electrodes. The addition of $Cu_2S$ to the positive electrode of Li-Al/FeS cells was found to improve performance, but cell failure was often caused by deposition of this copper in the separator.

In February 1978, ANL awarded Eagle-Picher a contract to develop, design, and fabricate the 40 kW-hr Mark IA battery. On the basis of previous test results from Li-Al/FeS and Li-Al/$FeS_2$ bicells, Eagle-Picher selected a Li-Al/FeS multiplate design (3 positive and 4 negative electrodes) for this battery (see Fig. 7). This cell is approximately 19.5 x 18 cm, with a thickness of 4 cm. This multiplate cell is essentially the same as the electrodes of three bicells incased in a single cell housing, with the appropriate changes in the positive and negative buses as well as in the feedthrough to accommodate the higher current flow.

Table 15. Performance Data of Prismatic Li-Al/FeS Bicells[1]

| Cells | No. of Cells | Ave. Specific Energy, W-hr/kg | Ave. Discharge Rate, hr | Ave. No. of Cycles Operated | Ave. % Decline In Specific Energy Per Cycle |
|---|---|---|---|---|---|
| Eagle-Picher | 5 | 52.0 | 8.0 | 676 | 0.05 |
| ANL | 20 | 53.3 | 3.8 | 240 | 0.06 |
| Gould | 7 | 55.8 | 7.6 | 107 | 0.05 |
| TOTAL | 32 | 53.7 | 5.3 | 279 | 0.06 |

Table 16. Performance Data of Prismatic Li-Al/$FeS_2$ Bicells[1]

| Cells | No. of Cells | Ave. Specific Energy, W-hr/kg | Ave. Discharge Rate, hr | Ave. No. of Cycles Operated | Ave. % Decline in Specific Energy Per Cycle |
|---|---|---|---|---|---|
| Eagle-Picher | 4 | 98.9 | 7 | 146 | 0.21 |
| Gould | 22 | 67.6 | 8 | 48 | 0.19 |
| TOTAL | 26 | 72.4 | 8 | 63 | 0.19 |

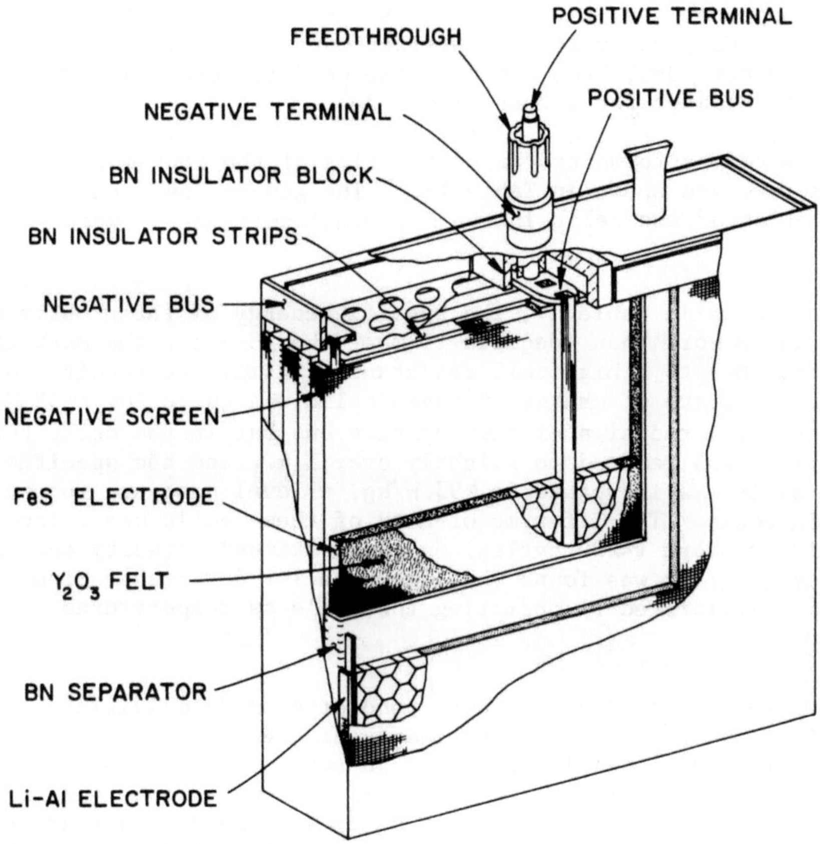


Fig. 7. Prismatic Multiplate Cell Design

In the cell development phase of the Mark IA program, Eagle-Picher fabricated four successive cell series--designated A, B, C and D--each consisting of 30 multiplate cells which were tested both at Eagle-Picher and ANL. The electrodes for these cells had many different designs, including the addition of 0 to 20 wt % $Cu_2S$ to the positive electrode, positive electrode loadings of 335 to 448 A-hr, and lithium to sulfur ratios of 0.89 to 1.29. Some of the other cell design variables included positive electrode current collector design and material (iron and nickel), bus and feedthrough designs, LiCl content of the electrolyte (eutectic and 67 mol % LiCl), and particle retainer materials ($Y_2O_3$ or $ZrO_2$ cloth on both electrodes, $Y_2O_3$ cloth on the positive electrode only, and no cloth retainer for either electrode).

Typical performance characteristics of the four multiplate-cell series are shown in Table 17.[1] The general performance characteristics of the cells in the different cell series were very similar.

As shown in Table 17, the specific energy of these cells was found to be more than adequate to meet the goal for the Mark IA battery. Due to a high cell resistance (∿2 mΩ) the specific power at the 50% state of charge of these cells was quite low (∿55 W/kg); however, by a redesign of the negative bus bar in the cell, this resistance was reduced to slightly over 1 mΩ, and the specific power as a result was increased to ∿95 W/kg, a level adequate to meet the Mark IA goals. The lifetime of many of these cells was found to be relatively short (<200 cycles) due to continued capacity decline with cycling; it was found that the capacity decline problem was greatly ameliorated by operating the cells as temperatures of 475-485°C.

Post-test examinations were conducted on 22 multiplate cells, operation had been terminated because of electrical short circuits in 12 cells and declining capacity in seven cells. The primary cause of cell failure was extrusion of the active material from the positive electrode; the secondary cause was the development of short circuits in the feedthrough. The active material in the negative electrode was found to be badly agglomerated in many of the cells, suggesting that the active material had not been properly utilized; this Li-Al agglomeration could, in part, be responsible for the capacity decline observed in the multiplate cells.

Based on the results of these cell tests and an engineering cell model developed by E. Gay of ANL,[2] Eagle-Picher and ANL selected the final design for the cells to be used for the Mark IA battery, and then Eagle-Picher began fabrication of these cells. The cell design was as follows: The Mark IA cell has dimensions of 19 x 18 x 3.9 cm and a weight of about 4 kg. The negative elec-

Table 17. Performance Data of Prismatic Multiplate Cells

| Cell Designation | No. of Cells | Ave. Specific Energy,[a] W-hr/kg | Ave. Discharge Rate, hr | Ave. No. of Cycles Operated | Ave. % Decline in Specific Energy Per Cycle |
|---|---|---|---|---|---|
| A-series | 7 | 95.4 | 5 | 97 | 0.21 |
| B-series | 18 | 100.5 | 4 | 100 | 0.23 |
| C-series | 13 | 102.4 | 4 | 111 | 0.32 |
| D-series | 24 | 97.2 | 4 | 119 | 0.24 |
| TOTAL | 62 | 99.0 | 4 | 109 | 0.25 |

[a] Determined during the first 24 cycles.

trodes have a theoretical capacity of 440 A-hr in dimensions of 16 x 17 x 0.57 or 0.27 cm. The positive electrodes consist of FeS with 15 wt % $Cu_2S$, and have a theoretical capacity of 410 A-hr in dimensions of 16 x 17 x 0.32 cm. Other cell components include AISI 1010 low-carbon steel current collectors, BN cloth separators, $Y_2O_3$ felt particle retainers on the positive electrodes and type 304 stainless steel screen (200 mesh) on both the positive and negative electrodes, a crimp-type feedthrough with a BN powder seal, and a cell container of AISI 1008 or 1010 low-carbon steel with a material thickness of 0.57 cm.

Approximately 250 multiplate cells were fabricated by Eagle-Picher for the Mark IA battery. Typical performance data obtained from the testing of these cells indicate specific energies at the 4 hr rate in the range of 95 to 100 W-hr/kg and specific powers of ∿95 W/kg at the 50% state of charge. Of these, 120 cells were selected for use in the Mark IA battery and 14 cells were selected as spares. Test results of this battery are given in the next section.

## V. BATTERY DEVELOPMENT

### A. Early Battery Development

In addition to the single-cell tests described above, ANL evaluated the performance of two to six Li/MS bicells connected in series or parallel during 1976-1977.[55] These small-scale battery tests were conducted to identify the problems associated with the development of Li/MS batteries.

Testing of the prismatic-bicell batteries proved that Li/MS batteries could be successfully operated.[55] Two Li-Al/FeS bicells in series operated for the longest period, 756 cycles (11,500 h), and six Li-Al/$FeS_2$ bicells in series obtained the highest specific energy, 87 W-hr/kg for the cells. These batteries had coulombic efficiencies of 87-99 percent, and energy efficiencies of 70-88 percent.

The results of these small-battery tests indicated the need for a suitable charging system. With normal bulk charging, the cell capacities in these batteries[55] were mismatched after several cycles because of the differing coulombic efficiencies of the cells and the limitations imposed in the bulk-charge cutoff voltage to protect the individual cells from overcharge. To avoid the problems associated with simply bulk charging, ANL constructed a charging system containing separate bulk and equalizer chargers that are monitored and controlled by a minicomputer system.[56] In this charger, bulk charging is performed at a constant current rate until one of the battery cells is detected to have reached a predetermined limit by the computer system. Bulk charging is then terminated and operation of equalizer charges is initiated. The individual cells are

Fig. 8. Six-Volt Battery.

then charged slowly to the predetermined cutoff voltage with the charge equalization circuit. For electric-vehicle applications, equalization would be required only occasionally, about once a week.

## B. The Development of the Mark IA Battery

Under the Mark IA contract, Eagle-Picher in 1978 constructed a battery containing five Mark IA-type cells connected in series (6 V) and housed in a thermally insulated case (shown in Fig. 8).[1,49] Heat-loss measurements on the 6-V battery indicated a value of 160 W. The 6-V battery was operated for ∿70 cycles (approximately 1000 hr) at ANL; it achieved a maximum specific energy of 73 W-hr/kg and an energy efficiency of about 75 percent. During a discharge cycle similar to that specified by the SAE J227 cycle C driving profile, this battery achieved a specific energy of ∿56 W-hr/kg, which is very near the goal of 60 W-hr/kg set for the Mark IA battery. An in-vehicle test was also conducted in which the 6-V battery was connected in series to a 144-V lead-acid battery in a Volkswagen Transporter Van. During this test, (1) there was no measurable

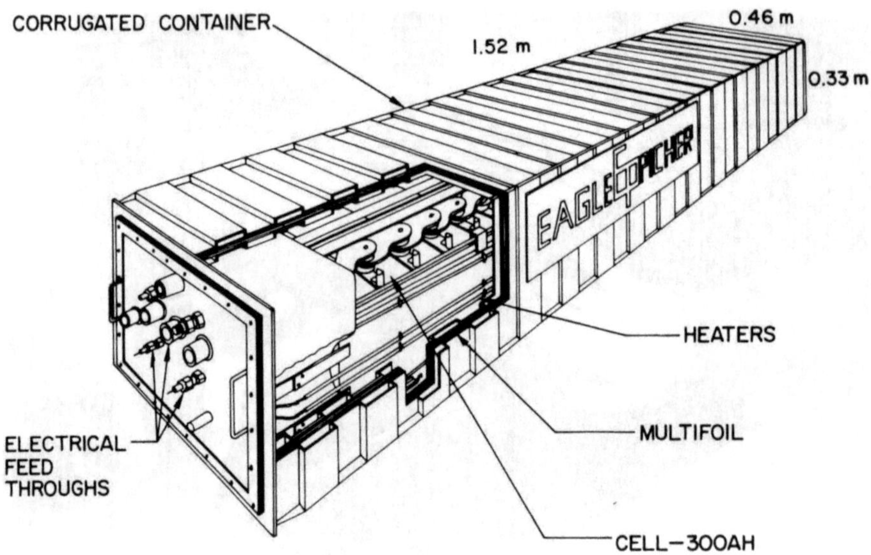


Fig. 9. 20-kW-hr Battery Module

inductive effect of the 6-V battery on the chopped current; (2) the module temperature decreased from 475 to 470°C during one hour of driving; and (3) road vibration and high-current peaks had no measurable effect on the electrical performance of the 6-V battery. Other than its relatively short cycle life, the 6-V battery performed as expected. The battery test was terminated because short circuits developed in the feedthroughs of two of the five cells.

The Mark IA battery, see Fig. 9, consisted of two 20 kW-hr modules; each module contained 60 multiplate cells, in two rows of 30 cells each, connected in series and housed in a thermally insulated case. The thermal case had two corrugated metal walls (Inconel 718) with multilayered foil in the evacuated annulus between the walls. The two module cases were fabricated by the Thermo Electron Corp. and the Budd Company under a subcontract from Eagle-Picher. A dynamic stress analysis gave the expected resonant frequencies of the system as well as the areas of maximum modal stress. The results indicated that structural deflections under operating conditions would be small and well within acceptable limits. The rectangular, corrugated cases were fabricated with some difficulty, primarily because of the nature and length of the welds involved. Cylindrical thermal cases appear to be more easily fabricated, but are less compatible with rectangular cells.

Upon completion of the Mark IA battery, it was delivered to Argonne for testing. The two modules weighed a total of ∿748 kg, and had a theoretical specific energy of 88.8 W-hr/kg and a theoretical voltage at full charge of 162 V. During the initial test, one of the two modules failed due to an internal short circuit. A failure analysis is in progress to determine the exact cause of the short circuit and to evaluate its effects. The other module was unaffected by the event.

## C. Stationary Energy Storage

The specific-energy and specific-power requirements for stationary-energy-storage (SES) battery cells are less demanding than those for the electric vehicle battery, but low cost (about 40 to $50/kW-h) and long life are essential; the ANL goals for the SES battery[57] are listed in Table 18. The size and voltage of the SES batteries being considered in the present conceptual designs by Rockwell International and ANL are in large measure dependent upon the ac-dc converter technology anticipated for the 1990's.[58] These large individual converters will need to operate at 2000 V for high

Table 18. Program Goals for Lithium/Metal Sulfide Stationary Energy-Storage Batteries

| Goal | BEST[a] 1983 | Demonstration 1987 |
|---|---|---|
| Battery Performance | | |
| Energy Output, kW-hr | 5,000 | 100,000 |
| Peak Power, kW | 1,500 | 25,000 |
| Sustained Power, kW | 1,000 | 10,000 |
| Cycle Life | 500-1,000 | 3,000 |
| Discharge Time, hr | 5 | 5-10 |
| Charge Time, hr | 10 | 10 |
| Cell Performance | | |
| Specific Energy, W-hr/kg | 60-80 | 60-150 |
| Specific Power, W/kg | 12-20 | 12-20 |
| Cell Cost, $kW-hr | 30-35[b] | 25-30[b] |

[a]Battery Energy Storage Test Facility. This facility, which is being contructed under joint sponsorship by the U.S. Department of Energy, the Electric Power Research Institute, and the Public Service Co. of New Jersey, will be used to test various types of batteries as load-leveling devices on an electric utility system.

[b]Projected cost for a production rate of 2000 MW-hr/y in 1979 dollars.

efficiency (*i.e.*, 95%) and will have to have power capabilities of about 10 MW. A 100 MW-hr energy-storage plant would thus consist of two converters and two 50 MW-hr battery units. A 50 MW-hr (10 MW) battery unit would consist of 10 sub-units, each housed in a separate thermally controlled enclosure, and contain between 400 and 20,000 cells (depending on cell size). A nominal dc potential of 2000 V would be achieved by interconnecting these sub-units so that some 1500 to 1700 cells would be in series. Depending on the cell size, the 50 MW-hr battery might consist of between one and 60 parallel strings. In one of the present conceptual designs, strings of cells are connected in parallel to form the 50 MW-hr battery. Such strings would require cells having an energy capacity of 2.5 kW-hr, which is about eight times larger than that of the Mark IA electric-vehicle cells discussed earlier.

The design and development of Li-Si/LiCl-KCl/FeS cells for SES systems are being carried out by the Energy Systems Group of Rockwell International.[2] The cell presently being developed for this application is a 2.5 kW-hr multiplate cell having 16 electrode pairs (23 x 23 cm) and an AlN powder separator.[2] Figure 10 is a photograph of the first such cell built by Rockwell International. Cells of this design have operated for 33 cycles (600 hr) and attained an energy storage capacity of 2.24 kW-hr. At the present time these cells are being redesigned to improve the performance characteristics of the AlN powder separator. The testing of large Li-Si/FeS cells at Rockwell International indicates that these cells have potential as SES devices. However, it may require a decade or more of experience with constructing these cells before a realistic assessment can be made of their economic potential.

## VI. COMMERCIALIZATION

To achieve successful commercialization of the lithium/metal sulfide battery, industry must be convinced that the battery is (1) technically feasible, (2) capable of meeting governmental requirements, and (3) economically viable.

In regard to the first condition, the Mark II tests of the electric-vehicle battery and the BEST facility tests of the stationary energy-storage battery are expected to establish the technical feasibility of the system.

The second condition involves the safety aspects of the battery, environmental and health considerations, and resource utilization. A preliminary assessment of the safety aspects of the lithium/metal sulfide battery has been conducted by the Budd Co.[59] In these tests, cells at operating temperature in air have been deliberately ruptured. The results of these tests confirmed theoretical predictions of chemical stability of the cell when exposed to air (no explosions

Fig. 10. Load-Leveling Cell (2.5 kW-hr)

or fires); in addition, little of the molten electrolyte escaped. These results are highly encouraging with respect to the safe deployment of lithium/metal sulfide batteries in electric vehicles. Institutional agencies with responsibilities in resource surveillance and management are, of course, interested in the effect of the mass-production of lithium/metal sulfide batteries on the world's lithium resources. The projected lithium requirements[2] for SES and electric-vehicle Li/MS batteries by the year 2000 is 70,000 MW-hr, which will require a substantial fraction of the known lithium reserves--about 15% of U.S. reserves (class A, B, and C) or about 5% of the world reserves (class A, B, and C). Although adequate domestic and world-wide reserves appear to be available up until the year 2000, the cost of the resource as a function of large volume production remains in question, and appears to be resolvable only as part of an actual commercial effort.

In regards to the third condition, the initial market to be penetrated by the lithium/metal sulfide batteries will probably be for limited markets such as postal vans, buses, minining vehicles and submarines.[2] Another near-term market of interest for the Li/MS battery is a high-performance battery for use in fork-lift trucks for the U.S. Army; Eagle-Picher Industries Inc. has been recently contracted to deliver several small battery modules to Fort Belvoir (MERADCOM) for test evaluation. In these near-term (1982-1990) markets the relatively high prices of the batteries should be offset by their favorable performance characteristics. A projection of the size of these near-term markets is shown in Table 19.[2] The van market after 1987, assuming that automated battery plants can produce Li/MS batteries at about $50/kW-hr, can be very large. An estimate has been made that this market could support 300,000 battery-powered vans per year, which would require 18,900 MW-hr/yr of plant capacity.

Table 19. Pilot Markets for the Lithium/Metal Sulfide Battery

| Price,[b] \$/kW-hr | Market Size,[a] MW-hr/yr | Markets | Year Initiated |
|---|---|---|---|
| 300 | 85 | Submersibles | 1982 |
| 200 | 119 | Postal vans | 1982 |
| 170 | 197 | Postal vans | 1982 |
| 150 | 441 | Postal vans, buses, mining vehicles | 1982 |
| 50 | 580 | Buses | 1987 |
| 50 | 18,900 | Vans | 1987 |
| 50 | 934 | Buses | 1990 |

[a]The market size shown assumes that the Li/MS battery captures the entire market. The effect of market-sharing because of competition from other advanced batteries will need to be considered as part of an overall business plan.

[b]Projected price in 1979 dollars.

## VIII CONCLUSIONS

Over the past seven years, considerable progress has been made in the development of Li/MS batteries. During this time, hundreds of engineering cells having different designs have been tested, some of these cells achieving specific energies >100 W-hr/kg, others operating for >1000 cycles. Development work on large-scale batteries is under way and appears very promising, even though problems were encountered in the first such battery (Mark IA). Although the present research and development program is oriented toward batteries for electric-vehicle propulsion and stationary energy storage, this technology has reached the stage where it is receiving consideration for use for other purposes where high specific power and energy are required and cost is a less important consideration. For example, batteries of this type are now being evaluated by the U. S. Army for forklift trucks and are being considered for aerospace, military, and communications applications.

ACKNOWLEDGMENTS

The author is grateful to L. Burris, D. S. Webster, D. Barney and R. K. Steunenberg for support and encouragement. The author would also like to acknowledge J. Harmon for his editorial assistance. This work was performed under the auspices of the U. S. Department of Energy.

REFERENCES

1. P. A. Nelson *et al.*, High Performance Batteries for Electric-Vehicle Propulsion and Stationary Energy Storage: Progress Report for the Period October 1978-March 1979, Argonne National Laboratory Report ANL-79-39 (May 1979).
2. P. A. Nelson *et al.*, High Performance Batteries for Electric-Vehicle Propulsion and Stationary Energy Storage: Progress Report for the Period October 1977-September 1978, Argonne National Laboratory Report ANL-78-94 (November 1978).
3. P. A. Nelson, A. A. Chilenskas, and R. K. Steunenberg, Proc. Fifth International Electric Vehicle Symposium (E.V. Council, New York), Paper No. 783104, (1978).
4. M. L. Kyle, E. J. Cairns, and D. S. Webster, Lithium/Sulfur Batteries for Off-Peak Energy Storage: A Preliminary Comparison of Energy Storage and Peak Power Generator Systems, E.R.D.A. Report No. ANL-7948, Argonne National Laboratory (1973).
5. P. A. Nelson and N. P. Yao, Proc. Amer. Power Conf. 38, 1381 (1976).
6. P. A. Nelson *et al.*, High-Performance Batteries for Off-Peak Energy Storage and Electric Vehicle Propulsion: Progress Report for the Period July-September 1976, Argonne National Laboratory ANL-76-98 (December 1976).
7. P. A. Nelson *et al.*, Development of Lithium/Metal Sulfide Batteries at Argonne National Laboratory: Summary Report for 1977, Argonne National Laboratory Report ANL-78-20 (March 1978).
8. N. P. Yao, L. A. Heredy, and R. C. Saunders, J. Electrochem. Soc., 118, 1039 (1971).
9. E. C. Gay, D. R. Vissers, F. J. Martino, and K. E. Anderson, J. Electrochem. Soc., 123, 1591 (1976).
10. E. J. Zietner, Jr. and J. S. Dunning, High Performance Lithium/Iron Disulfide Cells, Proc. 13th IECEC, pp. 697-701 (1978).

11. L. R. McCoy, S. Sudar, L. A. Heredy and J. C. Hall, Lithium Silicon-Iron Sulfide Load Leveling and Electric Vehicle Batteries, Proc. 13th LECEC, pp. 702-708 (1978).
12. D. R. Vissers, Z. Tomczuk and R. K. Steunenberg, J. Electrochem. Soc. 121, 665 (1974).
13. J. P. Mathers, C. W. Boquist, and T. W. Olszanski, J. Electrochem. Soc., 125, 1913 (1978).
14. J. P. Mathers and T. W. Olszanski, J. Electrochem. Soc., 124, 1149 (1977).
15. K. M. Myles, F. C. Mrazek, J. A. Smaga, and J. L. Settle, Materials Development in Lithium/Iron Sulfide Battery Program at Argonne National Laboratory, Proc. Symp. and Workshop on Adv. Battery Research and Design, Argonne National Laboratory Report ANL-76-8, p. B-50 (March 1976).
16. E. R. Van Artsdalen and I. S. Yaffe, J. Phys. Chem. 59, 118 (1955).
17. J. A. Plambeck, J. Chem. Eng. Data 12, 77 (1967).
18. H. Shimotake, M. L. Kyle, V. Maroni, and E. J. Cairns, Proc. 1st Int. Electric Vehicle Symp., Electric Vehicle Council, p. 392, New York (1969).
19. M. L. Kyle, H. Shimotake, R. K. Steunenberg, F. J. Martino, R. Rubischko, and E. J. Cairns, 6th IECEC, p. 80, New York (1971).
20. J. R. Selman, D. K. DeNuccio, C. C. Sy, and R. K. Steunenberg, J. Electrochem. Soc. 124, 1160-1164 (August 1977).
21. R. A. Sharma and R. N. Seefurth, J. Electrochem. Soc. 123, 1763 (1976).
22. F. A. Shunk, Constitution of Binary Alloys, Second Supplement, p. 480, McGraw-Hill Book Co., New York (1969).
23. S. C. Lai, J. Electrochem. Soc. 123, 1196 (1976).
24. D. R. Vissers, W. Frost, K. E. Anderson, and M. F. Roche, Argonne National Laboratory, unpublished data (1976).
25. E. C. Gay, R. K. Steunenberg, J. E. Battles, and E. J. Cairns, Proc. 8th IECEC p. 9-, AIAA, New York (1973).
26. W. Giggenbach, Inor. Chem. 10, 1308 (1971).
27. D. M. Gruen, R. L. McBeth, and A. J. Zielen, J. Am. Chem. Soc., 93, 6691 (1971).
28. L. A. Heredy *et al.*, Metal Sulfide Electrodes for Secondary Lithium Batteries: New Uses of Sulfur, J. R. West, Ed., pp. 203-215, ADV. Chem. Ser. 140, ACS, Washington (1975).
29. R. C. Weast, S. M. Selby, and C. D. Hodgman, Eds., Handbook of Chemistry and Physics, 45th Ed., The Chemical Rubber Co., Cleveland (1964).
30. E. J. Cairns and J. S. Dunning, High-Temperature Batteries, Proc. Symp. and Workshop on Adv. Battery Research and Design, Argonne National Laboratory Report ANL-76-8, p. A-81.
31 K. C. Mills, Thermodynamic Data for Inorganic Sulfides, Selenides and Tellurides, Butterworths, London (1974).
32. A. E. Martin, Z. Tomczuk, and M. F. Roche, Argonne National Laboratory Report ANL-78-94, p. 167.

33. Z. Tomczuk, M. F. Roche, and A. E. Martin, in Argonne National Laboratory Report ANL-79-39, p. 66.
34. B. Tani, American Mineralogist, 62, 819 (1977).
35. F. C. Mrazek and J. E. Battles, High Performance Batteries for Off-Peak Energy Storage and Electric Vehicle Propulsion: Progress Report for the period July to December 1974, Eds., P. A. Nelson *et al.*, Argonne National Laboratory Report ANL-75-1, p. 91 (July 1975).
36. M. L. Saboungi, J. J. Marr, and M. Blander, J. Electrochem. Soc., 125, 1567 (1968).
37. Z. Tomczuk and A. E. Martin, Argonne National Laboratory, unpublished data (1979).
38. A. E. Martin and Z. Tomczuk, High Performance Batteries for Stationary Energy Storage and Electric Vehicle Propulsion: Progress Report for the Period October-December 1976, Eds., P. A. Nelson *et al.*, ANL-17-17, pp. 45-46 (April 1977).
39. M. L. Saboungi and A. E. Martin, Extended Abstracts of the Electrochem. Soc. Meeting, Pittsburgh, PA, October 15-20, 1978, Vol. 78-2, p. 919 (1978).
40. D. R. Vissers, K. E. Anderson, C. K. Ho, and H. Shimotake, Proc. of the Symp. on Battery Design and Optimization, Ed., S. Gross, Electrochem. Soc. 79-1, 416 (1979).
41. J. Braunstein and C. E. Vallet, J. Electrochem. Soc., 126, 960 (1979).
42. John Newman and Richard Pollard, unpublished work, Univ. of California, Berkeley (1979).
43. Z. Tomczuk, A. E. Martin, and R. K. Steunenberg, The Chemistry of the FeS Electrode of Li/FeS Cells, Extended Abstract No. 47 of Electrochemical Society Fall Meeting, Las Vegas, Nev., October 17-22, 1976.
44. J. E. Battles, J. A. Smaga, and K. M. Myles, Metallurgical Trans. 9A, 183 (1978).
45. J. E. Battles, Materials for High Temperature Li-Al/$FeS_x$ Secondary Batteries, Critical Materials Problems in Energy Production, Ed., C. Stein, Academic Press, New York (1978).
46. D. R. Stull and H. Prophet, JANAF Thermochemical Tables, 2nd Ed., NSRDS-NBS-37, U. S. National Bureau of Standards, June 1971.
47. C. E. Wichs and F. E. Block, Thermodynamic Properties of 65 Elements--Their Oxides, Halides, Carbides and Nitrides, U. S. Bureau of Mines Bull. 605 (1963).
48. R. M. Yonco, E. Veleckis, and V. A. Maroni, J. Nucl. Mater. 57, 317 (1975).
49. R. Hudson and K. Gentry, The Design, Development, and Fabrication of a 40 kW-hr Lithium/Iron Sulfide Electric Vehicle Battery, Extended Abstract No. 91, Electrochemical Society, Fall Meeting, Pittsburgh, PA, October 15-20, 1978.
50. F. J. Martino, T. D. Kaun, H. Shimotake, and E. C. Gay, Proc. 13th IECEC, pp. 709-716 (1978).

51. E. C. Gay, W. E. Miller, R. F. Malecha, and R. C. Elliott, Proc. 13th IECEC, pp. 6900696 (1978).
52. K. M. Myles and J. L. Settle, in Argonne National Laboratory Report ANL-77-17, pp. 31-33.
53. P. A. Nelson *et al.*, Development of Lithium/Metal Sulfide Batteries at Argonne National Laboratory: Summary Report for 1975, Argonne National Laboratory Report ANL-76-45, (March 1976).
54. P. A. Nelson et. al., Development of Lithium/Metal Sulfide Batteries at Argonne National Laboratory: Summary Report for 1974, Argonne National Laboratory Report ANL-75-20, (March 1975).
55. V. M. Kolba, G. W. Redding, and J. L. Hamilton, High Performance Batteries for Stationary Energy Storage and Electric Vehicle Propulsion: Progress Report for January-March 1977, Ed., P. A. Nelson *et al.*, Argonne National Laboratory Report ANL-77-35, pp. 24-28 (June 1977).
56. W. Deluca, A. Chilenskas, and F. Hornstra, To be published in the Proc. of the 14th IECEC Meeting, Boston, MA (1979).
57. P. A. Nelson *et al.*, Development of Lithium/Metal Sulfide Batteries at Argonne National Laboratory: Summary Report for 1978, Argonne National Laboratory Report ANL-79-64 (july 1979).
58. S. M. Zivi, I. Pollach, H. Kacinskas, A. A. Chilenskas, D. L. Barney, S. Sudar, I. Goldstein, and W. Grieve, Battery Engineering Problems in Designing An Electrical Load Leveling Plant for Lithium/Iron Sulfide Cells, To be published in the Proceedings of the 14th IECEC, Boston, MA (1979).
59. A. A. Chilenskas, Argonne National Laboratory Report ANL-77-75, p. 12.

# SOLID ELECTROLYTES

Robert A. Huggins

Department of Materials Science & Engineering
Stanford University
Stanford, California 94305

## A. INTRODUCTION

This paper will discuss a number of aspects of solid electrolytes that are relevant to their potential use in practical advanced batteries. It is intended to serve as a starting point for the considerations of the solid electrolyte study group, rather than as an exposition of the current state of knowledge in this area.

Although materials that have properties such that they can be considered for use as solid electrolytes have been known for a long time, and were even used for practical devices before the end of the century, the possibility of the use of such materials as important components in advanced types of batteries arose only recently.

While they are not the subject of this Institute, we should recognize that there are a number of other applications in which solid electrolytes play important roles. These include their use as sensors for specific species (eg. oxygen) in gases, or (eg. fluorine) in liquids. Solid electrolytes have been shown to be useful in heterogeneous catalytic systems. They are being investigated as possible components in new types of solid state display systems, and a number of practical devices have been demonstrated, such as timers and switches, which depend upon their special properties. In addition, there are materials and systems which are presently technologically dormant, but which can have important scientific uses. Each of these applications imposes different requirements, and thus the optimization of different combinations of properties.

When discussing the use, or possible use, of solid electrolytes in advanced batteries, we should realize that the list of materials currently utilized in commercial batteries or in systems receiving a significant amount of developmental effort is really very short - the sodium beta alumina family, and various modifications of lithium iodide, and a sodium borate-based glass.

Naturally, the matters of assessing recent progress and considering areas in need of additional research are quite different if one considers the optimization of established materials than they are in the more "wide open" field of the exploration of other new possibilities. This talk will discuss each of these briefly, but is certainly not intended to be complete in either.

## B. CHARACTERISTICS GENERALLY REQIURED OF SOLID ELECTROLYTES

### Introduction

With regard to their use in advanced batteries, solid electrolytes can be employed for 3 different functions. First, they can be used strictly as electrolytes, separating either liquid or solid electrode systems. An example of this is the well known sodium-polysulfide cell involving a solid beta alumina electrolyte. Second, they can be used as ion-transparent physical separators in cells in which the primary electrolyte is a liquid. An example of this is the work presently going on to investigate the use of phosphate-doped lithium orthosilicates as either powder or sheet separators in cells utilizing molten $LiC\ell$-$KC\ell$ salt electrolytes in which lithium is the electroactive species. Third, solid electrolytes can be used as ion-transporting constituents in solid electrode structures. This was demonstrated several years ago in the case of ambient temperature silver-conducting cells in which modest quantities of an electrolyte was mixed with the electrode reactant materials in order to physically distribute the interfacial electrochemical reaction over a greater area.

While each type of application has somewhat different requirements they have a number of important common features. Some of these are as follows:

1. High ionic conductivity
2. Selectivity of ionic transport
3. Negligible electronic transport
4. Stability with respect to thermal decomposition
5. Stability with respect to electrochemical decomposition
6. Stability with respect to reaction with species in the environment
7. Ease of fabrication (at satisfactory cost)
8. Suitable mechanical and other related properties (e.g., thermal expansion coefficient that matches other constituents)
9. Readily available and low-cost ingredients.

10. Adequate safety (e.g., must not be poisonous).

Basic Questions Relating to Fast Ionic Conduction

I think that it is fair to say that we still do not understand satisfactorily why certain materials exhibit fast ionic conduction, nor do we fully understand the physical processes involved in the motion of ions under such conditions. Misunderstanding about such matters is rampant in the current literature, and it is quite reasonable to question whether there is really anything fundamentally different in such materials, as sometimes implied by use of the term "superionic conductors," or whether they merely exhibit extreme values of "normal" behavior [1].

However, it is now quite apparent that in the case of crystalline materials which exhibit fast ionic conduction this behavior is directly related to special characteristics of their crystal structures. Some time ago initial steps were taken to develop an explicit structure-dependent model for the transport of ions through specific crystal structures [2-5].

This approach has been able to show semiquantitatively the relationship between the details of the crystallographic structure, such as the configuration of the other ions in the lattice along and near crystallographic tunnels, upon the minimum energy path which should be preferred by species moving through the structure. This minimum energy path often does not follow the centerline between assumed crystallographic sites. Furthermore, it has been demonstrated that the locus of the minimum energy path should vary with the size of the mobile species, as well as the dimensions, charge distributions and other characteristics of the structure through which it moves. These variables also influence the energy profile along the path, and thus the activation enthalpy. Although this approach has only been applied to a relatively few crystal structures, it seems to provide a good deal of insight concerning the important parameters influencing the kinetics of transport through such structures.

Stability Requirements

While the potential utility of solid electrolytes surely depends to first order upon the magnitude and selectivity of the ionic conductivity, the practical utilization of such materials also requires that they meet the relevant stability requirements. As mentioned earlier, this means that they must be stable with respect to thermal decomposition, electrochemical decomposition, or reaction with other species in their environments. In addition, they must be utilized under conditions in which ionic, rather than electronic, species conduct most of the charge.

Stated another way, such materials must be utilized within their appropriate thermodynamic and electrolytic stability windows.

We can summarize this situation by saying that the practical utilization of materials as solid electrolytes is often limited by restricted ranges of temperature, pressure and chemical potentials over which they are thermodynamically stable and conduct charge primarily by the motion of ionic species. This matter has received relatively little attention to date, despite its obvious practical importance, as pointed out recently [6].

An example of such a limitation is the thermal decomposition of $H_3OC\ell O_4$, a very good proton conductor at ambient temperatures [7], which gives off water at about 50° C. On the other hand, some solid electrolytes go through crystallographic changes which cause their properties to be attractive over a limited range of temperature. In such cases a thermally-controlled crystallographic change is important rather than thermal decomposition. A well-known example of this is the sudden change in the ionic conductivity at the α/β phase transformation in AgI at 146° C.

Contrary to the assumptions which one often finds in the literature crystallographic phase changes do not always result in increased ionic conductivity above the transformation temperature. Examples are the α/β transformation in $Na_2WO_4$ and $\alpha_2/\alpha$ transformation in $Na_2MoO_4$ [8-10].

The question of the influence of chemical parameters and tempera ture upon the fraction of the charge transport which is carried by ionic species (ionic transference number) has been studied most in anionic conductors; for example, oxides with the fluorite structure: $ZrO_2$, $ThO_2$ and $CeO_2$, and solid solutions based upon them. Relatively little attention has been given to this matter in the case of cationi conductors.

There has been a great deal of interest in the possible development of batteries based upon the transport of lithium or sodium in which the negative electrode is the alkali metal itself. This desire flies in the face of the fact that elemental lithium is a very highly reducing material, and it is not easy to find solid electrolytes whic are thermodynamically stable in its presence.

One strategy is to use a binary intermediate phase which is the most lithium-rich in its family at the temperature of operation. Such a phase will obviously be stable in the presence of lithium, or the relevant lithium-rich terminal solid or liquid solution. This does not always mean, of course, that this phase will be a good ionic conductor or that it will even have a satisfactorily large value of ionic transference number An example of the latter is the fact that molten alkali halide salts dissolve appreciable quantities of their respective alkali metals. Since these salts are highly ionic, the excess alkali metal is present in the form of cations and electrons; the latter impart appreciable electronic conductivity to the liquid solution. This means that self discharge may be an important problem in a practical battery involving the use of such salts under conditions of high alkali metal activity.

The question of the thermodynamic stability of ternary and other multicomponent solid electrolytes is more difficult to visualize. A very useful approach to such matters involves consideration of the relevant phase diagram, and evaluation of the thermodynamic parameters involved in the various phase fields adjacent to the solid electrolyte phase. Most useful are data concerning the invariant equilibria (eg three-phase regions in ternary systems). One can often attain very useful information by this means. As an example, it was recently found that the lithium-conducting phase $LiA\ell C\ell_4$ is stable over the range from 1.68 to 4.36 volts versus lithium at 25° C [11].

This general matter of electrolyte stability windows was recently discussed in [6]. Table I includes data on the stability windows of a number of solid electrolytes.

A technique that can be employed in some cases in order to avoid problems of electrolyte stability is to use a second electrolyte between the two phases that are not mutually stable. An example of this is found in what is sometimes called the $Na/SbC\ell_3$ battery system, which can be represented by the cell

$$Na/Na^+ \text{ beta alumina}/NaA\ell C\ell_4/SbC\ell_3, Sb$$

and which operates above the melting points of both the sodium and the aluminochloride salt. The overall reaction of this cell can be simply thought of as the displacement reaction of sodium with $SbC\ell_3$, producing solid antimony and liquid $NaC\ell$. The molten alkali metal chloride electrolyte permeates the solid right hand electrode, providing a large reaction area. This molten salt electrolyte is not stable in the presence of elemental sodium, however. This problem, plus the practical difficulty of having both a molten electrode and a liquid electrolyte, are both solved by introducing a membrane of beta alumina solid electrolyte between them.

A similar phenomenon occurs in a number of other systems by the formation of a thin protective solid electrolyte layer as a result of a reaction between the electrode material and the electrolyte. Examples and probable solid electrolyte reaction products are listed in Table II.

TABLE I

Some Electrolyte Stability Windows

| Material | Li-Rich E vs. Li | Li-Poor E vs. Li | Temp. | Ref |
|---|---|---|---|---|
| LiI | 0 | 2.79 | 25°C | 12 |
| $Li_3N$ | 0 | 0.44 | 25°C | 13,14 |
| LiCℓ | 0 | 3.98 | 25°C | 12 |
| $LiAℓCℓ_4$ | 1.68 | 4.36 | 25°C | 11 |
| $LiAℓCℓ_4$ | 1.70 | 4.20 | 135°C | 11 |
| $Li_9N_2Cℓ_3$ | 0 | > 2.5 | 100°C | 15,16 |
| $Li_{11}N_3Cℓ_2$ | 0 | > 1.8 | 322°C | 16 |
| $Li_6NBr_3$ | 0 | > 1.3 | 176°C | 16,17 |
| $Li_{13}N_4Br$(LT) | 0 | > 1.3 | 146°C | 16,17 |
| $Li_{13}N_4Br$(HT) | 0 | > 0.65 | 300°C | 16,17 |
| $Li_5N\ I_2$ (LT) | 0 | > 1.9 | 98°C | 16 |
| $Li_5N\ I_2$(HT) | 0 | > 1.6 | 287°C | 16 |
| $Li_{6.67}N_{1.89}I$ | 0 | > 0.9 | 313°C | 16 |
| $Li_{9.11}N_{2.7}\ I$ | 0 | > 0.9 | 316°C | 16 |

TABLE II

Thin Film Solid Electrolytes
Formed by Electrode-Electrolyte Reactions

| System | Probable Solid Electrolyte Layer |
|---|---|
| $Li/H_2O$ | LiOH |
| Li/propylene carbonate | $Li_2CO_3$ |
| Li/thionyl chloride | $LiC\ell$ |
| $Li/SO_2$ | $Li_2S_2O_4$ |

## C. COMMENTS ON ESTABLISHED MATERIALS

### The Beta Alumina Family

At least eleven significant programs are currently being undertaken to develop sodium-polysulfide battery systems that employ one or another of the beta aluminas as a solid electrolyte. Nevertheless, there are still many areas in which insufficient understanding now exists, and in which additional research of a basic nature could be profitably undertaken.

Although very favorable properties can be obtained from beta alumina solid electrolytes, the situation is complicated by the fact that there are two different phases of interest. One of these, called β alumina, has a crystal structure with a unit cell containing two spinel blocks, separated by the highly-conductive bridging layer. This 2-block phase exists with an $A\ell_2O_3/Na_2O$ molar ratio in the range 9-11. The other phase, β" alumina, has a 3-block structure, producing a somewhat different environment for the mobile ions in the intermediate bridging layers. The β" phase is apparently not thermodynamically stable over about 1500° C unless it is stabilized by the incorporation of about 2% $Li_2O$ or MgO (or both).

Materials with the β" structure can be produced in polycrystalline form with a sodium ion conductivity some 3 to 5 times as great as with β structure at about 300° C. However, the choice is not so simple, for a number of other factors are also important with regard to both fabrication processes and the properties of the final product. The pure β phase requires sintering over 1650° C, whereas β" is sintered at about 1550° C.

Mixed microstructures resulting from incomplete conversion to the β" phase can produce inferior values of conductivity (eg. a factor of 4 or 5 worse than fully converted material), and in some cases it has been found that both the conductivity and the mechanical properties can be irreversibly degraded by water vapor. Materials containing either the β phase or fully converted β" have been found to be much less water-sensitive; properties can generally be recovere by annealing in air at about 650° C.

Different strategies are being followed in the various development programs, some work with β, some with β" , and some prefer two-phase microstructures. Fabrication can involve pre-reaction of the starting ingredients to form the desired phase followed by isostatic pressing (on some other mechanical densification process) and sintering, or it can be done by reactive phase sintering, in which phase conversion and final densification take place concurrently. Both the final composition, which can be affected by the volatization of $Na_2O$, and microstructural features such as grain size and shape, can influence the final properties. Interestingly, grain size plays a much more important role with respect to mechanical properties than to ionic conductivity.

Since the beta alumina family is now being employed in developmental programs, the types of problems that deserve attention from the research community are somewhat different from those which are more appropriate in more exploratory areas. While efforts to better understand the transport mechanism are surely worthwhile, work on problems such as solid solution effects upon both stability and trans port, sintering phenomena, particularly in the presence of a liquid phase in a multicomponent system, and microstructural generation and control would surely have a greater short-to-intermediate term impact

Sodium Borate-Based Glass

The second approach that is currently being followed in a develo mental program aimed at the commercialization of a battery based upon the sodium-polysulfide electrochemical couple employs a sodium borate-based glass instead of beta alumina. The ionic conductivity of the glass is much lower than that of the beta aluminas (about $4 \times 10^{-5}$ $ohm^{-1}cm^{-1}$ at 300° C), and a very different cell design is used, employing a very large number of fine hollow glass fibers (approx. 13,000 fibers for a 40 ampere-hour cell). Typical dimensions of these very small glass tubes are a 70 micron outer diameter, 10 micron wall thickness, and 10 cm length. This provides a very large electrolyte area, and thus compensates for the low conductivity The current density through the tube walls is typically about 2 $ma/cm^2$ for cells designed for relatively high power.

Current cells are designed so that about 90% of the total cell resistance is due to the ionic resistivity of the glass at 300° C, the total cell resistance causing a 5% change in cell voltage during both charge and discharge.

The activation enthalpy for ionic conductivity in glasses typically has a rather high value, so the resistance of such a cell is quite temperature-dependent; the resistance of these glasses drops about 50% if the temperature increases 25° C.

The cells now being developed use a glass electrolyte composition that is based on $Na_2O \cdot 2\ B_2O_3$, to which about 0.16 mole NaCℓ and 0.2 mole $SiO_2$ have been added. Silicate-based glasses are not sufficiently stable at high sodium activities, but the presence of a small amount of $SiO_2$ has a large influence upon the vitreous-crystalline transition, preventing devitrification near the operating temperature. The addition of NaCℓ increases the mobile sodium concentration, and thus the ionic conductivity. Similar effects have been found in other glasses, as will be mentioned later.

As is the case with all of the high performance battery development programs, longer cell lifetime is a significant problem. Reproducibility needs to be improved and failure modes better understood. In these cells fiber breakage near the tube sheet has sometimes been a problem, and may be related to small amounts of impurities, such as oxygen or hydroxide in the sodium, which apparently can cause corrosion of the glass.

While remaining practical problems that are presently recognized will probably be overcome during the development program, there are several areas in which further research could be very advantageous. Since the total cell resistance is dominated by the ionic resistance of the glass, both the output voltage and the energy loss through heat generation are strongly dependent upon this factor. Improvements in conductivity without deleterious changes in chemical stability or mechanical and thermal properties would be very welcome. The influence of composition in these several areas is also only very poorly understood, not only in the case of borate glasses, but glasses in general.

Lithium Iodide-Based Materials

Cells are now available commercially in which a solid electrolyte based upon lithium iodide is being used. As shown in Table I, this material is stable over a rather wide range of lithium activities. It has been found that the ionic conductivity can be greatly enhanced by the presence of a substantial amount of aluminum oxide introduced in the form of very fine particles.

The matter of the mechanism of this enhancement effect certainly deserves consideration from the research community. Similar, but smaller, effects seem to have been found in other systems. Questions involving the possible influence of internal 2-phase interfacial area and of modest amounts of water, hydrates, or hydroxide are intriguing, as well as having important practical ramifications.

## D. COMMENTS ON SEVERAL OTHER MATERIALS GROUPS

### Introduction

There is a substantial amount of exploratory work being undertaken in a number of laboratories in which a wide variety of materials are being evaluated as potential candidates for use as solid electrolytes. There will be no attempt here to either inventory or evaluate all this work. Instead, I shall briefly discuss a few areas that seem to me to provide some very interesting possibilities, and which might benefit by more attention.

### The Sodium-Zirconium-Silicon-Phosphorous Oxide Family

There is a good measure of present interest in sodium-conducting solid electrolytes of the general formula $Na_{1+x}Zr_2Si_xP_{3-x}O_{12}$ (sometimes called NASICON). The sodium ion conductivity of these materials is very high at elevated temperatures[18]. It varies with composition, apparently reaching a maximum (at 300° C) when $x \approx 2$ [19].

Although work on this material is still quite recent, it appears to be very interesting from a practical point of view because its conductivity is comparable to that of the best of the beta aluminas at the operating temperature of sodium-polysulfide cells. Furthermore, it can apparently be sintered at about 1230° C, whereas the β" aluminas must be processed at 1500° C. In addition, NASICON appears to be relatively insensitive to the presence of water or water vapor.

These materials may well compete with the beta aluminas in certain practical applications. Because of the large number of components present, compositional control, which is evidently quite important, is difficult. Many structure- and composition-related questions should be addressed in this system.

### Proton Conductors

Proton transport has been known to occur in a number of material for a considerable time [20,21]. However, interest in the use of such materials as solid electrolytes in batteries and similar purpose has arisen in the last few years primarily as a result of the discove of the high values of proton conductivity in three groups of material $H_3OClO_4$ [7], the tetrahydrate of hydrogen uranyl phosphate [$HUO_2PO_4 \cdot 4$I and related materials [22-27] and in the β" alumina structure [28-32].

The reported conductivity value of $4 \times 10^{-3}$ $ohm^{-1}$ $cm^{-1}$ at 295 K in these materials is high enough to make it worthwhile to consider their use in practical devices.

Experiments were also recently reported on proton conduction in materials of the montimorillinite clay family [33]. Quite high values of protonic conductivity were found (1 - 4 $\times 10^{-4}$ $ohm^{-1}$ $cm^{-1}$ at 290 K. However, these materials, while also having structures with water layers, are different from the uranyl phosphate family in several respects. As is also the case with the NAFION family of polymeric materials, the clays do not have a high degree of selectivity. Several alkali metal ions, as well as protons have been found to exhibit rapid transport within them [34]. The amount of water included in the structure, as well as between particles, can be varied appreciably, and has a pronounced effect upon the conductivity.

High values of protonic conductivity have also been reported in a family of solid heteropoly acids of molybdenum and tungsten (eg. $H_3PMo_{12}O_{40} \cdot 29\ H_2O$) [35-38]. These materials can be thought of as having structures consisting of interpenetrating diamond lattices, with large $(H_3 \cdot 29\ H_2O)^{+3}$ cations and $(PMo_{12}O_{40})^{-3}$ anions.

Another interesting approach involves organic sulfates, such as triethylene diamene sulfate and hexamethylene tetramine sulfate [40], although the ambient temperature conductivities of these materials is somewhat lower than the others.

Since proton-conducting solid electrolytes potentially have a number of practical applications, a considerable amount of further effort in this area certainly seems warranted. As pointed out earlier, questions relating to the ranges of chemical and thermal stability can often play a significant role. For application in water-containing environments both the water activity and the pH are surely important. Especially interesting for fuel cells, for example, would be proton-conducting materials that could be utilized in non-aqueous environments at elevated temperatures.

It is not an easy task to find such materials, however. It will probably require a structure containing structural units such as $H_3O^+$, $NH_4^+$, $H_2O$ or $NH_3$ which are in sufficiently close proximity to allow proton hopping and yet are quite tightly bound to other parts of the total structure.

Some of the available data concerning proton conduction in solid electrolytes are included in Table III.

TABLE III

Values of Proton Conductivity in Some Solid Electrolytes

| Material | Conductivity ($ohm^{-1}$ $cm^{-1}$) | Temp. (K) | Ref. |
|---|---|---|---|
| $H_3PMo_{12}O_{40} \cdot 29\ H_2O$ | $1.8 \times 10^{-1}$ | 298 | 35-38 |
| $H_3PW_{12}O_{40} \cdot 29\ H_2O$ | $1.7 \times 10^{-1}$ | 298 | 35-38 |
| $HUO_2PO_4 \cdot 4\ H_2O$ | $4 \times 10^{-3}$ | 290 | 22 |
| $HUO_2AsO_4 \cdot 4\ H_2O$ | $6 \times 10^{-3}$ | 310 | 24 |
| $H_8UO_2(IO_6)_2 \cdot 4\ H_2O$ | $7 \times 10^{-3}$ | 293 | 26 |
| $H_3OC\ell O_4$ | $3 \times 10^{-4}$ | 298 | 7 |
| $Sb_2O_5 \cdot 4\ H_2O$ | $3 \times 10^{-4}$ | 293 | 39 |
| $SnO_2 \cdot 3\ H_2O$ | $2 \times 10^{-4}$ | 293 | 39 |
| $ZrO_2 \cdot 2.3\ H_2O$ | $3 \times 10^{-5}$ | 293 | 39 |
| $H^+$ - Montmorillonite | $1\text{-}4 \times 10^{-4}$ | 290 | 33 |
| H-Aℓ-Montmorillonite | $10^{-4} - 10^{-2}$ | 290 | 33 |
| Hydrated hydronium β" alumina | $5 \times 10^{-3}$ | 298 | 31 |
| $NH_4^+$ - β alumina | $\sim 10^{-3}$ | 523 | 32 |
| $NH_4^+/H(H_2O)_x^+$ β" alumina | $\sim 10^{-4}$ | 298 | 32 |
| $C_6H_{12}N_2 \cdot 1.5\ H_2SO_4$ | $\sim 10^{-6}$ | 298 | 40 |
| $C_6H_{12}N_4 \cdot 2\ H_2SO_4$ | $\sim 10^{-7}$ | 298 | 40 |

Cationic Conduction in Noncrystalline Materials

Ionic conduction has also been recognized for many years in inorganic glasses, such as the alkali silicates, and reviews can be found in various places (eg. [41]]). This area is now getting greatly increased attention for two reasons. One of these is the impetus given by the developmental effort involving the use of fine hollow fibers of sodium borate-based glass as the solid electrolyte in a sodium-polysulfide cell that was mentioned earlier. The other is the considerable volume of interesting recent work on amorphous materials, both inorganic and organic, which show quite high values of cationic conductivity.

It has been known for some time that the ionic conductivity of alkali-containing oxide glasses increases, and the activation enthalpy decreases, as the alkali oxide concentration is increased. However, the conductivity is not a linear function of the alkali metal ion (presumably the mobile species) concentration. This has lead to the concept that some of the alkali metal ions are mobile, while others are "trapped" in the structure and can undergo local, but not long range, motion.

Recent work has provided significantly more insight into this phenomenon, which is leading to the design of new compositions with appreciably greater values of conductivity. One important observation was that the conductivity is proportional to the square root of the alkali oxide activity in the alkali silicates [42-45]. This leads to the concept that the alkali metal ions exist in the structure in such a way that one can think in terms of an association-dissociation equilibrium of the alkali oxide in the glass solvent, with the charge carried by only the dissociated cations. This approach is analagous to that used in liquid electrolytic solutions.

An obvious extension of these ideas is to seek to increase the alkali oxide activity (concentration) to enhance the conductivity. There are limits as to how far one can go in this direction, however, before devitrification becomes a problem due to a reduction in the size of the anionic groups in the structure. This problem can be circumvented, and the concentration of undissociated alkali metal ions increased by the addition of alkali halides. In such cases one can assume that neither the alkali metal ion nor the halide ion becomes bound into the macromolecular anionic structure, which consists primarily of corner-shared tetrahedral groups. Quite large concentrations of alkali halide and other simple salts can be dissolved in these glasses, and since the cationic and anionic species are highly dissociated, the conductivity can be increased substantially, about two orders of magnitude for lithium and sodium-containing oxide glasses, and four orders of magnitude for silver-based glasses. As is the case for analagous liquid solutions, the degree of dissociation, and thus the contribution to the conductivity, is greater the larger the anion with which the mobile cation is associated. It has also been found that sulfide analogs of the oxide glasses have ionic conductivities about an order of magnitude higher.

This general approach has been followed recently in a number of laboratories, including work in Grenoble on alkali silicates (eg. [46]) and phosphates, as well as several sulfide systems, in Bordeaux on lithium borate systems [47,48], some of which are related to the crystalline phase boracite ($Li_4B_7O_{12}C\ell$) which was first recognized by Jeitschko and co-workers [49–51] to be a fast lithium ion conductor, in Besancon on alkali and silver phosphates [52–54], and also in Japan [55]. Work on other silver-containing glasses has also been underway in Italy and Scotland [56,57] and on various lithium-containing glasses in the United States [58–60].

A different class of noncrystalline solids is also beginning to receive attention in which the major constituents in the structure are organic polymers and alkali metal salts. Work has been reported on the properties of poly-ether-based systems, including poly(ethylene oxide) and poly(propylene oxide) containing simple salts [61–63], and poly-organo-phosphonates in which the anion is inserted in the macromolecular chain [64].

Because of the possibility of the fabrication of noncrystalline solids into simple shapes, such as thin films or tubes, and the lack of grain boundaries and local anisotropy there seems to be ample justification for the expectation that this general area will receive considerably more attention in the future.

## Materials Based Upon the Lithium Orthosilicate and Orthophosphate Structures

High values of lithium ionic conductivity at temperatures comparable to those currently used in the lithium - based molten chloride salt cells and the sodium - polysulfide solid electrolyte cells have been reported in the last several years [66–70] in two families of ternary and quaternary lithium oxides. These are based upon the lithium orthosilicate and the lithium orthophosphate structures, which can both be viewed as consisting of isolated, rather than corner - sharing, tetrahedral anionic groups, between which the mobile lithium ions percolate.

Especially interesting have been investigations of solid solution effects in these structures. For example, it has been shown that several other anionic groups can replace a significant fraction of the $SiO_4^{-4}$ tetrahedra in the orthosilicates. Experiments have been performed on the effects of the presence of $AlO_4^{-5}$ [69], $PO_4^{-3}$ [66–68], $VO_4^{-3}$ [70], and $SO_4^{-2}$ [69] anions upon the lithium ionic conductivity in solid solutions with this structure over substantial ranges of both composition and and temperature. The inclusion of such aliovalent anionic groups causes appreciable changes in the concentration of lithium ions in the orthosilicate structure - decreasing concentrations when less negative anions are present, and increasing concentrations when more negative anions are substituted. One would therefore expect that such species in solid solution would influence the ionic conductivity primarily by changing the pre-exponential factor in the normal Arrhenius relation between the conductivity and the temperature, with opposite effects in the case of more positively and more negatively charged anionic groups. This type of effect has, indeed, been observed in the case of solid solutions based upon the lithium orthophosphate structure [69].

Experimental data, however, show quite a different effect in the case of the orthosilicate - based solid solutions. In all cases investigated to date, the presence of aliovalent anionic groups increases the the lithium ionic conductivity, and does so be influencing the activation enthalpy, rather than the pre-exponential factor.

These can be large effects, producing increases in the ionic conductivity of a number of orders of magnitude. However, I believe that it is fair to say that the reasons for these solid solution effects, and for the quite different behavior in materials with different structures, are not yet even qualitatively understood. As a result, very little guidance is available for the search for other such materials that might have equal or better properties. The desirability of employing solid solutions as solid electrolytes in terms of tailoring other properties as well as the ionic conductivity is inherently very great, and certainly calls for more efforts in this area.

## E. CLOSING COMMENTS

This paper has discusses only a few of the features of solid electrolytes that may be critical to their possible application in practical technological devices, such as batteries. Other properties will surely also be important, if not dominant, in specific cases.

The optimum mix of activities is quite different in the case of materials and configurations that are already established as deserving of developmental attention from that which is characteristic of more exploratory research efforts in which new materials are being sought, or from work aimed primarily at the achievement of better understanding of the physical phenomena giving rise to the properties which have already been observed. This area of science and developing technology is both exciting and potentially important. Go to it.

ACKNOWLEDGEMENTS

The author is pleased to acknowledge the many important theoretical and experimental contributions in this area by his students and colleagues at Stanford. Special thanks are due I. D. Raistrick, B. A. Boukamp, W. Weppner, and J. L. Souquet. Much of the financial support of work in this area at Stanford comes from the Department of Energy under Contract EC-77-S-02-4506.

REFERENCES

1. R. A. Huggins and A. Rabenau, Mat. Res. Bull. 13, 1315 (1978)

2. W. F. Flygare and R. A. Huggins, J. Phys. Chem. Solids 34, 1199 (1973).

3. O. B. Ajayi, Ph.D. Dissertation, Stanford University (1975).

4. O. B. Ajayi, L. E. Nagel, I. D. Raistrick and R. A. Huggins, J. Phys. Chem. Solids 37, 167 (1976).

5. R. A. Huggins, "Crystal Structures and Fast Ionic Conduction," in Solid Electrolytes, ed. by P. Hagenmuller and W. van Gool, Academic Press (1978), p. 27.

6. R. A. Huggins, "Evaluation of Properties Related to the Application of Fast Ionic Transport in Solid Electrolytes and Mixed Conductors," Proceedings of International Conference on EAst Ion Transport in Solids, Elsevier North-Holland (1979).

7. A. Potier and D. Rousselet, J. Chim. Phys. 70, 873 (1973).

8. P. H. Bottelberghs, "Anomalous Electrical Conductivity in $Na_2WO_4$ in Fast Ion Transport in Solids, ed. by W. van Gool, North-Holland (1973), p. 637.

9. P. H. Bottelberghs and E. Everts, J. Solid State Chem. 14, 342 (1975).

10. P. H. Bottelberghs and F. R. van Buren, J. Solid State Chem. 13, 182 (1975).

11. W. Weppner and R. A. Huggins, "Thermodynamic Properties of the Solid Electrolyte $LiAlCl_4$," Proceedings of International Conference on East Ion Transport in Solids, Elsevier North-Holland (1979).

12. I. Barin and O. Knacke, Thermochemical Properties of Inorganic Substances, Springer, Berlin (1973).

13. R. M. Yonco, E. Valeckis and V. A. Maroni, J. Nucl. Mat. 57, 317 (1977)

14. A. Bonomi, M. Hadate and C. Gentaz, J. Electrochem. Soc. 124, 982 (1977).

15. P. Hartwig, W. Weppner and W. Wichelhaus, Mat. Res. Bull. 14, 493 (1979).

16. P. Hartwig, W. Weppner and W. Wichelhaus, "Fast Ionic Conduction in Solid Lithium Nitride Halides," Proceedings of International Conference on Fast Ion Transport in Solids, Elsevier North Holland (1979).

17. P. Hartwig, W. Weppner, W. Wichelhaus and A. Rabenau, Solid State Communications, in press.

18. J. B. Goodenough, H. Y-P Hong and J. A. Kafalas, Mat. Res. Bull. 11, 203 (1976).

19. J. A. Kafalas and R. J. Cava, "Effect of Pressure and Composition on Fast $Na^+$-Ion Transport in the System $Na_{1+x}Zr_2Si_xP_{3-x}O_{12}$," Proceedings of International Conference on Fast Ion Transport in Solids, Elsevier North-Holland (1979).

20. J. Bruinink, J. Appl. Electrochem 2, 239 (1972).

21. L. Glasser, Chem. Rev. 75, 21 (1975).

22. M. G. Shilton and A. T. Howe, Mat. Res. Bull 12, 701 (1977).

23. P. E. Childs, A. T. Howe and M. G. Shilton, J. Power Sources 3, 105 (1978).

24. A. T. Howe, M. G. Shilton and P. E. Childs, "Fast Proton Conduction in Hydrogen Uranyl Phosphate and Arsenate," presented at Second International Meeting on Solid Electrolytes, St. Andrews, Scotland, September 1978.

25. T. K. Halstead, T. C. Jones, R. C. T. Slade and P. E. Childs, "Hydrogen Mobility Studies in Solid Electrolyte and Electrode Materials Using NMR," presented at Second International Meeting on Solid Electrolytes, St. Andrews, Scotland, September 1978.

26. M. G. Shilton and A. T. Howe, "Evidence for Proton Conductivity in Compounds in the Hydrogen-Uranyl-Periodate-Water System," Proceedings of International conference on Fast Ion Transport in Solids, Elsevier North-Holland (1979).

27. A. T. Howe and M. G. Shilton, J. Solid State Chem. 28 (1979).

28. G. C. Farrington and J. L. Briant, presented at Electrochemical Society Meeting, Atlanta, October, 1977.

29. G. C. Farrington, J. L. Briant and H. S. Story, "Conductance and Spectroscopy of Protonic Beta and Beta'' Aluminas," presented at Second International Meeting on Solid Electrolytes, St. Andrews, Scotland (1978).

30. Ph. Colomban, A. Kahn and J. P. Boilot, "Protonic Conductors with the β and β" $A\ell_2O_3$ Structure," presented at Second International Meeting on Solid Electrolytes, St. Andrews, Scotland (1978).

31. G. C. Farrington and J. L. Briant, Mat. Res. Bull. 13, 763 (1978)

32. G. C. Farrington and J. L. Briant, "Chemistry and Conductivity of Protonic Beta Aluminas," Proceedings of International Conference on Fast Ionic Transport in Solids, Elsevier North-Holland (1979).

33. S. H. Sheffield and A. T. Howe, Mat. Res. Bull 14, 929 (1979).

34. R. Calvet and J. Mamy, C. R. Acad. Sci. Paris 2720, 1251 (1971).

35. O. Nakamura, T. Kodama, I. Ogino and Y. Miyake, Japan Kokai Patent 76106683 (1976).

36. O. Nakamura et al., U. S. Patent 4025036 (1977).

37. O. Nakamura et al., Japan Patent 899328 (1978).

38. O. Nakamura et al., Chemistry Letters 17 (1979).

39. W. A. England, M. G. Cross, A. Hamnett, P. J. Wiseman and J. B. Goodenough, presented at Faraday Division Electrochemistry Group Meeting, Oxford, March, 1979.

40. T. Takahashi, S. Tanase, O. Yamamoto and S. Yamauchi, J. Solid State Chem 17, 353 (1976).

41. B. S. Dunn and J. D. Mackenzie, "Structure and Ionic Properties of Glasses," in International Review of Science, Physical Chemistry Series Two, Vol. 6, ed. by J. O'M. Bockris, Butterworths (1976), p. 231.

42. J. L. Souquet and D. Ravaine, Proceedings of 24th Meeting of the ISE, Eindhoven (1973).

43. D. Ravaine, These, Inst. Nat. Polytechnique, Grenoble (1976).

44. D. Ravaine and J. L. Souquet, Phys. Chem. Glasses 18, 27 (1977).

45. D. Ravaine and J. L. Souquet, "Ionic Conductive Glasses," in

Solid Electrolytes ed. by P. Hagenmuller and W. van Gool, Academic Press (1978), p. 277.

46. A. Kone, B. Barrau, J. L. Souquet and M. Ribes, Mat. Res. Bull. 14, 393 (1979).

47. A. Levasseur, B. Cales , J.-M. Reau and P. Hagenmuller, Mat. Res. Bull 13, 205 (1978).

48. A. Levasseur, J. -C. Brethous, J.-M. Reau and P. Hagenmuller, Mat. Res. Bull. 14, 921 (1979).

49. W. Jeitschko, Acta Cryst. B28, 60 (1972)

50. W. Jeitschko and T. A. Bither, Z. Naturforsch. 27b, 1423 (1972).

51. W. Jeitschko, T. A. Bither and P. E. Bierstedt, Act. Cryst. B33, 2767 (1977).

52. J. P. Malugani, A. Wasniewski, M. Doreau and G. Robert, C. R. Acad. Sci. Paris 284C, 99 (1977).

53. J.-C. Reggiani, J.-P Malugani and J. Bernard, J. Chim. Phys. 75, 849 (1978).

54. J.-P. Malugani and G. Robert, Mat. Res. Bul. 14, 1075 (1979).

55. J. Kuwamo, T. Isoda and M. Kato, Denki Kagaku 45, 104 (1977).

56. M. Lazzari, B. Scrosati and C. A. Vicent, Electrochim. Acta 22, 51 (1977).

57. M. Lazzari, B. Scrosati and C. A. Vicent, "Vitreous Solid Electrolytes," Proceedings of International Conference on Fast Ion Transport in Solids, Elsevier North-Holland (1979).

58. S. I. Smedley and C. A. Angell, Solid State Comm. 27, 21 (1978).

59. C. A. Angell and L. Boehm, "Ionic Conductivity in Lithium Oxide-Fluoride Glasses," Proceedings of Intenational Conference on Fast Ion Transport in Solids, Elsevier North-Holland (1979).

60. A. M. Glass and K. Nassau, "High Alkali Ion Conductivity in Rapidly Quenched Oxides Containing No Network Formers," Prodeedings of International Conference on Fast Ion Transport in Solids, Elsevier North-Holland (1979).

61. P. V. Wright, British Polymer Jour. 7, 319 (1975).

62. M. Armand, J. M. Chabagno and M. Duclot, "Polymeric Solid Electrolytes," presented at Second International Meeting on Solid Electrolytes, St. Andrews (1978).

63. M. B. Armand, J. M. Chabagno and M. Duclot, "Poly-Ethers as Solid Electrolytes," Proceedings of International Conference on Fast Ion Transport in Solids, Elsevier North-Holland (1979).

65. D. Chapin, H. Cheradame, J. M. Latour and J. L. Souquet, Proceedings of European Symposium on Electric Phenomena in Polymer Science, Pisa (1978).

66. Y-W. Hu, I. Raistrick and R. A. Huggins, Mat. Res. Bull. 11, 1227 (1976).

67. R. A. Huggins, Electrochimica Acta 22, 773 (1977).

68. Y-W. Hu, I. D. Raistrick and R. A. Huggins, J. Electrochem. Soc. 124, 1240 (1977).

69. Y-W. Hu, Ph.D. Dissertation, Stanford University (1979).

70. R. F. Aspandiar, unpublished results, Stanford University.

MOLTEN SALT ELECTROLYTES IN SECONDARY BATTERIES

Gleb Mamantov

Department of Chemistry
University of Tennessee
Knoxville, TN 37916

Introduction

In this paper I intend to present an introduction to molten salt electrolytes; the emphasis will be on melts of interest in rechargeable batteries.

Molten salts are predominantly ionic liquids; however, for some inorganic compounds, such as $AlCl_3$, the electrical conductivity decreases sharply upon melting indicating the formation of a molecular liquid ($Al_2Cl_6$). A few physical properties of some typical single component melts are given in Table 1. Most data were taken from a very recent compilation by Janz et al. (1) except for the data for NaBr and NaOH which were taken from the Molten Salts Handbook (2). It is apparent that the melting points are generally quite high (in excess of 600°C) except for $NaNO_3$, NaOH and $AlCl_3$. Molten nitrates are oxidizing solvents and are unsuitable for alkali metal anodes and other strong reducing agents. Molten hydroxides are quite corrosive and have not been of interest in battery applications. Note that the specific conductivity of all single component melts shown (except for $AlCl_3$) is in the range 1-6 $ohm^{-1}$ $cm^{-1}$; the specific conductivity of $AlCl_3$ is only an order of magnitude higher than that of very pure water.

The high melting points of most pure inorganic salts have resulted in the use of binary or ternary molten salt compositions; some physical properties of several common molten salt solvents (1) are given in Table 2. The somewhat lower conductivities (compared to molten alkali halides) of molten chloroaluminates, polysulfides and carbonates, and the high viscosities of molten

TABLE I

Some Physical Properties of Single Component Melts

| | $NaCl$ | $LiCl$ | $KCl$ | $NaF$ | $NaBr$ | $NaOH$ | $Na_2S$ | $NaNO_3$ | $Na_2CO_3$ | $Na_2SO_4$ | $MgCl_2$ | $CaCl_2$ | $AlCl_3$ |
|---|---|---|---|---|---|---|---|---|---|---|---|---|---|
| M.p. (°C) | 800±2 | 610±2 | 770±1 | 995±3 | 747 | 318 | 1170±10 | 307±1 | 858±1 | 884±2 | 714±4 | 782±5 | 192±1 (@ 2.3 atm.) |
| Density ($g/cm^3$) | 1.542 (@827°) | 1.490 (@637°) | 1.512 (@797°) | 1.928 (@1027°) | 2.320 (@774°) | 1.772 (@345°) | -- | 1.866 (@337°) | 1.959 (@887°) | 2.058 (@907°) | 1.668 (@747°) | 2.070 (@807°) | 1.231 (@217°) |
| Specific conductivity ($ohm^{-1}\ cm^{-1}$) | 3.687 (@827°) | 5.854 (@637°) | 2.229 (@797°) | 5.019 (@1027°) | 2.976 (@774°) | 2.332 (@345°) | -- | 1.107 (@337°) | 2.978 (@887°) | 2.333 (@907°) | 1.077 (@747°) | 2.147 (@807°) | $7.45 \times 10^{-7}$ (@217°) |
| Viscosity (cp) | 0.986 (@827°) | 1.377 (@637°) | 1.020 (@797°) | 1.78 (@1027°) | 1.26 (@800°) | 2.83 (@400°) | -- | 2.57 (@337°) | 3.40 (@887°) | 8.53 (@967°) | 2.04 (@747°) | 2.90 (@807°) | 0.271 (@217°) |

TABLE II

Some Physical Properties of Several Molten Salt Solvents

| | LiCl-KCl | NaCl-KCl | $CaCl_2$-NaCl | $AlCl_3$-NaCl | $Na_2S_x$ | LiF-LiCl-LiBr | $Li_2CO_3$-$Na_2CO_3$-$K_2CO_3$ |
|---|---|---|---|---|---|---|---|
| Liquidus temperature (°C) | 355 (eutectic, 58.5 mol % LiCl) | 685 (equimolar) | 500 (eutectic, 48 mol % NaCl) | 107 (eutectic, 36.8 mol % NaCl) ∿155 (equimolar) | 230 ($Na_2S_3$) 255 ($Na_2S_5$) | 445 (eutectic, 22 mol % LiF, 21 mol % LiCl, 47 mol % LiBr) | 397 (eutectic, 43.5 mol % $Li_2CO_3$, 31.5 mol % $Na_2CO_3$, 25.0 mol % $K_2CO_3$) |
| Density ($g/cm^3$) | 1.646 (for 60 mol % LiCl @ 447°) | 1.571 (@ 717°) | 1.855 (for 49.1 mol % NaCl @ 787°) | 1.691 (for 48 mol % NaCl @ 177°) 1.644 (for 38.2 mol % NaCl @ 177°) | 1.888 (for $Na_2S_3$ @ 327°) 1.876 (for $Na_2S_{4.8}$ @ 327°) | 2.19 (for eutectic @ 500°) | 2.110 (for eutectic @ 467°) |
| Specific conductivity ($ohm^{-1}$ $cm^{-1}$) | 1.615 (for 58.8 mol % LiCl @ 457°) | 2.396 (for 51.2 mol % NaCl @ 717°) | 1.183 (for 51.8 mol % NaCl @ 557°) | 0.462 (for 50 mol % NaCl @ 187°) 0.197 (for 31 mol % NaCl @ 187°) | 0.561 (for $Na_2S_3$ @ 327°) 0.296 (for $Na_2S_{5.1}$ @ 327°) | -------- | 0.516 (for eutectic @ 457°) |
| Viscosity (cp) | 1.46 (for 60 mol % LiCl @ 617°) | 1.58 (for 51.2 mol % NaCl @ 727°) | 4.43 (for 50 mol % NaCl @ 657°) | 2.645 (for 50 mol % NaCl @ 187°) 3.246 (for 31.1 mol % NaCl @ 187°) | 24.87 (for $Na_2S_{3.1}$ @ 327°) 19.37 (for $Na_2S_{5.2}$ @ 347°) | -------- | 5.402 (for eutectic @ 487°) |

polysulfides should be noted.

The electrochemical behavior of numerous solute species in several molten salt solvents has been reviewed previously by Fung and Mamantov (3) and by Plambeck (4); earlier reviews, employing a different arrangement, are also available (5-8). The discussion below will emphasize solvents and solute species of interest to rechargeable molten salt batteries.

## LiCl-KCl Eutectic

The molten salt solvent that has been used most widely for electrochemical and spectroscopic studies, as well as in the Li/Al-FeS (or $FeS_2$) battery, is the LiCl-KCl eutectic. It has a relatively low liquidus temperature for an alkali chloride melt; Pyrex glass can be used as the container material provided that moisture is completely removed (in the presence of $H_2O$ hydrolysis will result in the formation of HCl and $OH^-$ ions which attack glass) and the temperature is kept below the softening point of Pyrex (> 550°C). Most of the results have been obtained at 450°C. The potential span accessible in this melt, ∿3.6V, is determined by alkali metal deposition and chlorine evolution. The solvent must be treated with gaseous HCl or $Cl_2$ to remove hydrolysis products ($O^{2-}$ and $OH^-$); this treatment is followed by purging with a dry inert gas (e.g. Ar) (9). Preelectrolysis can be used to remove heavy metal impurities and traces of HCl and $Cl_2$.

The reference electrode that has been most commonly used in this solvent is based on the reversible Pt(II)/Pt couple; the standard potential of the Pt(II)/Pt(0) couple has been defined to be 0.000 V in this melt (10). Complications may result from the formation of the passivating $K_2PtCl_6$ layer and the anodic oxide films (in the presence of oxide ions) (4). Such complications are not present in the case of the Ag(I)/Ag couple, also used as the reference electrode in this solvent. The use of the $Cl_2(g)/Cl^-$ couple as the reference electrode is less convenient than that of the metal ion/metal couples.

The electrochemistry of oxygen (or water) derived species is controversial (4). The possible species involved, in addition to $O^{2-}$ and $OH^-$ ions, are peroxide, $O_2^{2-}$, superoxide, $O_2^-$, as well as "adsorbed oxides" which are dependent on the nature of the electrode material. The container material (e.g. Pyrex) probably has an effect on the equilibria involving these various species (such an effect was observed in studies using nitrate melts (11)).

The standard potential of the $Cl_2/Cl^-$ couple (+0.322±0.002V) was established using a carbon rod (Pt and Au are attacked); the standard states correspond to 1 atm pressure of $Cl_2$ and the activity of $Cl^-$ions corresponding to the eutectic. The solubility of

Table 3

EMF Series in LiCl-KCl Eutectic at 450°C

| Couple | $E^\circ_M$(Pt), V | $E^\circ_m$(Pt), V | $E^\circ_X$(Pt), V | Precision, V |
|---|---|---|---|---|
| Li(I)/Li(0) | −3.304 | −3.320 | −3.410 | 0.002 |
| Na(I)/Na(0) | −3.25 | −3.23 | −3.14 | 0.008 |
| Mg(II)/Mg(0) | −2.580 | −2.580 | −2.580 | 0.002 |
| Al(III)/Al(0) | −1.762 | −1.767 | −1.797 | 0.009 |
| Zn(II)/Zn(0) | −1.566 | −1.566 | −1.566 | 0.002 |
| Cr(II)/Cr(0) | −1.425 | −1.425 | −1.425 | 0.003 |
| Fe(II)/Fe(0) | −1.172 | −1.172 | −1.172 | 0.005 |
| Se(l),C/$Se_x^{2-}$ | −1.141 | −1.172 | −1.252 | 0.002[a] |
| S(l),C/$S_x^{2-}$ | −1.008 | −1.039 | −1.219 | 0.002[a] |
| Cu(I)/Cu(0) | −0.957 | −0.941 | −0.851 | 0.004 |
| Ni(II)/Ni(0) | −0.795 | −0.795 | −0.795 | 0.002 |
| Ag(I)/Ag(0) | −0.743 | −0.727 | −0.637 | 0.002 |
| HCl(g)/$H_2$(g),Pt | −0.694 | −0.710 | −0.800 | 0.005 |
| Pt(II)/Pt(0) | 0.000 | 0.000 | 0.000 | 0.002 |
| Cu(II)/Cu(I) | +0.061 | +0.045 | −0.045 | 0.002 |
| Fe(III)/Fe(II) | +0.086 | +0.070 | −0.020 | 0.003 |
| Au(I)/Au(0) | +0.205 | +0.221 | +0.311 | 0.008 |
| $Cl_2$(g),C/$Cl^-$ | +0.322 | +0.306 | +0.216 | 0.002 |

[a] extrapolated

chlorine has been reported to be $1.54 \times 10^{-4}$ $\underline{M}$ at 450°C (12); spectroscopic evidence for the $Cl_3^-$ ion has been obtained (13).

Sulfur, which boils at 445°C, can be reduced reversibly to sulfide at a carbon electrode; the standard potential of the S($\ell$)/$S^{2-}$ couple has been reported as -1.008±0.002V at 450°C (an extrapolated value from the temperature range 400-440°C (4)). In the presence of both sulfur and sulfide colored polysulfide species, such as $S_2^-$ and $S_3^-$, are formed (4,14). Oxidation of sulfur at ∿0V results in the volatile $S_2Cl_2$. The reduction of selenium to selenide, the formation of polyselenide species, and the oxidation of selenium at potentials similar to those for sulfur have also been observed (4).

The electrochemistry of iron in this melt is of importance to the Li/Al-FeS battery. Iron can be oxidized reversibly to Fe(II) (E° for this couple is -1.172±0.005V) which can be oxidized further to Fe(III) (E° for the Fe(III)/Fe(II) couple is +0.086±0.003V) which is a powerful oxidizing agent in this melt.

Lithium metal (m.p. 179°C) is somewhat soluble in this melt resulting in some electronic conductivity and the evolution of gaseous potassium (vapor pressure of potassium at 450°C is 0.7 torr (15)). Lithium also forms stable intermetallic compounds or alloys with a number of metals including aluminum and silicon. Such alloys have been found to be more useful than pure lithium as negative electrodes in rechargeable molten salt batteries. An abbreviated EMF series for this melt is shown in Table 3 (4).

## Polysulfide Melts

Sodium polysulfide melts are employed in the sodium/sulfur battery. In contrast to sodium sulfide which has a very high melting point (1170°C), the melting points of the higher polysulfides ($Na_2S_3$, $Na_2S_4$, $Na_2S_5$) are in the range 230°-294°C (1). In the initial discharge reaction sulfur is reduced at a carbon electrode to $Na_2S_5$; this material is not soluble in sulfur, resulting in two liquid phases throughout ∿60% of discharge. During the rest of the discharge process $Na_2S_5$ is transformed to $Na_2S_3$. The electrochemistry of sulfur-containing species in these melts has been reviewed by Tischer and Ludwig (16). According to them the undiluted sodium polysulfide melts in the 300-400°C range contain mainly $S_2^{2-}$, $S_4^{2-}$, $S_5^{2-}$ and some $S_6^{2-}$, the relative amounts of each depending on the overall composition. No free elemental sulfur is believed to be present in the melt; evidence for the disproportionation of $S_3^{2-}$ into $S_2^{2-}$ and $S_4^{2-}$ has been obtained. Tischer and Ludwig also conclude that the concentration of ion radicals in the undiluted polysulfide melt is extremely small. The electrochemical processes involved are complex; for the reduction of $Na_2S_5$ the following reactions have been proposed (16):

$$S_5^{2-} + e^- = S_5^{3-}$$

$$S_5^{3-} + e^- = S_2^{2-} + S_3^{2-}$$

$$2S_3^{2-} \longrightarrow S_2^{2-} + S_4^{2-}$$

## Chloroaluminate Melts

Molten chloroaluminates ($AlCl_3$-MCl mixtures where M is an alkali metal) have low liquidus temperatures and reasonable conductivities (see Table 2). The acid-base properties of these melts (determined primarily by the $AlCl_3$/MCl ratio) and the factors affecting the chemistry of solutes have been reasonably well characterized (17). $AlCl_4^-$ is the predominant anion in the equimolar $AlCl_3$-NaCl melt. For more acidic ($AlCl_3$-rich) melts the $Al_2Cl_7^-$ ion becomes more abundant.

Molten chloroaluminates have been used as solvents in the Al/$Cl_2$ (18,19), Na/$Cl_2$ (20), Al/S (21,22) cells as well as in cells utilizing a metal halide as the active cathode material and either aluminum or sodium as the anode (23-26). The use of sodium as the anode is made possible by the compatibility of β- (or β"-) alumina with chloroaluminate melts (20,24). The use of aluminum as the anode is potentially attractive since this metal is quite electropositive, relatively inexpensive and has a low equivalent weight. Problems caused by dendrite formation during charging and by passivation phenomena (formation of a poorly conducting $Al_2Cl_6$ layer at high current densities) during the dissolution of aluminum can be attacked by adding selected contaminants (27) and by using proper melt compositions (28).

The most developed battery using chloroaluminates is based on the Na/$SbCl_3$ system for which an open circuit voltage of 2.83V and a theoretical energy density of 825Wh/kg have been reported (20,26). In the discharge process Sb(III) is reduced to metallic antimony; the cathode compartment becomes more basic (NaCl-rich) during this process. These reactions are reversed during the charge process. Although an operating life of ∿500 cycles has been achieved, problems were encountered caused primarily by the ceramic tube cracking and the corrosion of the molybdenum current collector used in the cathode compartment (25). In addition, studies have shown that this system will be significantly higher in cost than the sodium-sulfur battery (25).

We have demonstrated the feasibility of using tetravalent sulfur as the active cathode material dissolved in an acidic chloroaluminate melt at temperatures 180-260°C (29-31). Such a cathode coupled with a sodium anode separated by β"-alumina results in a rechargeable cell with an OCV of 4.2V. The overall reaction at the positive electrode upon discharge is

$$SCl_4 + 4Na^+ + 4e^- = S + 4\ NaCl$$

where S(IV) is written as $SCl_4$ in order to simplify the stoichiometry ($SCl_4$ is unstable above -30°C (32)). Sulfur is present predominantly as $S_8$ (33); $Cl^-$ ions react with $Al_2Cl_7^-$ ions to form additional $AlCl_4^-$ ions. Sulfur can be reduced further to sulfide present mainly as AlSCl or $AlSCl_2^-$ (17). The reduction of S(IV) to sulfur proceeds through several intermediates which include S(I), $S_8^{2+}$ and some radical species (34). The calculated energy density for the reaction

$$SCl_4 + 4Na = S + 4\ NaCl$$

in a chloroaluminate melt which has the initial mole ratio of total Al(III) to Na(I) of 70/30, is 325 Wh/kg (this value includes the weight of the solvent but not the weight of the separator); this parameter is increased to 419 Wh/kg using the overall reaction

$$SCl_4 + 6\ Na = Na_2S + 4\ NaCl$$

and assuming that the reduction of sulfur to sulfide occurs in melts saturated with NaCl (35). Measured energy densities for discharge to sulfur have exceeded 250 Wh/kg (at ∿15 mA/cm$^2$); the corresponding energy efficiency and utilization of S(IV) are >85%. Cycle life of >100 cycles has been demonstrated. Most of the cell failures to date have been caused by the formation of cracks in the β"-alumina tubes or in the β"/α-alumina seal (35).

A corresponding Se(IV) cathode in chloroaluminate melts has been described by Marassi et al. (36).

Other Low-Melting Electrolytes

The use of ternary chloroaluminate compositions, such as $AlCl_3$-LiCl-KCl or $AlCl_3$-NaCl-KCl, results in melts with lower liquidus temperatures as compared to $AlCl_3$-NaCl (for the binary eutectic composition containing 61.4 mol % $AlCl_3$ the liquidus temperature is 107.2°C (37)). The values reported in the literature differ considerably because of supercooling of the melts. For the composition $AlCl_3$-NaCl-KCl (66-20-14 mol %) the liquidus temperature of 93°C appears to be reliable (4,38). It should be noted that electrolytes which are liquid over a narrow composition range, may not be satisfactory for rechargeable battery applications because of composition gradients produced by the electrode reactions (39).

Much lower liquidus temperatures may be obtained with

pyridinium halide-containing melts, such as the $AlCl_3$-1-ethyl-pyridinium bromide (2/1 molar ratio) melt which has been used at room temperatures (40-42). Alkyl pyridinium chloride-$AlCl_3$ mixtures (43,44) are under investigation for possible battery applications (42). The specific conductivities of these melts are considerably lower than those of $AlCl_3$-alkali chloride melts.

Acknowledgement

I would like to acknowledge the support of the Department of Energy, Contract EY-76-S-05-5053, and the Air Force Aero Propulsion Laboratory, Contract F33615-78-C-2075.

References

1. G. J. Janz, C. B. Allen, N. P. Bansal, R. M. Murphy and R. P. T. Tomkins, "Physical Properties Data Compilations Relevant to Energy Storage. II. Molten Salts," NSRDS-NBS 61, Part II, April 1979.

2. G. J. Janz, "Molten Salts Handbook," Academic Press, 1967.

3. K. W. Fung and G. Mamantov, in Advances in Molten Salt Chemistry, Vol. 2, J. Braunstein, G. Mamantov, and G. P. Smith, eds., Plenum Press, 1973, pp. 199-254.

4. J. A. Plambeck, "Fused Salt Systems," Vol. 10, Encyclopedia of Electrochemistry of Elements, A. J. Bard, ed., M. Dekker, 1976.

5. C. H. Liu, K. E. Johnson and H. A. Laitinen, in "Molten Salt Chemistry," M. Blander, ed., Interscience, 1964, pp. 681-733.

6. H. A. Laitinen and R. A. Osteryoung, in "Fused Salts," B. R. Sundheim, ed., McGraw-Hill, 1964, pp. 255-300.

7. Yu. K. Delimarskii and B. F. Markov, "Electrochemistry of Fused Salts," Sigma Press, 1961.

8. D. Inman, A. D. Graves and R. S. Sethi, in "Electrochemistry," Vol. 1, The Chemical Society, London, 1970; D. Inman, A. D. Graves and A. A. Nobile, in "Electrochemistry," Vol. 2, The Chemical Society, London, 1972; D. Inman, J. E. Bowling, D. G. Lovering and S. H. White, in "Electrochemistry," Vol. 4, The Chemical Society, London, 1974.

9. Melt purification and other experimental aspects are discussed in more detail by K. W. Fung and G. Mamantov, in Wilson & Wilson's "Comprehensive Analytical Chemistry," Vol. III, G. Svehla, ed., Elsevier Publishing Co., 1975, pp. 305-370.

10. H. A. Laitinen and C. H. Liu, J. Amer. Chem. Soc., 80, 1015 (1958).

11. J. Jordan, W. B. McCarthy and P. G. Zambonin, in "Molten Salts; Characterization and Analysis," G. Mamantov, ed., M. Dekker, 1969, pp. 575-592.

12. J. D. Van Norman and R. J. Tivers, in "Molten Salts: Characterization and Analysis," G. Mamantov, ed., M. Dekker, 1969, pp. 509-527.

13. J. Greenberg and B. R. Sundheim, J. Chem. Phys., 29, 1029 (1958).

14. D. M. Gruen, R. L. McBeth and A. J. Zielen, J. Amer. Chem. Soc., 93, 6691 (1971).

15. R. N. Seefurth and R. A. Sharma, J. Electrochem. Soc., 122, 1049 (1975).

16. R. P. Tischer and F. A. Ludwig, in "Advances in Electrochemistry and Electrochemical Engineering," Vol. 10, H. Gerischer and C. W. Tobias, eds., John Wiley & Sons, 1977, pp. 391-482.

17. G. Mamantov and R. A. Osteryoung, in "Characterization of Solutes in Non-Aqueous Solvents," G. Mamantov, ed., Plenum Press, 1978, pp. 223-249.

18. J. Giner and G. L. Holleck, NASA Report CR 1541, March 1970.

19. G. D. Brabson, A. A. Fannin, Jr., L. A. King and D. W. Seegmiller, Abstract #26, Electrochemical Society Extended Abstracts, Spring Meeting, Chicago, Illinois, May 13-18, 1973.

20. J. J. Werth, U.S. Patent 3,847,667, Nov. 12, 1974; U.S. Patent 3,877,984, April 15, 1975.

21. L. Redey, I. Porubszky and I. Molner, 9th International Power Sources Symposium, Brighton, England, September 17-19, 1974.

22. J. Greenberg, U.S. Patent 3,573,986 (1971); 3,635,765 (1972).

23. G. D. Brabson, J. K. Erbacher, L. A. King, and D. W. Seegmiller, FJSRL Technical Report 76-0002, January 1976.

24. J. J. Werth, EPRI Report EM-230, December 1975.

25. A. M. Chreitzberg, EPRI Report EM-751, April 1978.

26. W. P. Sholette, I. S. Klein, and J. Werth, in "Proceedings of the Symposium on Load Leveling," N. P. Yao and J. R. Selman, eds., The Electrochemical Society Symposium Series, Vol. 77-4, Princeton, N. J., (1977), pp. 306-317.

27. R. C. Howie and D. W. MacMillan, J. Appl. Electrochem., 2, 217 (1972).

28. B. Gilbert, D. L. Brotherton and G. Mamantov, J. Electrochem. Soc., 121, 773 (1974).

29. G. Mamantov, R. Marassi and J. Q. Chambers, U.S. Patent 3,966,491, June 29, 1976.

30. G. Mamantov and R. Marassi, U.S. Patent 4,063,005, December 13, 1977.

31. G. Mamantov, R. Marassi, J. P. Wiaux, S. E. Springer and E. J. Frazer, in "Proceedings of the Symposium on Load Leveling," N. P. Yao and J. R. Selman, eds., The Electrochemical Society Symposium Series, Vol. 77-4, Princeton, N. J., (1977), pp. 379-383.

32. J. C. Bailar, H. J. Emeleus, R. Nyholm, and A. F. Trotman-Dickenson, eds., "Comprehensive Inorganic Chemistry," Vol. 2, Pergamon Press, 1973, p. 857.

33. R. Huglen, F. W. Poulsen, G. Mamantov, R. Marassi and G. M. Begun, Inorg. Nucl. Chem. Letters, 14, 167 (1978).

34. R. Marassi, G. Mamantov, M. Matsunaga, S. E. Springer, and J. P. Wiaux, J. Electrochem. Soc., 126, 231 (1979).

35. Y. Ogata, R. Marassi and G. Mamantov, unpublished work.

36. R. Marassi, D. Calasanzio and G. Mamantov, Proceedings of NATO Institute on Advanced Batteries (in press).

37. E. M. Levin, J. F. Kinney, R. D. Wells and J. T. Benedict, J. Res. Nat. Bur. Stand., 78A, 505 (1974).

38. R. Midorikawa, J. Electrochem. Soc. Japan, 23, 127 (1955).

39. C. E. Vallet and J. Braunstein, J. Electrochem. Soc., 125, 1193 (1978).

40. H. L. Chum, V. R. Koch, L. L. Miller and R. A. Osteryoung, J. Amer. Chem. Soc., 97, 3264 (1975).

41. V. R. Koch, L. L. Miller and R. A. Osteryoung, J. Amer. Chem. Soc., 98, 5277 (1976).

42. C. L. Hussey, J. C. Nardi, L. A. King, and J. K. Erbacher, J. Electrochem. Soc., 124, 1451 (1977); C. L. Hussey, private communication.

43. J. Robinson and R. A. Osteryoung, J. Amer. Chem. Soc., 101, 323 (1979).

44. R. J. Gale and R. A. Osteryoung, Inorg. Chem., 18, 1603 (1979).

# THE RECHARGING OF THE LITHIUM ELECTRODE IN ORGANIC ELECTROLYTES

S. B. Brummer, V. R. Koch
and R. D. Rauh

EIC Corporation
55 Chapel Street
Newton, MA 02158

## ABSTRACT

The state of the art of the secondary lithium electrode in organic solvent based electrolytes at ambient temperature is reviewed.

## INTRODUCTION AND PURPOSES - THE PROBLEM

We have chosen to restrict the topic "organic electrolytes" to the deposition and redissolution of Li in electrolytes based on organic solvents. Interest in this field is growing as there is a need for an ambient temperature rechargeable Li cell. Indeed it is now possible to purchase a rechargeable Li cell, albeit of rather limited capabilities (1). Because of the limited amount of good recent data that have been published, the main focus of our review is on the work we have carried out at EIC.

The Li/organic electrolyte field has developed since the thesis of Harris in 1958 (2). The majority of a large body of work since then has been concerned, among all the alkali and alkaline earth metals, with Li the lightest metal. Why is it then, more than 20 years later, that so little is known about how to recharge the Li electrode? One may point out, for example, the very limited body of data that was available for a recent review (3). The answer is that it turned out to be difficult to cycle Li (or any of the other alkali metals for that matter), and attention turned to the development of primary cells. This development has been highly successful, at least technically, and batteries based on $Li/I_2$, $Li/MnO_2$, $Li/V_2O_5$, $Li/Ag_2CrO_4$, $Li/CF_x$, $Li/SO_2$ and $Li/SOCl_2$ are commercially available. The cell based on $SOCl_2$ has extraordinary performance characteristics (e.g., >300 Whr/lb at ∼C/30 days) and exemplifies the problems which beset the recharging of Li in both inorganic and organic solvents.

In this cell, $SOCl_2$ is both the electrolytic solvent and the cell depolarizer. The open circuit potential is ∿3.6V. On a carbon cathode, $SOCl_2$ will reduce at up to 20 $mA/cm^2$. Yet Li will remain shiny when it is boiled in $SOCl_2$ (4). The metal is not unattacked, however, and a thin film of LiCl grows on its surface (5). Dey has carried out a detailed study of the growth of these films as a function of temperature and solution composition (6). The crucial finding in the present connection is that the LiCl film appears to be present under all experimental conditions. It is not our purpose to speculate about whether the LiCl observed on the surface is the primary passivating film or is only a reaction product. The important point is that it is always present.

Similarly, it is now generally agreed that the stability of propylene carbonate (PC) - the most popular of the organic Li-battery aprotic solvents - in contact with Li, which is considerable (7), results from formation of a film, probably of $Li_2CO_3$ (8,9). It was shown quite early on that in PC, amongst other solvents, Li can be plated with 100% efficiency (10,11). But it cannot be stripped as efficiently, particularly after the deposit has been allowed to stand in contact with the solution (10,11). Selim and Bro (11) showed that although plated Li becomes less accessible to electrostripping during stand, much of it is still in the form of metal. But this metal is electrically isolated from the conductive substrate.

The model in Fig. 1, based on our earlier publication (12), will account for the above results: Li is deposited with 100% faradaic efficiency: The surfaces of the grains of the deposit react with the solution (including some impurities) at a high rate, then with solvent and/or salt at a lower rate. The Li grains become partially under-cut and some of them become partially or completely isolated from the substrate. Subsequent stripping is not then 100% efficient, and some undischarged metal obscures the surface. This type of intergrannular attack continues during stand, albeit more slowly. The residual undischarged Li results in an irregular surface, which distorts further the morphology of subsequent plates. Thus, cycle-by-cycle, the deposit becomes increasingly dendritic, and the stripping efficiency declines rapidly over a small number of cycles and becomes essentially zero (12-14).

This would not be serious if solvents could be found which are inherently non-reactive towards Li. But it seems unlikely that many solvents could be found which would be thermodynamically stable, particularly when we consider that to sustain electrolyte solubility and conductance the solvent must be polar. The best we could hope for is that the solvent would be kinetically hindered from reacting with Li, but without the need for the presence of protective films. This is by no means impossible. For example in our laboratory we have

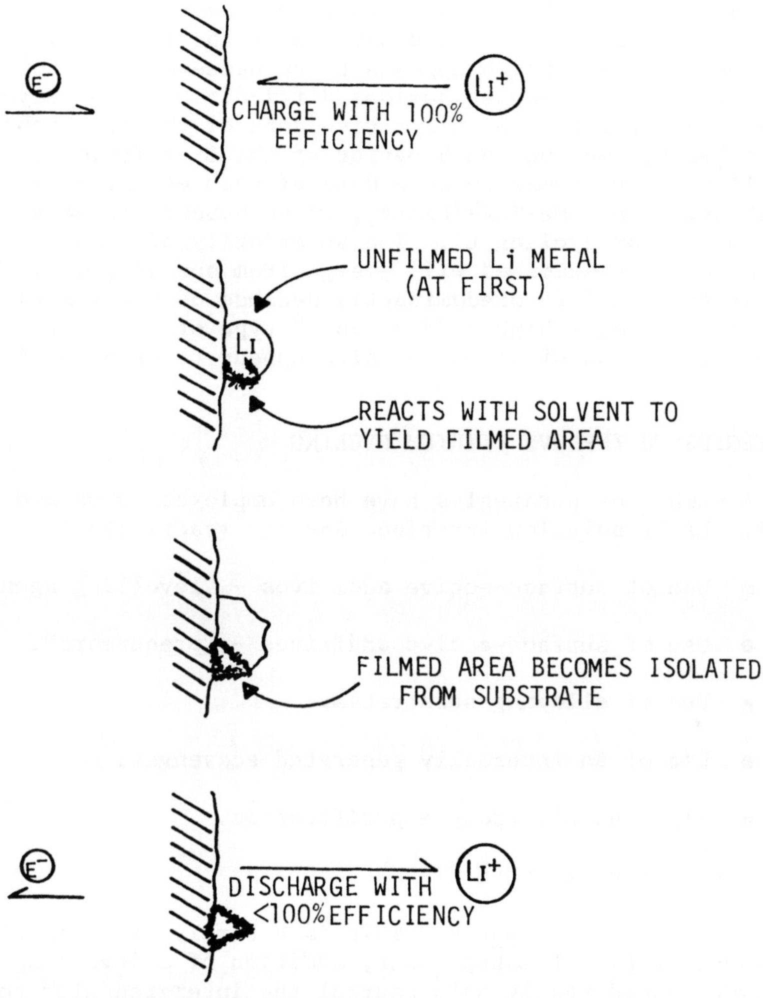


Fig. 1. Mechanism for isolation of plated Li

observed that 1M $Br_2$ in PC attacks Li only at the rate of ∿1 $mA/cm^2$ (15), even though the LiBr reaction product appears to be much too soluble to form a protective film. We must be careful though to appreciate that the $Li_2CO_3$ film on the metal from the Li-PC reaction may be partially protective.

In the main, though, it appears necessary for us to live with the presence of some films on the surface of the metal. The problem is that the very film which appears to be necessary for the metal's stability on stand causes serious difficulties with recycling. In the succeeding sections, we will review some of the approaches that have been taken to improve the behavior of the Li-solution interface. Then we will briefly summarize some data of Koch et al. at EIC on 2-methyltetrahydrofuran (2Me-THF)/$LiAsF_6$, which appears to be a superior electrolyte for cycling Li. The superiority of $LiAsF_6$ with every solvent we have examined will emerge from our body of work. We will suggest that this is predominantly because of the reactivity of the salt with Li, forming a highly "desirable" type of film. Finally, we will suggest the areas of research which appear to be promising at this time.

## STRATEGIES TO IMPROVE LITHIUM CYCLING

A number of strategies have been employed by us and others to modify the Li-solution interface and its reactivity:

- Use of surface-active additives - "levelling agents".
- Use of surface-active additives - "precursors".
- Use of alloying substrates.
- Use of an internally generated scavenger.
- Rigorous electrolyte purification.

We discuss these in turn.

• "Levelling Agents". This is a barely explored field but, as the model of Fig. 1 makes clear, addition of a levelling or brightening agent could really help control the intergrannular corrosion process that is so damaging. There are very little data in this area. One report (15) claims that in PC-based electrolytes, the addition of surface active materials has some beneficial effects on Li plate morphology, albeit at impractically low (≤0.1 $mA/cm^2$) current densities (16).

Some improvement in Li adherence was noted in early work due to addition of Rhodamine B sodium salt and disodium fluorescein to PC/$LiAlCl_4$ (17). Besenhard and Eichinger (3) also note some claims in the patent literature. This area obviously merits more attention.

● "Precursors". The passivation of Li towards highly oxidizing media is frequently observed and, as we have pointed out, is an intrinsic feature of batteries such as that based on $SOCl_2$. We have attempted to take advantage of this type of passivation phenomenon to modify the Li-solution film. The concept is to use a compound, such as $SOCl_2$ or $SO_2$, which will be able to compete for the surface with the solvent, and will form a "desirable" type of film. "Desirable" might mean thin and $Li^+$-ion conductive but impermeable to solvent molecules:

$$\text{Li} + \text{Pr} \xrightarrow{k_1} \text{LiPr}' \text{ (insoluble film)} \qquad (1)$$

$$\text{Li} + \text{solvent} \xrightarrow{k_2} \text{Li-solvent}' \text{ (insoluble film)} \qquad (2)$$

Here $k_1 > k_2$ and Pr is the "precursor" of the desired film, LiPr'.

An extensive investigation of this phenomenon was carried out in PC/1M $LiClO_4$ and tetrahydrofuran (THF) (12,18,19). Previous work had also shown some value for this concept in methyl acetate (MA) (20). Precursors (generally at 0.1M) were chosen which were likely to form passivating films on Li. The rationale for the choices is given in ref. 19 (p. 54 et seq.) The compounds investigated as precursors were: $CS_2$, $PSCl_3$, $POBr_3$, $PNBr_2$, $POCl_3$, $MoOCl_4$, $CH_3NO_2$, $VOCl_2$, $CO_2$, $N_2O$, and $SO_2$. The cycling regime was $i_p = i_s = 2.5$ mA/cm$^2$; $Q_p = Q_s = 10$ coul/cm$^2$ (p ≡ plating; s ≡ stripping). In general, the pure solvent had been treated with activated molecular sieves before addition of the supporting electrolyte. Electrolytes were treated again with molecular sieves following the addition of the precursor. The base line system, PC/1M $LiClO_4$ treated with molecular sieves before and after addition of the $LiClO_4$, gave an average cycling efficiency of 40%. The effect of additives on this efficiency is summarized in Table 1.

$CS_2$ led to a deterioration in average cycling efficiency, $\bar{E}$ being 37%. The stripping curves were generally sloping anodically, and the dual plateau that we usually observe at low current densities when the stripping progresses from electroplated to massive Li was not evident. A second cell containing $CS_2$ was allowed to stand overnight before cycling was begun. The working electrode became highly polarized during the first stripping cycle and second plating cycle, and "failed" on the second stripping cycle. Disassembly of this cell revealed that the Li electrode was still intact, unlike the former case in which the electrode was fully stripped. It is clear that the formation of a resistive film on overnight stand gave rise to the poor cycling behavior of the second cell.

Addition of $PSCl_3$ to the 1M $LiClO_4$ in PC electrolyte resulted in a slight improvement in Li cycling efficiency. The initial stripping cycle did show a clear anodic overshoot, not present without the precursor, indicating a film-forming reaction between $PSCl_3$ and the Li.

Table 1

Effects of Additives on the Cycling Efficiency of Li on a 5-mil Li Foil Substrate. Base Electrolyte is PC/1M $LiClO_4$. Following Addition of Additive, the Electrolyte was Stirred with Activated Molecular Sieves

| Additive | Additive Concentration (M) | Average Efficiency % | Comments |
|---|---|---|---|
| $CS_2$ | 0.1 | 37.0 | |
| " | 0.1 | Polarized | Cycled after 12 hr stand |
| $PSCl_3$ | 0.1 | 70.7 | |
| " | 0.1 | 63.7 | |
| $PSBr_3$ | 0.1 | 84.5 | |
| $POBr_3$ | 0.1 | 63.7 | |
| " | 0.01 | 85.1 | |
| $PNBr_2$ | 0.1 | 31.7 | |
| " | 0.1 | 51.6 | Electrolyte treated with Li |
| $POCl_3$ | 0.1 | 34.0 | |
| " | 0.1 | Polarized | |
| " | 0.01 | 48.8 | |
| $CH_3NO_2$ | 0.1 | 34.0 | |
| $MoOCl_4$ | 0.1 | 63.7 | |
| $VOCl_2$ | 0.01 | 59.0 | |
| $CO_2$ | sat. (∿0.3) | 34.4 | Gas evolved during cycling |
| " | ∿0.1 | 64.4 | |
| $N_2O$ | 0.11 | 74.1 | |
| $SO_2$ | 0.11 | 34.0 | |
| | 3.6 | 79.8 | |

$PSBr_3$ gave rise to a significant improvement in Li cycling efficiency, yielding an average of 84.5%. Also, as in the case of $PSCl_3$, a definite anodic overshoot was present at the beginning of the first stripping cycle. The two-stepped stripping curves were not as evident in this case as $PSCl_3$, the stripping curves being more sloped, as with $CS_2$.

$POBr_3$ gave rise to a very resistive coating on the Li electrode. The polarization of the first stripping cycle was not anomalous, and the anodic overshoot was quite small. After a short delay, the 10 coul/cm$^2$ plating cycle was begun, and this was accompanied by a polarization of ∿2.5V. Subsequent plating and stripping cycles also showed unusually large polarizations (∿3V). The final value of $\bar{E}$ was 63.7%; the Li working electrode was completely stripped after being cycled to failure. Further improvement was achieved by decreasing the $POBr_3$ concentration to 0.01M, the average efficiency being 85.1%. Here, the observed polarizations were not abnormally high.

Analogous P-containing additives gave only moderate improvements in the cycling efficiency. $PNBr_2$ yielded only 31.7%, which improved to 51.6% when the electrolyte was treated with Li. $POCl_3$, like $POBr_3$, causes low efficiencies at the 0.1M level, but efficiencies improve at the 0.01M level.

The remaining oxychlorides - $MoOCl_4$ and $VOCl_2$ - showed moderate improvements in the cycling efficiency over the base line. With 0.1M $VOCl_2$, Li could not be plated or stripped, as the film was too resistive. Lowering the $VOCl_2$ concentration to 0.01M allowed a cycling efficiency of 59.0%.

$CO_2$ is only moderately soluble in PC/1M $LiClO_4$ (ca. 0.1-0.3M). A saturated solution, made by slowly bubbling $CO_2$ through the electrolyte for several hours, gave erratic plating and stripping curves: Gas bubbling from solution resulted in noisy chronopotentiograms, and polarizations were high. The Li cycling efficiency was only 34.4%. A somewhat less concentrated solution (0.05-0.1M) behaved more normally, and yielded an efficiency of 64.4%.

$N_2O$ and $SO_2$ also gave improved cycling efficiencies. The $SO_2$ had to be present in a high concentration: Li cycled in 0.11M and 3.6M solutions yielded efficiencies of 34% and 79.8%, respectively. In 0.11M $N_2O$, a cycling efficiency of 74.1% was obtained.

$CH_3NO_2$, which improved cycle efficiency and cycle life on a Ni substrate (12), gave very poor results (34%) on a Li substrate.

It appears, then, that many of these compounds can act as precursors and form films on the Li surface in preference to PC itself. This is encouraging. There seems to be a maximum improvement, however. In this particular experiment, we have only limited data but the best we can achieve appears to be an average cycling efficiency of $\sim$85%. This suggests, perhaps, that PC adsorption on certain crystal faces still controls some basic aspect of the morphology of the deposit.

In limited investigations, we have also found precursor effects in MA/$LiClO_4$ (20) and THF (21).

In 1M $LiAsF_6$, the cycling efficiency in this experiment was 85.2%, in striking agreement with the best results for precursors. Addition of 0.01M $POBr_3$, an excellent additive with $LiClO_4$, reduced $\bar{E}$ to 75.8%. These results suggest that $AsF_6^-$ and $POBr_3$ compete for the same sites and act in the same way.

We have invariably found that $LiAsF_6$ is the best supporting electrolyte for plating Li, for example, in PC (13) or MA (14) or THF (25) or 2Me-THF (22). We suggest that this is partly because of the high purity of the starting material but not principally. The major effect of $LiAsF_6$, we believe, is because it forms a "desirable" film on the Li surface. It is acting as a precursor. In support of this idea, we note that brown films are formed on Li in presence of $LiAsF_6$ in THF (23) and 2Me-THF (24). This film, which in THF has the composition $(\text{-As-O-As-})_n$ + LiF (23), obviously results from reaction of the anion with Li (and the solvent).

● Alloying Substrates. It is well known that as-plated Li will alloy with many metals. The most complete listing of such effects appears to be that of Dey (26), but there are many earlier reports of good cycling efficiency with alloying substrates and thin deposits of Li (27-30). A particularly popular alloying substrate is Al (31,32).

Alloying can function in two ways: the more anodic potential of the alloy lessens the driving force for solution reduction; dissolution of the Li deposit into the alloying substrate elminates the possibility of undercutting of the deposit by reaction with the solution. The difficulty is that this effect is only observed for very thin plates. To obtain plates of practical thickness, a high area substrate is necessary. The problem then is that the volume changes during formation and dissolution of the alloy tend to destroy the substrate. A cell incorporating a Li-Al anode is commercially available (1). We believe that the merit of this approach to demonstrate a practical plate (say 100+ cycles at 15-30 $mAhr/cm^2$) is yet to be demonstrated.

● Scavenger. In this concept, we add a material to the solution (the "reduced scavenger") which is normally inert (e.g., LiBr), but which can be converted into a "scavenger (e.g., $Br_2$) which will attack encapsulated Li. The scavenger redox potential must be above the potential of the positive electrode. Hence, the reduced scavenger ($Br^-$) will be the normal (non-corrosive) form. Hence, the self-discharge rate on stand will be very low. With a positive-limited battery, the scavenger could be generated on overcharge, or on an auxiliary electrode. A better alternative would be to employ a negative-limited configuration, and to overdischarge the system. This would generate the scavenger (e.g., $Br_2$) right at the Li-electrode substrate. This scavenger would attack and redissolve the encapsulated Li. Since there would be no other Li metal available, we would anticipate good efficiency.

There is evidence from the literature (33,34) that this is a viable approach: Weininger et al. explored the $Li/Br_2$ cell and achieved ∿1800, ∿3 $coul/cm^2$ charges and discharges. This is by far the largest number of cycles that has been reported for Li in organic electrolytes. The reason, we believe, is that the dissolved $Br_2$ positive kept the Li-electrode substrate free of encapsulated Li.

Criteria for selection of scavengers are:

- They (and where there are $Li^+$ salts of their reduction products) must be soluble.

- The scavenger redox potential must be above the potential of the positive electrode.

- The scavenger redox process must be simple.

Possible scavenger couples for this purpose are $LiI/I_2$, $LiBr/Br_2$, chloranil, fluoranil, Li polysulfide and ferrocene/ferrocinium.

We have carried out some preliminary investigations of this concept in MA ($Br^-/Br_2$) and THF ($I^-/I_2$ and $S^{-2}/S_n^{-2}$), with some promising results (19). For example, Fig. 2 shows some data for 1 coul/$cm^2$ of Li onto Ta from 0.5M $LiAsF_6$ + 0.5M Li in THF. The first 10 cycles, without any $I_2$ generation, show only modest efficiency, with a steady cycle-by-cycle decline. After the 10th cycle, 1 coul/$cm^2$ of $I_2$ was generated. Cycling efficiency improves from <30% to >90%. Subsequently, over the next 10 cycles, it declines, as before. Then $I_2$ was generated again, and subsequently between every cycle. The improved resistance to decline in cycling efficiency is manifest. At the time of this investigation, we were not aware of the intrinsic reactivity between Li and the solvents, and a further investigation in a more stable medium appears appropriate.

- Electrolyte Purification. The basic problem in recycling Li is reactivity. To assess the intrinsic reactivity of the solvent, it is necessary to remove electrolytic impurities as far as possible. We have worked very extensively with PC and THF.

Since PC is one of the organic solvents known to be most stable (kinetically) towards Li, we first employed it in a detailed investigation of desiccant effects on electrolyte purification (13). Prior to our work, there was only one publication addressing itself to the analysis and purification of PC (35). It was demonstrated that, even after careful treatment with molecular sieves followed by fractional distillation, small amounts of water and organic impurities remained. Our work took a different tack in that the combination of anhydrous alumina and galvanostatic preelectrolysis techniques afforded $LiClO_4$/PC electrolyte which provided the best Li cycling efficiencies obtained at that time (13). Since then, others have extolled the use of alumina in electrolyte purification schemes over molecular sieves (36).

More recently, we completed a study which provided the most impurity-free THF possible (23). Thus, THF was refluxed and subsequently distilled off benzophenone ketyl ($\phi_2CO^{-\cdot}$, $Na^+$), a material which scavenges $H_2O$, protic organics, and reactive gases such as $O_2$ and $CO_2$. As will be described in the following section, this highly purified solvent reacted with Li at a faster rate than either unpurified or marginally purified THF.

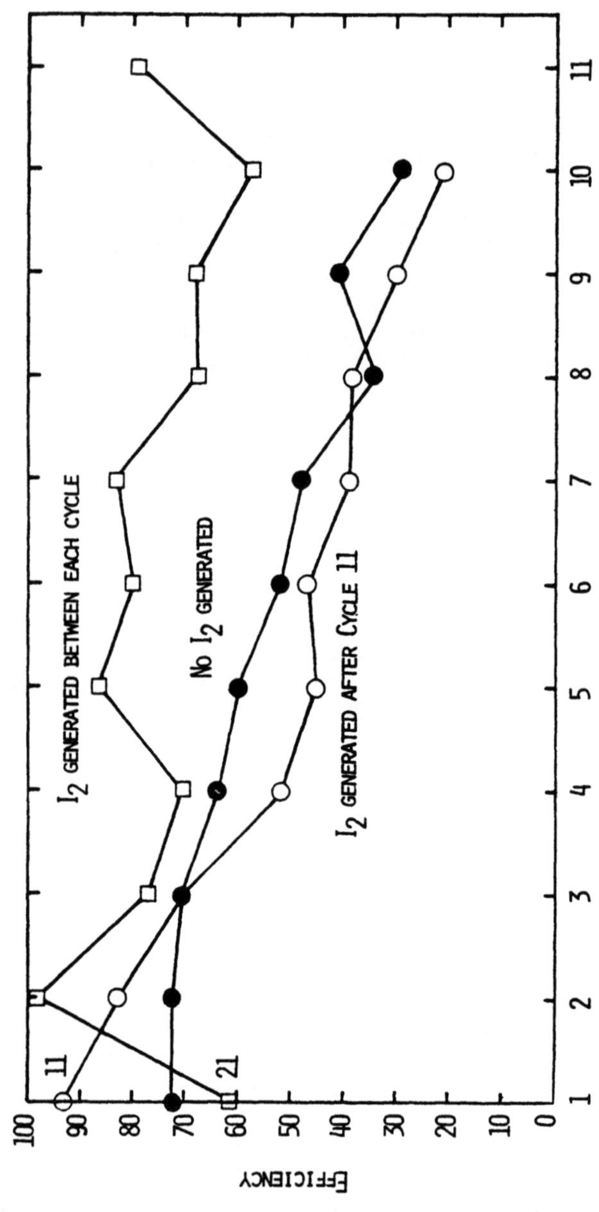


Fig. 2. Effect of producing an $I_2$ scavenger on the cycling efficiency of Li on Ta in $LiAsF_6$, LiI/THF, ($Q_p$ = 1.0 coul/$cm^2$).

We have concluded that with the solvents we have explored stringent purification is a necessary but insufficient condition for obtaining high Li cycling efficiencies. The key to achieving adequate cycling efficiencies for practical batteries involves the nature of the solvent molecule itself.

MODIFICATION OF THE SOLVENT - 2-METHYLTETRAHYDROFURAN/$LiAsF_6$ A SUPERIOR ELECTROLYTE

Our most recent approach to electrolyte development has been to choose solvents expected to manifest low reactivity towards Li, and then refine electrolytes based on these solvents via stringent purification procedures. In selecting a solvent, one may be guided by Selim and Bro (11) who point out that polar solvents are likely to be reactive with Li due to the presence of a molecular dipole. The dipole results from an unequal distribution of electron density about a carbon-heteroatom bond. This facilitates electron transfer from Li onto that bond. More recently, a thermodynamic analysis of the possible direct interaction between Li and a variety of aprotic solvents demonstrated that all may be reduced by Li (37). We suspected, therefore, that cyclic ethers afforded the best chance of retarding Li/solvent reactivity, since the C-O bond is far less polar than C=O (MA, methyl formate and PC) or S=O (dimethylsulfite and dimethylsulfoxide) -- other solvents in which the Li electrode has been cycled. Thus THF/Li reactivity was investigated in depth (23,25). We found that while $LiAsF_6$/THF electrolytes outperformed $LiAsF_6$/PC and MA in terms of Li electrode cycling efficiency, THF media were nonetheless too reactive for use in a practical secondary battery. Because of this finding, we went on to investigate the reactivity of substituted THF with Li. On the basis of our findings, we reported that 2Me-THF/$LiAsF_6$ is a superior electrolyte for cycling Li (25,39).

On the basis of isolated reaction products and earlier mechanistic work, we postulated a reduction mechanism involving an initial transfer of an electron from Li to the lowest unfilled molecular orbital (LUMO) centered on the oxygen atom of THF (23).

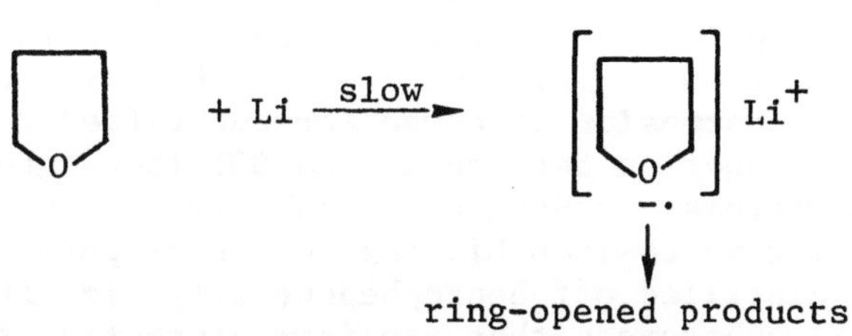


(3)

If equation 3 accurately describes the rate-determining step in the reduction of THF by Li, one can envisage modifying reaction kinetics by raising the activation energy of the slow step. This may be accomplished by perturbing the energy of the LUMO upward. Thus, locating an electron donating group, e.g., alkyl, in the 2-position adjacent to the oxygen atom, raises the activation energy required to form the anion-radical by localizing additional electron density on the oxygen atom.

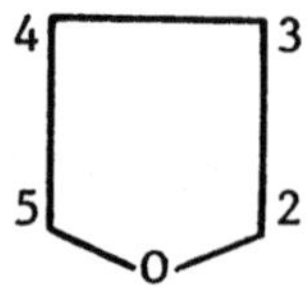

1

I

Consider the numbering scheme of the oxolane ring shown in I. On the basis of the LUMO argument, 2-methyl and 2,5-dimethyl-THF are predicted to be more stable to Li than THF. 3-methyl-THF, on the other hand, should be about as stable as THF, since the inductive effect of the methyl group is rapidly attenuated through saturated C-C bonds (38). To test this hypothesis, a series of methylated-THFs were obtained and exposed to Li under static and dynamic conditions. Static tests involved the incubation of Li foil with electrolyte at 71°C. The onset of Li-electrolyte reaction visually manifests itself in terms of corrosion on the Li foil and concurrent yellowing of the electrolyte. Dynamic conditions were achieved by galvanostatically cycling Li to and from Li and Ni substrates at 25°C. A fresh Li surface, therefore, came into contact with electrolyte on every cycle. Li-electrolyte reactivity was noted in terms of a loss of cycling efficiency (Li stripped/Li plated) with increasing cycle number.

Tables 2 and 3 summarize the results of the static test (40). It is readily apparent that those solvents and electrolytes comprising 2-methyl-(2-Me) or 2,5-dimethyl-THF (2,5-di-Me-THF) are markedly superior to THF or 3-methyl-THF (3-Me-THF) on exposure to Li at elevated temperature. We see that purified 2-Me-THF itself, as well as electrolyte prepared from it, require months before any sign of reaction with Li is noticeable. In fact, the ampoule containing Li and 2-Me-THF off benzophenone ketyl was deliberately opened after 10 months of storage prior to the onset of reaction. The UV spectrum of this solvent was essentially super-imposable with freshly distilled 2-Me-THF. The Li corrosion observed for unpurified 2-Me-THF was found to be due to the reaction between Li and BHT (butylated hydroxytoluene) added to the commercial product as a stabilizer. On the other hand, THF is quite reactive towards Li, regardless of purification procedure. Solvent distilled off benzophenone ketyl is effectively free of $H_2O$ and $O_2$ -- substances that can form protective films on Li (25). Yet even this extremely pure THF reacts with Li. Electrolyte prepared from THF is more stable, reacting after 25 days of storage. Possibly, impurities introduced in the electrolyte form protective films or, as we speculated earlier, the salt itself can form a protective film on the metal. The major reduction product in the reaction of THF with Li was found to be n-butanol from hydrolysis of lithium n-butoxide (23).

Table 2

The Onset of Li Reaction with Cyclic Ethers at 71°C

| Purification Procedure | Time, days | | | |
|---|---|---|---|---|
| | THF | 2-Me-THF | 3-Me-THF | 2,5-di-Me-THF |
| None | 1(3)[a] | 1(2) | 1(1) | 1(1) |
| A[b] | 4(7) | 10 Mo. | 1(1) | 1(2) |
| Benzophenone ketyl | 3(2) | >10 Mo | - | - |

[a]Observable Li corrosion after 1 day; observable solvent coloration after 3 days.

[b]Solvent passed through activated alumina.

Table 3

The Onset of Li Reaction with Cyclic Ether Based Electrolytes at 71°C

| Purification Procedure | 1M $LiAsF_6$/ THF | 1M $LiAsF_6$/ 2-Me-THF | 1M $LiClO_4$/ 2-Me-THF | 1M $LiAsF_6$/ 3-Me-THF | 1M $LiAsF_6$/ 2,5-di-Me-THF |
|---|---|---|---|---|---|
| None | 2(16)[a] | - | - | - | - |
| A[b] | 25(28) | 13 Mo | - | - | - |
| DAPA[c] | 4(7) | 10 Mo | 6 Mo | 10(4) | 12 Mo |

[a]Observable Li corrosion after 2 days; observable electrolyte coloration after 16 days.

bSolvent passed through alumina.

[c]Solvent distilled; passed through alumina; salt added in the cold followed by prelectrolysis; electrolyte passed through alumina.

A more stringent test of electrolyte inertness involves the dynami cycling of Li to and from a conducting substrate. As indicated earlie good cycling efficiency can only be achieved if the electrolyte is effectively inert to Li. This requires an absence of reactive impurities as well as the chemical compatibility of salt and solvent towards Li. In Fig. 3, the cycling efficiencies of THF and 2-Me-THF electrolytes (1M $LiAsF_6$) are presented as a function of cycle number for 1.1 coul/cm$^2$ Li on Ni plates. Although 50-100 coul/cm$^2$ of Li would normaly be cycled in a practical cell, thin plates are useful in rapidly assess ing the quality of a given electrolyte. Cycling of thin plates many times exacerbates conditions leading to inefficiencies and electrode failure. On the 10th cycle run in THF-based electrolyte, 80% of the Li plated is encapsulated by films and lost to anodic dissolution. By comparison, only 7% of Li plated from 2-Me-THF based electrolyte is electrically isolated. This electrolyte nevertheless does degrade with cycle number, as evidenced by the slow decay in cycling efficiency. Either reactive impurities and/or a very slow reaction of 2-Me-THF with Li yield products which perturb the morphology of subsequent plates.

Our experimental work with cyclic ether electrolytes attemps to mimic the charge and discharge characteristics of the secondary Li electrode in a practical battery. Thus, a key experiment employed by us to evaluate electrolyte stability involved cycling Li to and from a Li rather than Ni substrate in a half-cell configuration. This was accomplished by plating a known amount of Li onto a Ni electrode, and then sequentially stripping and plating a lesser charge of Li. The amount of excess Li determines, in part, the number of "100%" cycles to be achieved. For example, a typical Li-on-Li cycling experiment consists of plating 4.5 coul/cm$^2$ Li onto a Ni electrode; 1.1 coul/cm$^2$ are then stripped leaving 3.4 coul/cm$^2$ of excess Li. Subsequent plating and stripping cycles employ 1.1 coul/cm$^2$. Were each cycle 100% efficient, the cell would cycle indefinitely with a 3.4 coul/cm$^2$ reserve of Li (Efficiency = Q stripped/Q plated). Of course, each stripping cycle is <100% efficient, which means that each strip cuts into the Li reserve, yielding an apparent "100%" cycle until the Ni substrate is reached. At this point the excess Li is exhausted and one may calculate the average efficiency per cycle $\bar{E}$ (Eq. 4):

$$\bar{E} = \frac{Q_s - \frac{Q_{ex}}{n}}{Q_s} \tag{4}$$

where $Q_s$ is the charge of Li stripped, $Q_{ex}$ is the amount of excess Li, and n is the number of "100%" cycles.

Li-on-Li cycling efficiencies for $LiAsF_6$/THF and 2-Me-THF were determined at three salt concentrations. The cycling results and conductivity measurements are summarized in Table 4 (40). Electrolytes containing 2-Me-THF significantly outperformed those prepared with THF, even though the conductivities of the former were much less

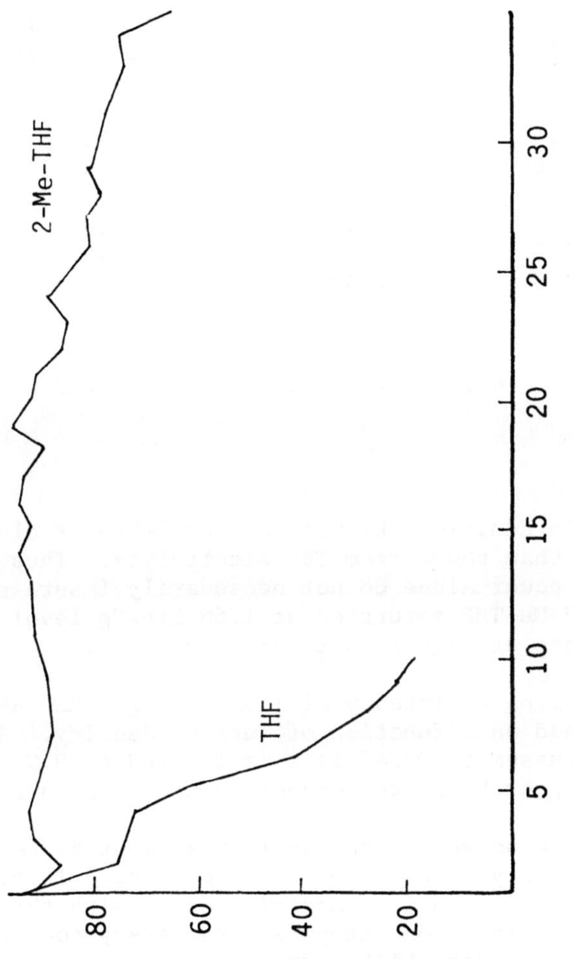


Fig. 3. Efficiencies of cycling Li on a Ni substrate. 1M $LiAsF_6$/cyclic ether; $i_p = i_s = 5$ $mA/cm^2$; $Q_p = 1.1$ $coul/cm^2$.

Table 4

Comparison of Conductivity and Cycling Efficiencies for 1.5M $LiAsF_6$/THF and 2-Me-THF Electrolytes at 25°C[a]

| | $[LiAsF_6]$ (mol/ℓ) | $K_{sp} \times 10^{-2}$ $(ohm\ cm)^{-1}$ | Number of "100%" Cycles | $\bar{E}$ (%) |
|---|---|---|---|---|
| THF | 1.0 | 1.37 | 18 | 83.3 |
| | 1.5 | 1.65 | 25 | 88.0 |
| | 2.0 | 1.45 | 17 | 82.4 |
| 2-Me-THF | 0.5 | 0.14 | 30 | 90.0 |
| | 1.0 | 0.30 | 45 | 93.3 |
| | 1.5 | 0.40 | 84 | 96.4 |

[a] $Q_{ex} = 3.4$ coul/cm$^2$; $Q_s = 1.1$ coul/cm$^2$; $i_p = i_s = 5$ mA/cm$^2$.

than the latter. In addition, Li plates from 2-Me-THF electrolyte were less dendritic than those from THF electrolyte. Thus, good condu tivity and throwing power alone do not necessarily insure regular Li plate morphology. 2-Me-THF saturates at 1.6M $LiAsF_6$ level, precluding Li cycling experiments at higher salt concentrations.

The average cycling efficiency of 1.5M $LiAsF_6$/2-Me-THF electrolyte was also assessed as a function of current density. As seen in Table 5 (22), $\bar{E}$ increases to 97.4% as i is lowered to 0.9 mA/cm$^2$, presumably due to better Li plate morphology at lower current densities

Another test of a solvent's suitability as a Li battery medium is to monitor its reactivity with plated Li on open circuit storage. This technique assesses the intrinsic reactivity of Li with the solvent, and the achievement of low reaction rates is a necessary condition for the implementation of a secondary Li battery.

One experiment consists of plating 1.1 coul/cm$^2$ onto a Ni substrate, and then switching the cell to OCV. After a predetermined length of time, the Li plate is electrostripped and the efficiency determined. Figure 4 compares data obtained from $LiAsF_6$-based THF and 2-Me-THF electrolytes. After 96 hours on OCV, almost 70% of Li plated from the 2-Me-THF electrolyte is seen to be electroaccessible. With THF however, all of the plated Li is isolated after 48 hours. These data translate into average isolation rate of 1.1 μA/cm$^2$ (2-Me-THF, 96 hrs)

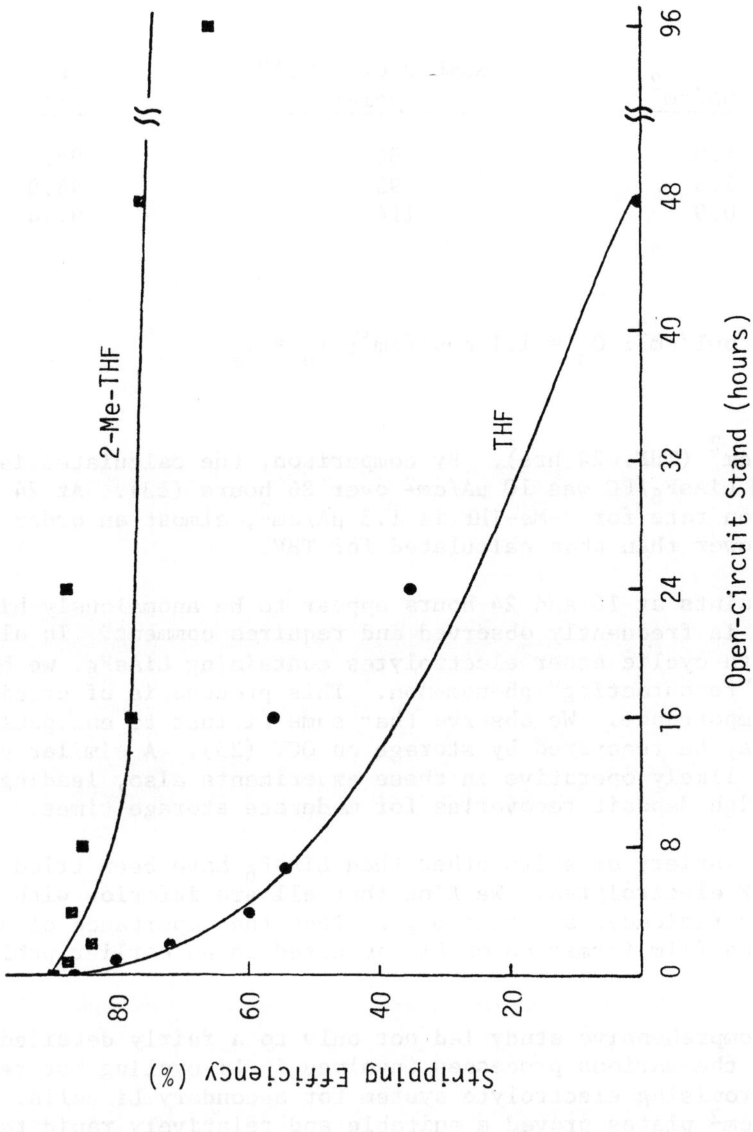


Fig. 4. Stripping efficiencies vs. time on open circuit storage. 1M $LiAsF_6$/cyclic ether; $i_p = i_s = 5$ mA/cm$^2$; $Q_p = 1.1$ coul/cm$^2$ (22).

Table 5

Variation of Average Cycling Efficiency with Current Density for 1.5M $LiAsF_6$/2-Me-THF[a]

| $i$, $mA/cm^2$ | Number of "100%" Cycles | $\bar{\bar{E}}$ (%) |
|---|---|---|
| 5.0 | 80 | 96.3 |
| 2.5 | 95 | 96.8 |
| 0.9 | 116 | 97.4 |

[a] $Q_{ex}$ = 3.4 coul/$cm^2$; $Q_s$ = 1.1 coul/$cm^2$; $i_p = i_s$.

and 8.3 $\mu A/cm^2$ (THF, 24 hrs). By comparison, the calculated isolation rate for 1M $LiAsF_6$/PC was 10 $\mu A/cm^2$ over 26 hours (13). At 24 hours, the isolation rate for 2-Me-THF is 1.3 $\mu A/cm^2$, almost an order of magnitude lower than that calculated for THF.

Data points at 16 and 24 hours appear to be anomalously high. This result is frequently observed and requires comment. In all of our work with cyclic ether electrolytes containing $LiAsF_6$, we have observed a "recontacting" phenomenon. This process is of crucial practical importance. We observe that some Li lost to encapsulation reactions may be recovered by storage on OCV (25). A similar phenomenon is most likely operative in these experiments also, leading to unusually high deposit recoveries for moderate storage times.

A wide variety of salts other than $LiAsF_6$ have been tried in THF and 2-Me-THF electrolytes. We find that all are inferior with respect to cycling efficiency, and this may reflect the importance of $AsF_6^-$-induced brown film formation on Li, as noted in an earlier publication (23).

This comprehensive study led not only to a fairly detailed understanding of the various processes involved in Li cycling but resulted in a very promising electrolyte system for secondary Li cells. Cyclin of ∿1 coul/$cm^2$ plates proved a suitable and relatively rapid test for initial evaluation and study of Li electrodes, but naturally does not duplicate the conditions characteristic for practical battery applications. We have, therefore, devoted considerable effort towards expanding our data to secondary Li electrodes of practical capacity for unit area. This study has resulted in actual battery anodes confirming in complete practical cells the results obtained with ∿1 coul/$cm^2$ plates.

## SUMMARY - WHERE DO WE GO FROM HERE?

As will be clear, our review has mostly focussed on summarizing results obtained within our own group. The major conclusions we would draw are:

(1) The common solvents PC, MA and THF are too reactive for use in a rechargeable Li cell.

(2) Modification of the cyclic ether structure, such as in 2-Me-THF, leads to lower reactivity.

(3) $LiAsF_6$ is by far a preferred electrolyte. It appears to function by reaction with Li, i.e., as a precursor.

(4) Both precursors (i.e, reactive compounds) and surface adsorption agents show some promise for improving Li cycling efficiency.

This assessment of the state of the art leads us to suggest the following research directions:

(1) Why is $LiAsF_6$ so outstanding? To what extent is this due to film formation or adsorption or throwing power effects? We believe that advancing our understanding of the mechanism of why $LiAsF_6$ works so well can lead to improvements. Studies of plate morphology and of the composition of the thin protective films on the metal surface are needed.

(2) The precursor concept has shown promise in quite reactive solvents. It seems likely that it should work better in less reactive media such as 2-Me-THF.

(3) The "re-contacting" phenomenon observed with deposits in cyclic ether solutions is a key to a practical rechargeable battery. The mechanisms of this re-contacting effect is obviously of great interest.

(4) How can the conductivity of solutions such as $LiAsF_6$ in 2-Me-THF be improved?

(5) Work is still needed to define solvents, other than those based on THF or tetrahydropyran, where small structural changes can lower reactivity.

## REFERENCES

1. Exxon Enterprises, Somerville, NJ, Preliminary Application Guide 0578-1.

2. W. S. Harris, Thesis UCRL 8381, University of California (1958).

3. J. O. Besenhard and G. Eichinger, J. Electroanal. Chem., 68, 1 (1976).

4. J. Auborn, K. French, S. Lieberman, V. Shah and A. Heller, J. Electrochem. Soc., 120, 1613 (1973).

5. A. N. Dey and C. R. Schlaiker, Proc. 26th Power Sources Symp., Atlantic City, NJ (1974).

6. A. N. Dey, Thin Solid Films, 43, 131 (1977).

7. R. Jasinski and S. Carroll, J. Electrochem. Soc., 117, 218 (1970)

8. A. N. Dey and B. P. Sullivan, J. Electrochem. Soc., 117, 222 (1970).

9. G. Eichinger, J. Electroanal. Chem., 74, 183 (1976).

10. J. E. Chilton, Jr., W. J. Conner, W. J. Cook and A. W. Holsinger: Lockheed Missiles and Space Co., Final Report on AG-33(615)-1195, AFAPL-TR-64-147 (1965).

11. R. Selim and P. Bro, J. Electrochem. Soc., 121, 1457 (1974).

12. R. D. Rauh and S. B. Brummer, Electrochimica Acta, 22, 75 (1977).

13. V. R. Koch and S. B. Brummer, Electrochimica Acta, 23, 55 (1978).

14. F. W. Dampier and S. B. Brummer, Electrochimica Acta, 22, 1339 (1977).

15. R. D. Rauh and S. B. Brummer, unpublished results.

16. J. Broadhead and F. A. Trumbore, 9th Intl. Power Sources Symposium Brighton, England (1974), Paper 41.

17. H. F. Baumann, J. E. Chilton, W. J. Conner and G. M. Cook, Lockhead Space and Missiles Co., Report AD 425 876 (1963).

18. R. D. Rauh, T. F. Reise and S. B. Brummer, unpublished data.

19. S. B. Brummer, F. W. Dampier, V. R. Koch, R. D. Rauh, T. F. Reise and J. H. Young, EIC Corporation, Final Report on Grant AER75-0377 (1978).

20. R. D. Rauh and S. B. Brummer, Electrochimica Acta, 22, 84 (1977).

21. R. D. Rauh, G. F. Pearson and S. B. Brummer, Proc. 12th IECEC Conference, Vol. 1, Washington, D. C. (1977); p.

22. V. R. Koch and J. H. Young, Science, 204, 499 (1979).

23. V. R. Koch, J. Electrochem. Soc., 126, 182 (1979).

24. V. R, Koch and J. H. Young, Extended Abstract No. 548, Spring Meeting of the Electrochemical Society, Seattle, WA (1978).

25. V. R. Koch and J. H. Young, J. Electrochem. Soc., 125, 1371 (1978).

26. A. N. Dey, J. Electrochem. Soc., 118, 1547 (1971).

27. J. P. Gabano, G. Lehmann, A. Gerbier and J. F. Laurent, Power Sources 3, Ed., D. H. Collins (Oriel Press, 1971) p. 297.

28. M. L. B. Rao and K. Hill, Extended Abstract No. 14, Fall Meeting of the Electrochemical Society (1965).

29. D. Sam and J. H. Ambrus, Extended Abstract No. 22, Fall Meeting of the Electrochemical Society (1974).

30. M. M. Nicholson, J. Electrochem. Soc., 121, 134 (1974).

31. J. O. Besenhard, J. Electroanal. Chem., 94, 77 (1978).

32. B. M. L. Rao, R. W. Francis and H. A. Christopher, J. Electrochem. Soc., 124, 1490 (1977).

33. J. L. Weininger and F. F. Holub, J. Electrochem. Soc., 117, 340 (1970).

34. J. L. Weininger and F. Secor, J. Electrochem. Soc., 121, 315 (1974).

35. R. J. Jasinski and S. Kirkland, Anal. Chem., 39, 1663 (1967).

36. F. A. Trumbore and J. J. Auborn, Extended Abstracts of the Electrochemical Society Meeting, Pittsburgh, PA (1978) p. 65.

37. I. A. Kedrinskii, S. V. Morozov, G. I. Sukhova and L. A. Sukhova, Soviet Electrochemistry, 12, 1094 (1977).

38. E. S. Gould, Mechanism and Structure in Organic Chemistry, (Rinehart and Winston, New York, NY, 1959) p. 203.

39. V. R. Koch, U.S. Patent 4,118,550 (1978).

40. J. L. Goldman, R. M. Mank, J. H. Young and V. R. Koch, In process of publication.

[illegible]

[illegible]

[illegible]

[illegible]

[illegible]

[illegible] and [illegible]

INTERCALATION ELECTRODES

M.B. Armand

Laboratoire d'Energétique Electrochimique
Université de Grenoble
B.P. 44 Domaine Universitaire
38401 - St-Martin-d'Hères France

INTRODUCTION

The need for more efficient electrical energy storage devices has prompted research on new electrode materials. In this view, intercalation compounds offer a promising route towards high energy density batteries.

Formally, an electrode is defined as an interface between an ionically conductive electrolyte ($M^+$ or $X^-$) and an electronic conductor :

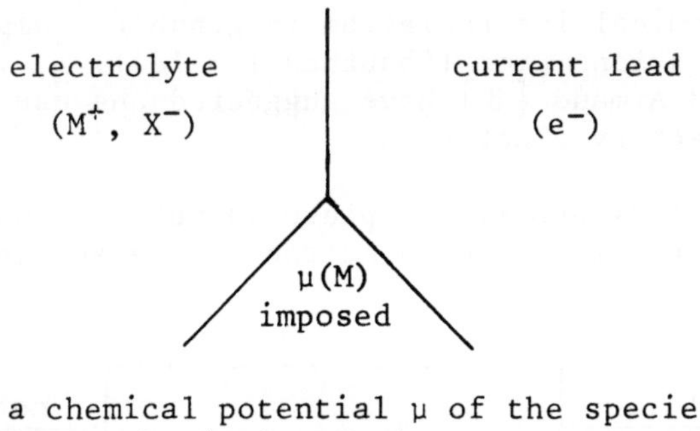


at the junction, a chemical potential $\mu$ of the species (M or X) is imposed for the electrochemical reaction :

$$\text{or} \left\{ \begin{array}{l} M \rightleftharpoons M^+ + e^- \\ \frac{1}{2} X_2 + e^- \rightleftharpoons X^- \end{array} \right.$$

In this ideal scheme, the electrode is a surface without any capacity. A capacity would be observed only if the surface electrode is expanded in volume, by interposing a material possessing the three properties of the junction :

- an electronic conductivity ($e^-$)
- an ionic conductivity ($M^+$ or $X^-$)
- a chemical potential ($\mu_M$ or $\mu_{X_2}$) imposed

as a consequence, the electrode material must have a host lattice structure, <H> and participates to the electrochemical reaction via an intercalation mechanism :

$$\underset{\text{electrolyte}}{xM^+} + \underset{\text{current lead}}{xe^-} + \underset{\text{electrode material}}{\langle H \rangle} \rightleftharpoons \underset{\text{intercalation compound}}{\langle M_xH \rangle} \quad (1)$$

or :

$$xX^- + \langle H \rangle \rightleftharpoons \langle X_xH \rangle + xe^-$$

The electrolyte and the current lead have in this case the sole function of carrying charged species, and do not enter the electrochemical process by dissolution and transport of neutral reactants.

An intercalation electrode can thus operate in totally solid-state system, the rate depending only on the transport properties in the bulk of the host lattice.

Though chemical intercalation in graphite compounds had been recognized for a long time (Shaufhautl - 1841) [1], only recently Steele [2] and Armand [3] have suggested the use of these compounds as battery electrodes.

There is an "a priori" complete symmetry between the intercalation of electropositive or electronegative species, schematized as :

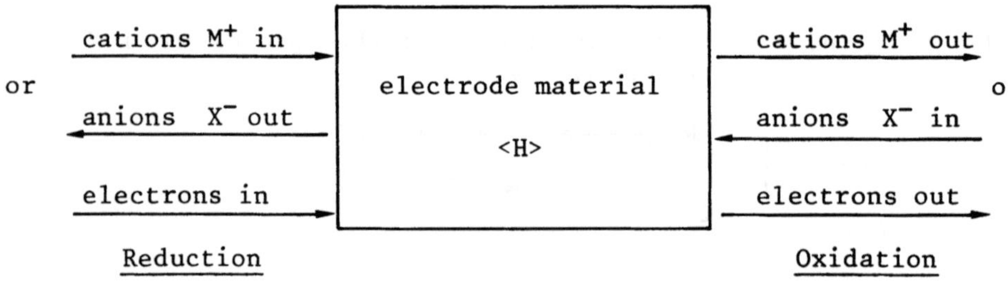

The intercalate acting as a source and a sink of either M or X. In reality, there are very few examples of an anionic intercalation mechanism (graphite "salts" $X_xC$), and we shall mainly restrict ourselves to the intercalation of monopositive species. Besides, lithium plays a dominant rôle, as a large majority of studies deal with compounds of this light metal.

## CLASSIFICATION OF INTERCALATION REACTIONS

The intercalation reaction is a dissolution of M (as $M^+ + e^-$) in the "solvent" <H>. The properties of the "solute" $M_xH$ vary depending on the host-guest interactions.

### 1) The perfectly non-stoichiometric compounds. Type I

These compounds are defined by the intercalation reaction (1):

$$xM^+ \quad + \quad xe^- \quad + \quad \langle H\rangle \rightleftharpoons \langle M_xH\rangle \qquad (1)$$

for which the x coefficient can vary continuously from 0 to a maximum value $x_{max}$ as :

$$0 < x < x_{max}$$

We define $\boxed{y = \frac{x}{x_{max}}}$ as the degree of occupancy. (2)

These materials constitute the "solid solution electrodes" as defined by Steele [4].

Properties : a type I compound is an ideal case for a reversible electrochemical reaction ; transport of matter occurs through a double flux of charged species ($M^+$ and $e^-$) in the bulk of the material whatever the concentration. This is schematized in Fig. 1.

Examples :

| | |
|---|---|
| $H_xMnO_2$ (γ) | $0 < x < 1$ |
| $Li_xTiS_2$ | $0 < x < 1$ |
| $Li_xNbSe_2$ | $0 < x < 1$ |

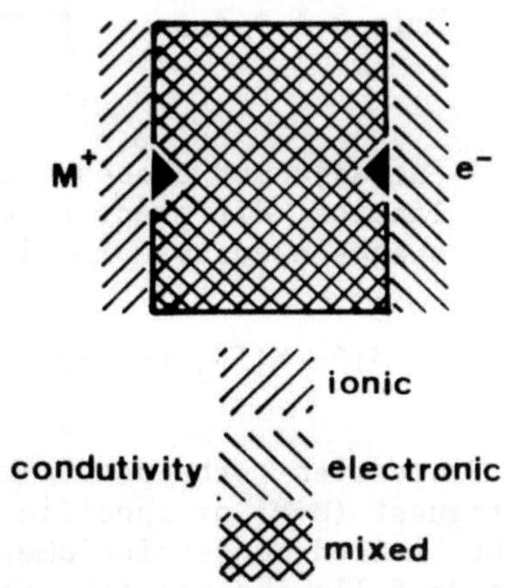


*Fig. 1. Type I compounds : schematization of the intercalation process.*

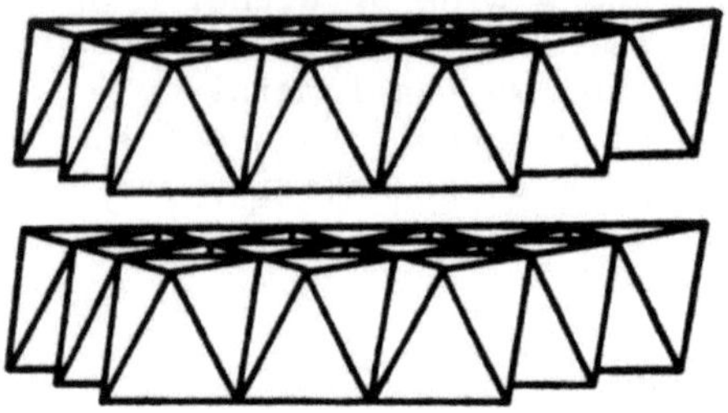

These materials are characterized by anisotropic structures providing space and preferential diffusion paths for small particles ($H^+$, $Li^+$) with weak guest-guest interactions. The lamellar structure of $TiS_2$, with a Van der Walls gap provides a good example Fig. 2.

*Fig. 2. The lamellar $TiS_2$ structure formed from $TiS_6$ octahedra*

2) <u>The pseudo two-phase compounds. Type II</u>

These compounds undergo the same electrochemical intercalation reaction written in eq. ① . However, the guest species (M) tend to form within the host phase ($M_xH$) an ordered sub-lattice with a limited departure from stoichiometry.

There are two possibilities :

- the threshold compounds, where intercalation reaction starts at a minimum concentration $x_{min}$

$x_{min} < x < x_{max}$

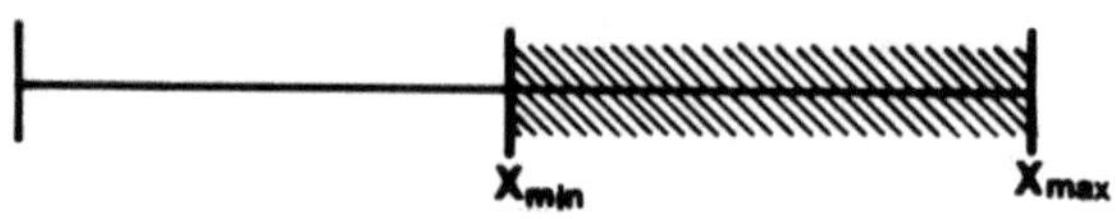


- the adjacent phases : as M species are injected in the <H> lattice, different superstructures appear with pseudo-phases bondary compositions

$0 < x_1 < .. x_j < x_{j+1} .. < x_{max}$

Properties : the pseudo-two-phase compounds result from strong guest-guest (M-M) or specific guest-host interactions. As a result of the limited non-stoichiometry domains, the intercalation proceeds following one (threshold compounds) or several (adjacent domains) reaction fronts.

Also, it becomes understandable that only one type of conductivity ($M^+$ or $e^-$) is needed in the starting material ( <H>), provided that the complementary conduction appears with the intercalation reaction and follows the reaction front(s). This is depicted in Fig.3., where the same symbolism as for Fig.1. has been kept.

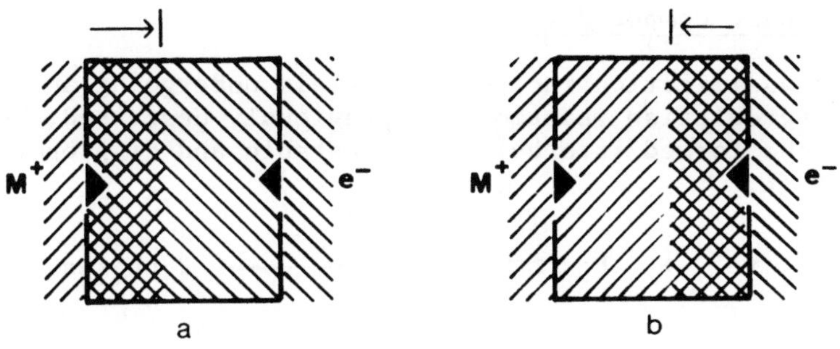


*Fig. 3. Type II compounds : propagation of the intercalation reaction front*
*a) appearance of the ionic conductivity*
*b) appearance of the electronic conductivity*

However, in this case, the electrochemical reaction is not evidently reversible ; the desintercalation process (i.e. the removal of the M species from the host lattice) can be hindered or even blocked by the disappearance of one type of conductivity at either the electrode-electrolyte or electrode-current lead interface. The intercalate $<M_xH>$ becomes thus electrochemically insulated, as shown schematically in Fig. 4.

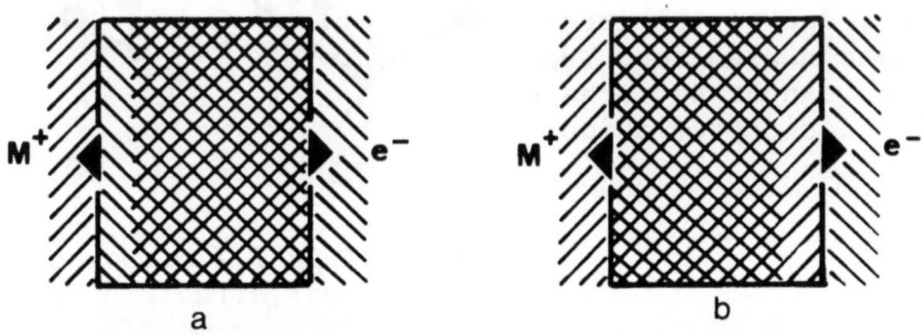


*Fig. 4. Type II compounds : formation of a blocking layer during the desintercalation*
*a) disappearance of the ionic conductivity*
*b) disappearance of the electronic conductivity*

The parameters affecting the formation of Type II versus Type I compounds shall be detailed with specific examples in the following chapter.

3) <u>The two-phase compounds. Type III</u>

This classification includes materials undergoing an electrochemical intercalation reaction, as previously described, but in this case, the formed intercalate is unstable and disproportionates, according to :

$$xLi^{+} + xe^{-} + <H> \rightleftharpoons \underset{\text{unstable}}{\overset{\text{Type I or II}}{<M_xH>}} \longrightarrow 2 \text{ phases}$$

Properties : the criteria for the progression of the electrochemical reaction are the same as for Type II compounds. In addition, the two types of conductivities needed can be separated in the resulting two phases. As for instance, the poly(carbon fluoride) <CF> whose structure is given Fig. 5. [5], is electrochemically active :

$$Li^{+} + e^{-} + <CF> \rightleftharpoons <LiCF> \downarrow <LiF> + <C>$$

The transient intercalate formed decomposes to electronically conducting carbon and highly defective LiF with appreciable ionic conductivity, allowing the progression of the reaction front.

However, due to the irreversible phase nucleation, Type III electrode cannot be recharged through a desintercalation process.

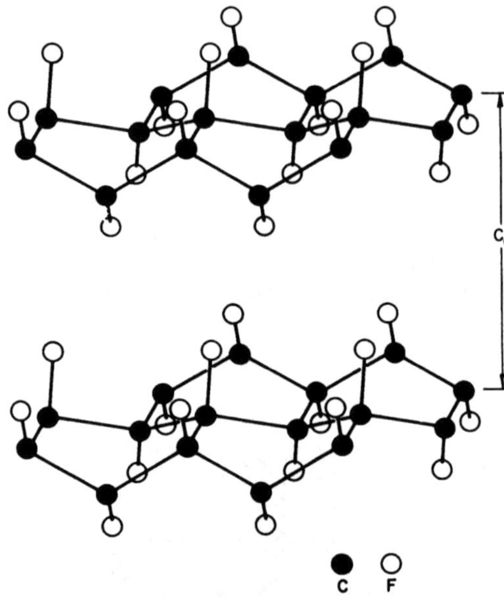


*Fig. 5. Type III compounds : the host structure of poly-(carbon-monofluoride) CF*

THERMODYNAMICS

An intercalation electrode, considered as solution of M in the host lattice <H> has thus a potential written as :

$$\phi = -\frac{1}{F}\,\mu(M)_{<H>} + \text{constant} \qquad (3)$$

where $\mu_{<H>}$ denotes the chemical potention of M in the <H> phase.

We are interested in the variation of the electrode potential as a function of the electrode composition x in $<M_xH>$

$$\phi = f(x) = f(y)$$

1) Non-stoichiometric Type I Compounds

a) Basic model

In the simplest ideal case, the potential of M is sum of the chemical potential contribution from the ions and the electrons :

$$\mu(M) = \mu(M^+) + \mu(e^-) \quad (4)$$

Ionic contribution : (the subscript i refers to the ionic species).

The host lattice $<H>$ keeps this integrety through the intercalation reaction, fixing the concentration of sites available for $M^+$ ions. A Fermi-Dirac statistic has to be used to write the chemical potential :

$$\mu_{M^+} = \mu_i^\circ + RT.\ln \frac{\text{occupied sites}}{\text{empty sites}}$$

or, if $y_i$ denotes the ionic site occupancy :

$$\mu_{M^+} = \mu_i^\circ + RT.\ln \frac{y_i}{1-y_i} \quad (5)$$

Electronic contribution : (the subscript e refers to the electrons).

The electrons are fermions, and are distributed in the lattice in a hand of energy width L.

We can express implicitly the total band occupancy $y_e$ as :

$$y_e = \frac{\int_{\mu_e^\circ}^{\mu_e^\circ+L} D(E).P(E).dE}{\int_{\mu_e^\circ}^{\mu_e^\circ+L} D(E).dE} \quad (6)$$

where : P(E) is the Fermi function
D(E) the density of states

This expression can be simplified with the hypothesis that the election band is narrow (L = 0 D(E) = Dirac distribution function), a logical assumption in the case of ionic crystals.

Eq. (6) thus becomes :

$$\mu(e) = \mu_e^\circ + RT.\ln \frac{y_e}{1-y_e} \qquad (7)$$

The electrode potential $\phi$ is thus compiled from eq. (3) (5) and (7) as :

$$\phi = \phi^\circ - \frac{RT}{F} \cdot \ln \frac{y_i}{1-y_i} \cdot \frac{y_e}{1-y_e} \qquad (8)$$

This equation implies two possibilities :

- one type of site is limiting

$$\text{or} \begin{cases} y_i = y = x/x_{max} & y_e \ll y_i \\ y_e = y = x/x_{max} & y_i \ll y_e \end{cases}$$

- both types of sites are simultaneously limiting

$$y_i = y_e = y = x/x_{max}$$

These two choices are simply expressed as :

$$\boxed{\phi = \phi^\circ - \frac{nRT}{F} \cdot \ln \frac{y}{1-y}} \qquad (9)$$

$$n = 1 \text{ or } 2$$

Fig. 6 shows the potential-composition ror n=1.

This curve corresponds to the experimental determination of the $H_xMnO_2$ electrode in Leclanché cells [6].

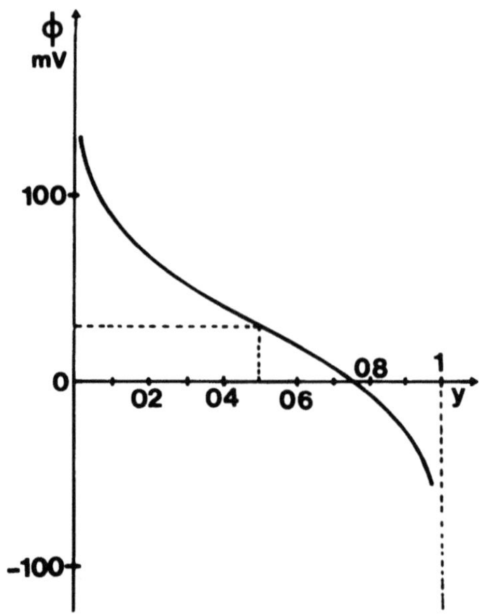


*Fig. 6. Basic model : shape of the potential-composition curve*

b) Interaction terms

The basic model take only into account the entropy term, the enthalpy being supposed constant. Interactions between guest species contribute to a new energy term proportional to the site occupancy (i.e. the concentration of interacting $M^+ - M^+$ ions), leading to :

$$\phi = \phi^{\circ} + Ky - \frac{nRT}{F} \cdot \ln \frac{y}{1-y} \qquad (10)$$

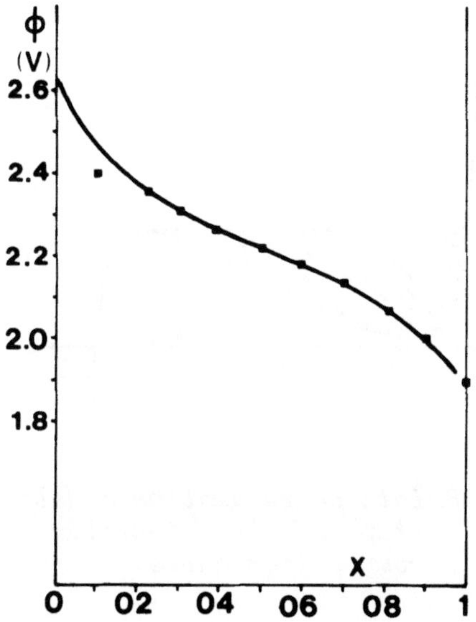


*Fig. 7. $Li_xTiS_2$ : potential-composition curve. Reference: $Li^{\circ}$. [7]*

However, the condition that no ordering can take place in the M sublattice (Type I) expresses that the potential is a smooth, decreasing function of the degree of occupancy :

$$\frac{\partial \phi}{\partial y} < 0 \qquad \forall y \qquad (11)$$

imposing

$$K < \frac{4nRT}{F} \qquad (12)$$

$0<K<\frac{4nRT}{F}$ : attractive interactions

$K < 0$ : repulsive

In Fig. 7., we have compared the experimental results from Whittingham[7] with eq. (9) , for a repulsive interaction term.

K = 0.22 eV and n = 2, showing a very good agreement with theory.

2) Pseudo two phases. Type II compounds

The condition expressed by eq. (11) is no longer valid as Type II compounds have narrow non-stoichiometry domains.

a) Lattice relaxation model

When the host lattice <H> contains a transition element, the electrons injected in the intercalation process are distributed in the empty d orbitals. The decrease in the formal oxidation state of the transition element results in a change of either the ionic radius or of the coordination shell symmetry (Jahn-Teller effect), inducing strains on the <H> lattice. This situation is expressed as a strong positive interaction term proportional to the number of intercalated species, as :

$$\phi = \phi^{\circ} + Ky - \frac{nRT}{F} \cdot \ln \frac{y}{1-y} \qquad (13)$$

with $K < \frac{4nRT}{F}$

This case is similar to the one presented in equation (9), but the larger value of the interaction term K leads to a minimum and a maximum of the curve $\phi = f(y)$.

The maximum corresponds to :

$$y_{\circ} = \frac{1 + \sqrt{1 - \frac{RT}{KF}}}{2} \qquad (14)$$

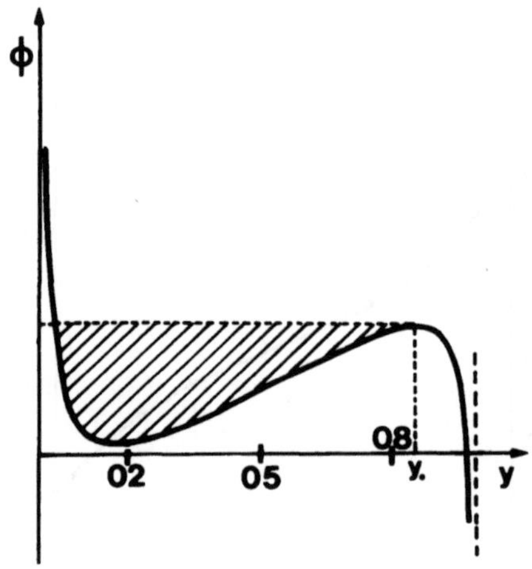


*Fig. 8. Lattice relaxation model: shape of the potential-composition curve.*

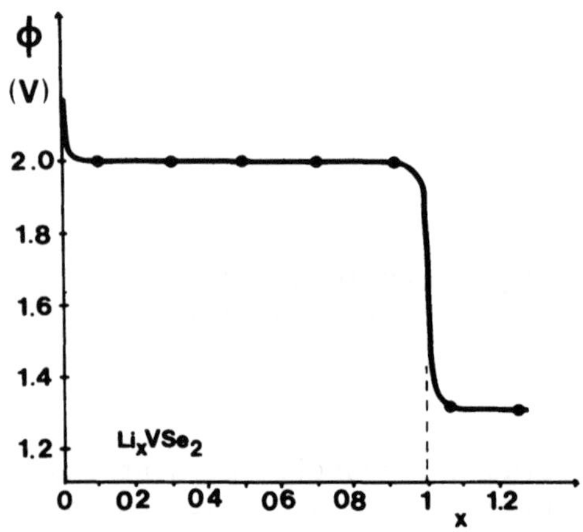


*Fig. 9. $Li_xVSe_2$ : potential composition curve. Reference : $Li^{\circ}$.* |8|

The corresponding curve is plotted in Fig. 8. It appears that the existence of a maximum of the potential $\phi(y_{\circ})$ implies that all the compositions in the hatched area are unstable ; a disproportionation reaction takes place in this "forbidden composition range" into a concentrated ($y = y_{\circ}$ $x = x_{\circ}$) phase and a diluted phase ($y \sim x = \varepsilon$).

$$\langle M_xH \rangle \rightarrow \langle M\varepsilon H \rangle + \langle M_{x_{\circ}}H \rangle$$

dilute phase     concentrated phase

The dilute phase having a negligible concentration, the host lattice behaves as a type II material with a threshold composition :

$$y \geqslant y_o \qquad (15)$$

As a consequence, the voltage-composition curve shows a plateau in the "forbidden composition range" due to the equilibrium of the two pseudo-phases.

In reality, the disproportionation reaction can take place with appreciable kinetics only when the driving faces are all positive, i.e. at the minimum of the $\mathfrak{A}(2)$ curve (Spingdal decomposition). For the same reason, the disproportionation for the desintercalation starts at the maximum of $\mathfrak{A}(2)$, leading to a charge-discharge hysteresis.

In Fig. 9., we have represented the experimental curve for $Li_xVSe_2$ (Murphy and al. [8]).

$VSe_2$ has a structure similar to that of $TiS_2$, but with a strong Jahn-Teller effect on the vanadium ion, a threshold composition $x \sim 1$ is observed instead of a non-stoichiometry domain.

b) The intercalation pressure

The lattice relaxation model is general for any type of host structure. In the special case of bi-dimensional compounds, with Van der Waals gaps, the structure expands parallel to the planes to accomodate the intercalated species.

A corrective term has to account for the energy needed to separate the host lattice planes.

If we suppose that the planes are perfectly rigid, the work per mole of <H> is constant and shared equally by the intercalated guest atoms.

The potential is then written as :

$$\phi = \phi^\circ - \frac{P}{y} - \frac{nRT}{F} \cdot \ln \frac{y}{1-y} \qquad (16)$$

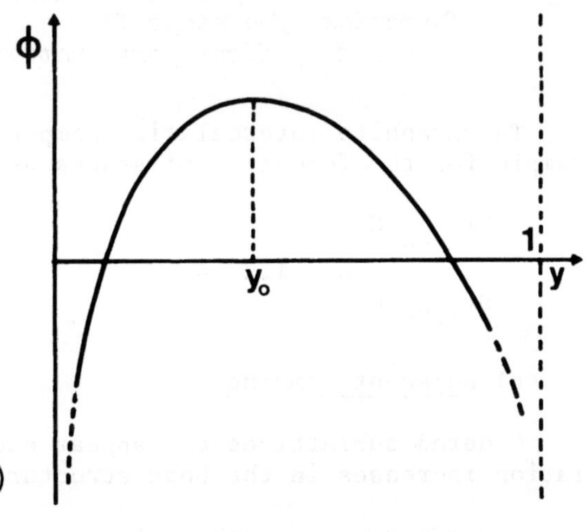


*Fig. 10. Intercalation pressure model: shape of the potential-composition curve.*

This function is represented on Fig. 10 ; it goes through a maximum for :

$$y_o = \frac{P_F}{P_F + nRT} \qquad (17)$$

and this value determines an optimal composition for the intercalate.

The leaner compounds corresponding to $y = y_o$ are thus unstable versus the disproportionation reaction :

$$\langle M_x H\rangle \longrightarrow \frac{x_o}{x}\langle M_{x_o} H\rangle + \langle H\rangle \qquad (18)$$

yielding the intercalate at the optimal composition $x = x_o$ and the pure phase <H>.

The ordered epitaxy of filled ($x = x_o$) and empty ($x = 0$) planes corresponds to the well-known formation of stages. This phenomenon is depicted in Fig. 11. for the formation of a stage II compound (i.e. the alternance of empty and filled planes).

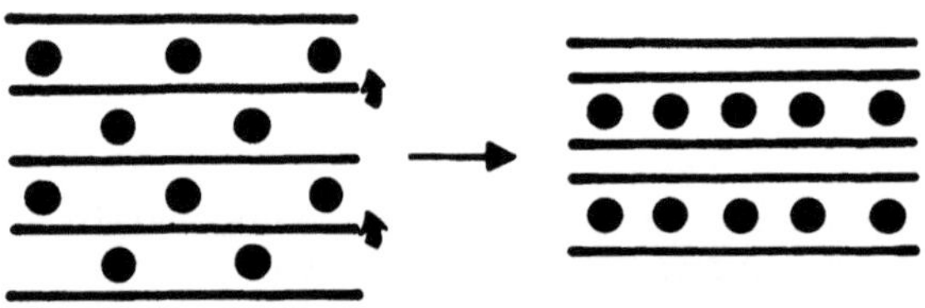

*Fig. 11. Intercalation pressure model: formation of a stage II compound by disproportionation*

The validity of the model is of course limited at the low x (y) values, as the host lattice planes are not perfectly rigid. This condition is expressed as :

$$\frac{P}{y} \not\rightarrow \infty \qquad y \longrightarrow 0 \qquad (19)$$

The graphite intercalation compounds provide a very good example for the formation of stages as in :

$$Li_{1/6n}C \qquad n = 1,2..4$$
$$Br_{1/8n}C$$

c) Adjacent domains

Ordered sublattices can appear successively as the M concentration increases in the host structure :

$0 < x < x_1$ phase 1

$x < x < x_j$ phase j

Each pseudo-phase corresponds to a different host-guest interaction. ($\phi°$ term)

$$\phi_j = \phi°_j - \frac{nRT}{F} \cdot \ln \frac{y_j}{1-y_j} \quad (20)$$

where $y_i$ represents the degree of occupancy of the j phase sites.

The pseudo-phases are in equilibrium :

$$\phi_j = \phi_{j+1}$$

We have thus implicitely access to the chemical potential as a function of the global occupancy y :

$$y = \frac{x}{x_{max}} = \frac{1}{x_{max}} \sum_j y_j (x_j - x_{j-1}) =$$

$$\frac{1}{x_{max}} \sum_j \frac{x_j - x_{j-1}}{1+\exp\frac{-F(\phi-\phi°_j)}{nRT}} \quad (21)$$

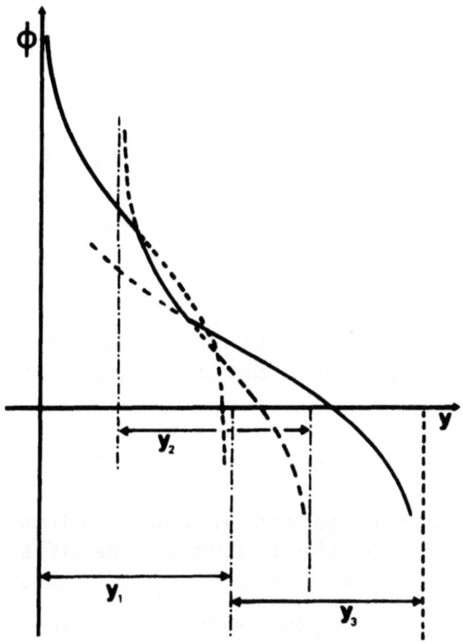


*Fig. 12. Adjacent domains: shape of the potential-composition curve.*

The corresponding curve $\phi = f(y)$ is drawn in Fig. 12. The pseudo-phases are filled successively, according to their site energy as shown by the knees of the curve ; the saturation of each subphase is obtained only at total lattice saturation.

In Fig. 13., the results obtained by Steele & Wynn [ 9 ] for $Na_xTiS_2$ have been reproduced.

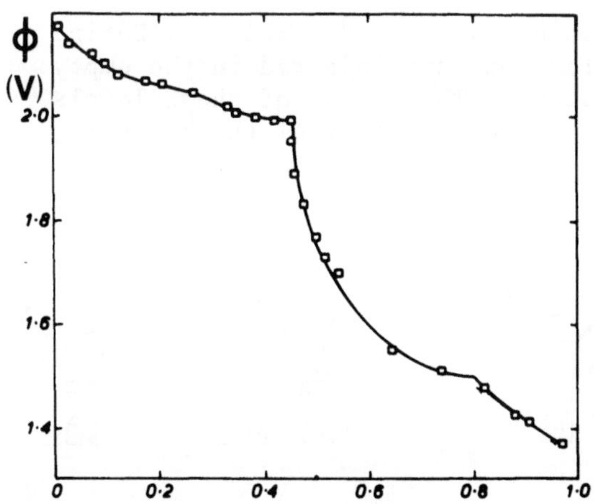


*Fig. 13. $Na_xTiS_2$ : potential-composition curve. Reference: $Na°$ [ 9 ]*

The larger size of the $Na^+$ ion (0.95 Å) as compared to $Li^+$ (0.65 Å), is favorable to a sublattice ordering with different site energies. Three domains exist for $Na_xTiS_2$ ; the heavier alkalis behave similarly.

3) Two-phase compounds. Type III

As defined, the type III electrodes function via the disproportionation of an intercalation compounds. This step is irreversible and the thermodynamic behaviour is imposed by the reversible intercalation reaction leading to the unstable compound of type I or II.

Potential composition curves for type III compounds fall into the categories presented above, limited at the composition corresponding to the maximum disproportionation rate.

Two factors are mainly responsible for the instability of the transient intercalate <$M_x$H>. We shall mention them qualitatively :

- variation of the coordination number or a large increase of the radius of one of the element constituting the <H> framework, leading to its destruction.

  In $H_xMnO_2$ $0<x<1$, we have a mixture of Mn(III) and Mn(IV) with similar ionic radii (0.5 Å)

  For $x>1$, the presence of the large Mn(II) (0.8 Å) results in the nucleation of $Mn(OH)_2$. This explains the non-rechargeability of the positive electrode in Leclanché cells.

- electronic levels incompatible with bond stability. During the intercalation, the electrons are injected in the empty orbitals of the host structure. The energy of these levels can be too high compared to the stability of the bonds in the structure.

A good example is given with the transition metal dichalcogenides, whose energy levels are shown in Fig. 14.

In $TiS_2$, with octahedral coordination, all the d orbitals are empty. During the intercalation, the electrons occupy the low-lying $t_{2g}$ level.

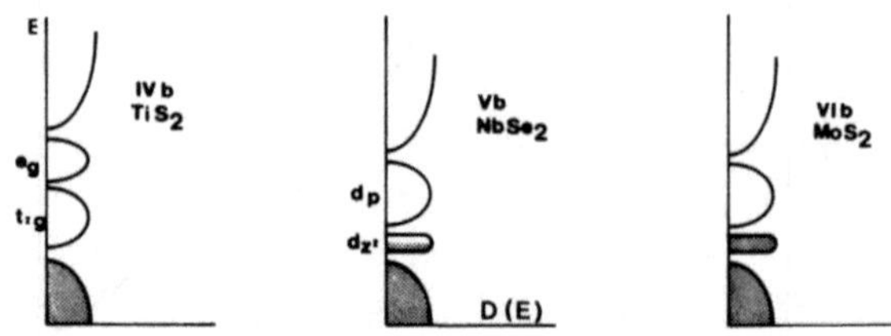


*Fig. 14. Electronic levels of the transition metals dichalcogenides. Energy as a function of the density of states D (E)*

For $NbSe_2$, with a prismatic coordination of Nb, the intercalation electron completes the filling of the $d_{z^2}$ level (1 d electron left).

Both $Li_xTiS_2$ and $Li_xNbSe_2$ are stable. $MoS_2$ with a filled $d_{z^2}$ band can only accept electrons in the higher dp level ; the Mo-S bonds becoming unstable at this electron activity ; the intercalate disproportionates according to :

$$<Li_xMoS_2> \qquad x/2<Li_2S> + <MoS_{2-x/2}>$$

A similar explaination holds for poly(carbon monofluoride). The "σ" C-F bond is cleaved by the electrons accomodated in the "π" anti-bonding level.

KINETICS

The rate of the electrochemical process in general limited by the diffusing species in the bulk of the host lattice.

As we have shown that the thermodynamics are different from that of classical solutions, we have to derive a new diffusion equation for these compounds.

Starting from Onsager's equation :

$$J = -c.u.\mathrm{grad}(\mu_M)^{<H>} = -c.u.\frac{\partial \mu}{\partial r} \quad \text{(one dimension)} \qquad (22)$$

where :

$J$ = flux

$c = c_o y$ = concentration

$r$ = space coordinate

$u$ = mobility

The mobility is proportional to the number of vacant sites for the diffusing species

$$u = u_o(1-y) \qquad (23)$$

with the chemical potential as in eq.(9), we have :

$$J = nRT.u_o.c_o.\frac{\partial y}{\partial r} = \boxed{D.c_o.\frac{\partial y}{\partial r}} \qquad (24)$$

Remarkably, the diffusion equation derived for particles obeying a Fermi-Dirac statistic is expressed in the same simple form as Fick's first law :

$$J = D \cdot \frac{\partial c}{\partial r}$$

The diffusion coefficient D keeps the same meaning for intercalation compounds and for solutions.

## CONCLUSIONS

The intercalation compounds are of growing importance in the field of solid-state chemistry, and we have simply emphasized here on the electrochemical aspect of there properties.

A rapidly increasing number of these compounds are known, as a wide variety of host structures appear favorable to intercalation, especially with the non-destructive lithium atoms. This metal keeps then with intercalation electrodes all its advantages already demonstrated for non-aqueous electro-chemistry (organic solvents and molten salts).

The intercalation compounds can be used as a positive electrode only, as in :

⊖ Li / $Li^+$ electrolyte / $<Li_xH>$ ⊕

An alternative is to use two different compounds as a positive and a negative electrode, in the so-called "rocking chair battery":

| | | | | |
|---|---|---|---|---|
| ⊖ $<Li_xH^1>$ | / | $Li^+$ electrolyte | / | $<Li_xH^2>$ ⊕ |
| low potential | | | | high potential |

the lithium ions being transferred from one side to the other, to avoid the yet unsolved problems encounterred with the rechargeability of the lithium electrode.

The lighter host structures have massic capacities much larger than that of classical electrode material. With the large potential difference allowed with lithium compounds, very high theoretical energy densities can be obtained :

| | | | |
|---|---|---|---|
| $TiS_2$ | $0 < x < 1$ | capacity | : 240 A.h/kg |
| | | energy density | : (Li/$TiS_2$ couple) 480 Wh/kg |
| $V_6O_{13}$ | $0 < x < 7.5$ | capacity | : 380 A.h/kg |
| | | energy density | : (Li/$V_6O_{13}$ couple) 800 Wh/kg [10] |

A question not yet clearly elucidated is the choice is the choice between Type I or Type II compounds for secondary batteries.

On one hand, Type I compounds can offer the guarantee of perfect reversibility with the same rate for the intercalation and the desintercalation process ; also, very high diffusion coefficients ($10^{-6} < D < 10^{-8}$ $cm^2/s$) [9] has been observed for bidimensional structures and allow to design relatively high-rate batteries (100 W/kg) for traction purposes. But necessarily, the emf of a cell using these materials varies markedly with the state of discharge, as shown by the potential-composition curves, a drawback for practical energy storage devices.

On the other hand, Type II compounds are much more widespread. The threshold-composition compounds exhibit a discharge plateau, and this advantage might compensate for the expected smaller rate capabilities due to narrower non-stoichiometry domains, and the already mentioned problems on recharge.

In this view, the host structures containing vanadium are very interesting, as they systematically show a limited voltage sloping due to the Jahn-Teller effect ($V_2O_5$, $VS_2$ , $V_6O_{13}$).

More informations on the thermodynamics and the kinetics of all sorts of intercalation compounds are needed before an optimal choice can be made.

REFERENCES

[1] P. Shaufhautl J. Prakt. Chem. 21 155 (1841).

[2] B.C.H. Steele "Fast transport in solids" W. Van Gool ed. North Holland Amsterdam 103 (1973).

[3] M.B. Armand *ibidem* 665 (1973).

[4] B.C.H. Steele "Trends in Electrochemistry" J.O'M. Bockris ed. Plenum Press New-York (1977).

[5] L.B. Ebert Ann. Rev. Materials Science 6 181 (1976).

[6] R.S. Johnson, W.C. Vosburgh, Trans. Electrochem.Soc 100 471 (1953).

[7] M.S. Whittingham, J. Electrochem. Soc. 123 3 315 (1976).

[8] D.W. Murphy, J.N. Carides, *ibidem* 126 3 349 (1979).

[9] D.A. Winn, J.M. Shemilt, B.C.H. Steele Mat. Res. Bull 11 559 (1976).

[10] D.W. Murphy, P.A. Christian, F.J. Di Salvo & P.N. Carides J. Electrochem. Soc. 126 3 497 (1979).

# INTERFACE PHENOMENA IN ADVANCED BATTERIES

Fritz G. Will

General Electric Corporate Research and Development
Schenectady, New York

## ABSTRACT

A variety of interfacial phenomena occurring in selected advanced batteries are reviewed. The effect of passive film formation, corrosion, reaction product layers, wetting and other factors on electrode and battery performance is discussed. The types of batteries considered are the sodium-sulfur, lithium/aluminum - ironsulfide, lithium-organic, zinc-chlorinehydrate and zinc-bromine systems.

## INTRODUCTION

Various types of batteries exhibit a multitude of different interfaces, giving rise to a great variety of interfacial phenomena involving metal dissolution, passivation, corrosion, electrocrystallization and - deposition, precipitation, adsorption, and wetting.

The "classical" batteries, such as lead-acid, nickel-ion and nickel-zinc, are predominantly characterized by phenomena at solid metal/or metal compound/aqueous solution interfaces. By comparison, "advanced" batteries, such as the sodium-sulfur, lithium/aluminum - ironsulfide, lithium-organic electrolyte, zinc-chlorinehydrate, zinc-bromine, hydrogen-halogen, redox, hydrogen-nickel, and metal-air systems, involve a much larger variety of interfaces whose reactions are significant for battery operation. Important types of interfaces include, in addition to those occurring in classical batteries, those between liquid metal/solid electrolyte, molten salt/solid metal or metal compound/organic electrolyte, gas/liquid, liquid/liquid, and gas/solid.

One of the salient interfacial considerations that dominate battery performance refers to the charge transfer process at the electrode/electrolyte interface and how the rate of this process is affected by the formation of passivating films and reaction products or corrosion layers. The fundamental aspects of the electrode/electrolyte interface and charge transfer across it are amply discussed in the relevant text books and articles. The following review will address major experimental observations and their interpretations on interfaces occurring in selected advanced batteries. As it appeared that the observed interfacial phenomena on different types of batteries are often specific to the particular battery, each battery will be considered separately. An alternate approach would have been to discuss generic problems of various batteries in common sections. Major emphasis will be placed on the sodium-sulfur battery because a considerably larger number of publications has addressed interface phenomena in this system than in other advanced batteries. Also considered are the lithium/aluminum - ironsulfide, lithium-organic, zinc-chlorinehydrate and zinc-bromine system. Batteries with gas diffusion electrodes, such as the nickel-hydrogen and metal air batteries, were not included in this review since they involve interface phenomena specific to chemisoption and electrocatalysis which have been amply discussed in the literature. The other advanced systems not considered here either share some of the major phenomena with the batteries discussed or else insufficient published information was available.

## SODIUM-SULFUR SYSTEM

The sodium-sulfur system employs a solid electrolyte to separate the two reactants, liquid sodium and liquid sulfur or polysulfide and is operated at temperatures of typically 350°C. It exhibits a number of interfaces each of which give rise to specific interfacial phenomena. The solid beta alumina electrolyte has interfaces with two liquid reactants: sodium on one side and sulfur or polysulfides on the other. Other important interfaces are those between porous conducting matrix and sulfur or polysulfide melt and between melt and current collector. Interfacial polarization and lack of good wetting are often associated with the sodium/beta alumina interface. Formation of insulating liquid sulfur and solid sulfide films at the interfaces with the beta alumina and the carbon felt have been held responsible for poor charge acceptance and lack of good high-rate discharge behavior. Furthermore, passive film formation or corrosion of the metal container are associated with the interface between many metals and the melt. A large number of studies in the past few years have addressed various aspects of these interfacial phenomena.

### Sodium/Beta Alumina Interface

The magnitude of interfacial polarization of the sodium/beta alumina interface is evident from a marked discrepance between d.c. and high frequency a.c. resistance[1] and depends upon surface roughness[1], temperature[1,2], thermal history[1,2] and beta alumina composition[3]. Interface

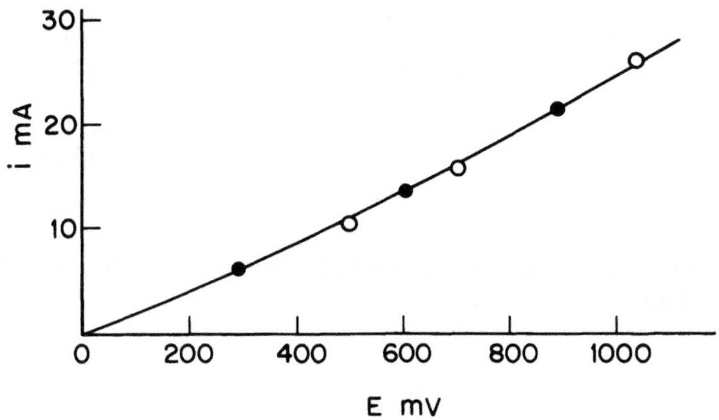


Fig. 1. Current-potential relation for "unwetted" disc at 150°.(0) Forward polarization; (●) reverse polarization.

polarization appears to be linked to the wetting behavior of sodium on beta alumina[2] and affected by protonated impurities such as water[4].

Armstrong et al.[1] determined the d.c. polarization curves and a.c. impedance of Na/Beta"/Na cells employing beta" alumina disks of approx-i mate composition $Na_2O:5\ Al_2O_3$ and < 5% porosity. The beta" alumina had been thoroughly outgassed at 400°C immediately prior to the measure ments. As-cut disks with rough surfaces showed quite different behavior from polished disks and the results in all cases depended strongly on the temperature of exposure to sodium. At temperatures below 250°C, as-cut disks gave rise to large cell polarizations, a.c. resistances approximately 10 times larger than attributable to the beta" alumina resistivity and anomalous complex impedance plots. The cell polarization curves for this case are shown in Figure 1 and the complex impedance plot in Figure 2. Ideal behavior for a cell of the type metal/electrolyte/metal would result in a semicircle in a complex impedance plot.

If the Na/Beta"/Na cell is kept at increasingly higher temperatures, about 250°C, the cell resistance decreases with time and approaches values predicted from the beta" alumina resistivity. Such a decrease with time for a temperature of 350°C is shown in Figure 3. Simultaneously, the cell polarization in Na/Beta"/Na cells decreases substantially. Once low a.c. impedance interface polarization has been achieved by heating the cell to ≥ 350°C, this state is preserved even if subsequent measure-

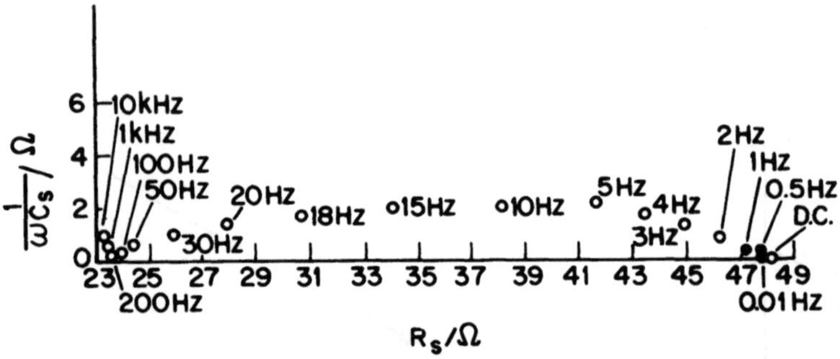


Fig. 2. Complex impedance of "unwetted" disc at 150°. (0) P.S.D.; (●) Lissajous figures.

ments are performed at lower temperatures. This is evident from comparing the cell polzarization curve of Figure 4 obtained at 200°C, for a cell exposed previously to 350°C with that of Figure 3 for a cell exposed to < 250°C.

On beta" alumina disks polished with 1 μ diamond paste, low d.c. polarization and a.c. impedances were obtained much more readily than on as-cut disks. Exposure of Na/beta"/Na cells to 150°C for 2 hours resulted in identical values of the d.c. and a.c. resistance and in a cell polarization

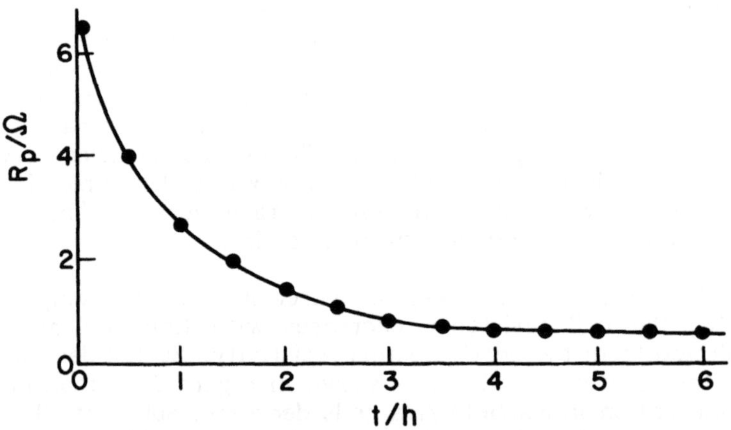


Fig. 3 Time dependence of $R_p$ at 350°.

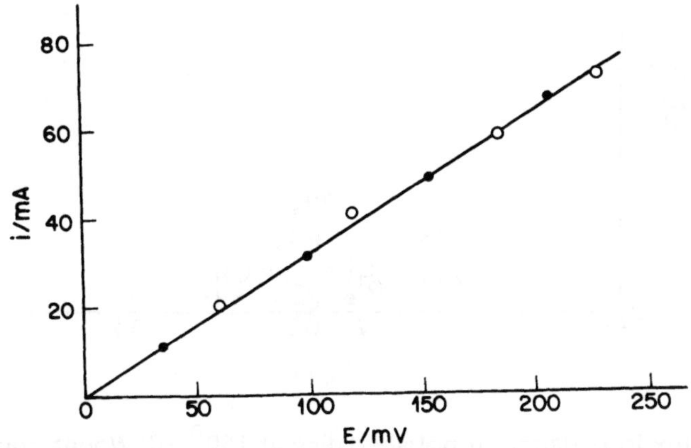


Fig. 4. Current-potential relation for "wetted" disc at 200$^{o}$. (0) Forward polarization; (●) reverse polarization.

identical to that predicted from the beta" bulk resistivity; this indicates that no interfacial polarizations were present in this case. Furthermore, the complex impedance plot of the polished disk, as shown in Figure 5 starts approximating the ideal behavior expected for a metal/electrolyte/metal system. From Figure 5, the charge transfer resistance and sodium exchange rate can be determined. The exchange current density between sodium and beta alumina at 150$^{o}$C was found to be in excess of 0.5 A/cm$^{2}$.

Armstrong et al. interpreted these findings in terms of intimate and rapid wetting of polished beta" alumina surfaces as opposed to slow wetting of rough surfaces. They envisioned wetting of the latter to occur only at the filamentary projections from the surface. This type of model required the assumption of a large number of small (100 to 1000 Å) filaments, between 15 and 50 μ long. Electron micrographs of the beta" alumina surfaces did not however, provide evidence for such large numbers of small protrusions. More studies on carefully characterized beta alumina of high density and different compositions seem to be needed for a full understanding of the possible differences in interfacial polarization between smooth and rough surfaces. It appears that different surface compositions produced by different surface treatments, such as cutting and polishing, might be an important aspect of the observed differences.

Gibson[2] studied the polarization and wetting behavior of the sodium/beta alumina interface on tubes consisting of a mixture of beta and beta" alumina. He substantiated Armstrong's et al. findings that the d.c. resistance of Na/beta/Na cells decreases substantially with time after filling the cells with sodium and that the rate of resistance decrease increases with increasing temperature. After heating to temperatures of 350$^{o}$C, Gibson found that the cell resistance attained a lower limit, essentially determined by the bulk beta alumina resistivity.

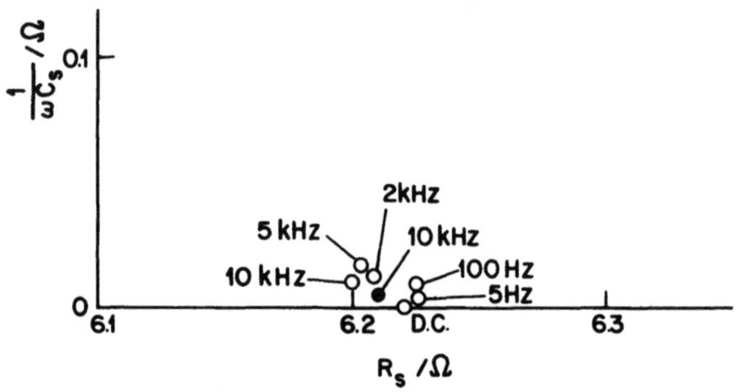


Fig. 5. Complex impedance of polished disc at 150°. (O) Wheatstone bridge; (●) P.S.D.

Gibson was also able to relate the polarization behavior to the degree of wetting of beta alumina by sodium. By rapidly quenching the cells from temperatures between 200 and 400°C to the freezing point of sodium and by applying a radiographic technique, the wetting angle of sodium on beta alumina was determined as a function of temperature. Figure 6 shows the effect of various temperatures applied for 24 hours and also the effect of time when the cell was kept at 300°C. Table I summarizes the contact angle measurements. A weakness of the technique lies in the need to freeze the sodium prior to contact angle measurement. This procedure could lead to a change of the contact angle.

Once beta alumina is wetted by applying a temperatures of 350°C or above the beta alumina remains wetted even at lower temperatures. This, too, is in agreement with Armstrong's et al. findings.

The important results of Gibson's work are that the interface polarization of the sodium/beta alumina interface corresponds to the degree of wetting by sodium and that heating of the ceramic in sodium to at least 350°C produces good wetting -- then maintained even at lower temperatures -- and negligible interface polarization. The mechanism involved in leading to good wetting was left open, but the author suggests that the wetting behavior might be related to hydration of the beta alumina. Such a correlation would be in agreement with studies of Will[4] on the effect of water on the d.c. and a.c. resistance of beta alumina carried out on the interface beta alumina/$Na^+$ (organic solution). In that study it was shown that water occlusion, accompanied by hydronium ion exchange, leads to increased interfacial polarization, an effect which could be reversed by removing the water at elevated temperatures.

Recently, Breiter et al.[3], by using reference electrodes in the beta alumina and in a sodium nitrate melt, determined the contributions of both the sodium/beta alumina and the melt/beta alumina interface separately. These authors found the interface polarization at the

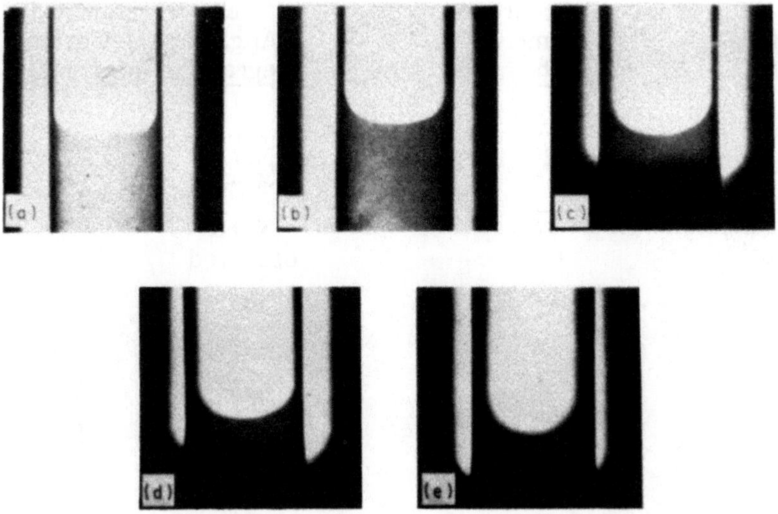

Fig. 6a. Wetting of beta alumina by sodium after 24 h at constant temperature. (a) 200°C, (b) 250°C, (c) 300°C, (d) 350°C, 400°C.

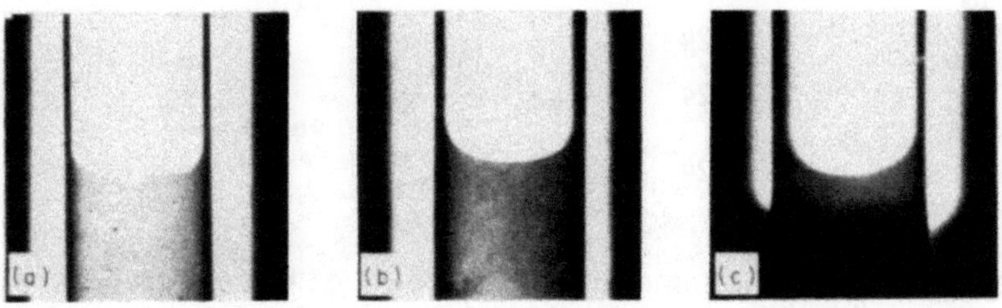

Fig. 6b. Wetting of beta alumina by sodium at 350°C. (a) 5 h soak, (b) 15 h soak, (c) 24 h soak.

Na/beta" interface was 2-½ times larger when sodium entered the beta alumina from the sodium (discharge) than when it exited into the sodium (charge). The asymmetry was found to be composition-dependent and disappeared completed on pure beta alumina compositions. Breiter et al. attributed the interface polarization and non-wetting of beta" alumina to the presence of an $Na_2O$ film on the beta" alumina. They believed that this film was produced by the interaction of water -- spuriously present in beta" -- with sodium at 320°C and dissolved in sodium by heating to temperatures above 350°C.

Table 1. Contact Angles of Sodium on Beta Alumina

| Temperature (°C) | Exposure Time (h) | Mean Angle (degrees) | Standard Deviation (degrees) |
|---|---|---|---|
| 200 | 24 | None observed | - |
| 250 | 24 | None observed | - |
| 300 | 6 | 45 | 4 |
| 300 | 24 | 39 | 8 |
| 350 | 5 | 29 | 4 |
| 350 | 15 | 21 | 5 |
| 350 | 24 | 23 | 6 |
| 350 | 48 | 21 | 4 |
| 400 | 24 | 22 | 4 |
| 400 | 48 | 21 | 3 |
| 350<br>Cooled to<br>200 | 24<br><br>24 | 21 | 3 |
| 400<br>Cooled to<br>200 | 24<br><br>24 | 24 | 4 |

A similar type of asymmetry or "anistropy" of the electrolyte resistance on the sodium/beta alumina interface has also been reported by other investigators[2,3,6]. However, the phenomenon was not addressed in detail and no mechanism was postulated.

Additional studies appear desirable to bring forward a true understanding of the asymmetry phenonemon.

Sulfur Electrode Interfaces

Lack of rechargeability of sodium-sulfur cells and increase of cell resistance with cycling are two often-observed phenomena[7-10] that have been attributed to the sulfur electrode. The "sulfur electrode" exhibits three important interfaces, namely, between the sulfur/polysulfide melt and (1) the beta alumina electrolyte, (2) the carbon felt or other porous conducting matrix and (3) the current collector, usually consisting of a metal container or a metal or graphite rod.

While different groups agree that the sulfur electrode is responsible for the observed effects, sufficient experimental data do not as yet exist to allow a unique interpretation. Hence, no commonly accepted mechanism has been advanced to explain the observed sulfur electrode behavior in actual sodium-sulfur cells.

Planar-Carbon/Melt Interface. Armstrong et al.[11] studied the vitreous carbon/polysulfide interface with voltammetric, pulse and rotating disk techniques. They found diffusion - controlled currents for the reduction of both $Na_2S_5$ and $Na_2S_3$ melts, however, with currents substantially higher in the former. The authors suggested that solid films were present on the carbon electrode when the cathodic current became diffusion - limited and that the film formed during $Na_2S_2$ reduction consisted of $Na_2S_3$.

In anodic current-voltage curves, the authors observed the occurrence of pronounced current maxima in both $Na_2S_3$ and $Na_2S_5$ melts at 350°C, but much less pronounced maxima at 305°C. Diffusion-control was found at potentials less anodic than corresponding to the current maximum but not for more anodic potentials. The observed anodic behavior was interpreted with the formation of a film of elemental sulfur on the carbon electrode. Figure 7 summarizes in diagrammatic form the various processes thought to occur during reduction and oxidation of polysulfide melts.

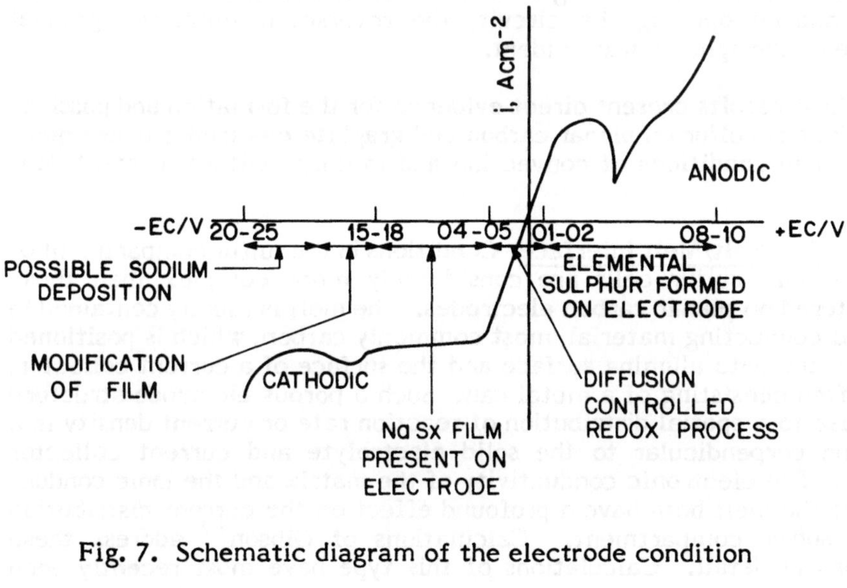


Fig. 7. Schematic diagram of the electrode condition

Brennan[12] investigated polysulfide reduction and oxidation on vitreous carbon and pyrolytic graphite, employing voltammetric and in-situ micro scopic techniques. He found pronounced reduction maxima in $Na_2S_4$ melts at 350°C with the maxima height proportional to the square root of the voltage sweep rate which he interpreted in terms of rate-limiting diffusion. Importantly, the voltage position of the maxima was found to be independent of sweep rate, namely at about 1.7 V. According to the phase diagram, solid $Na_2S_2$ should form at about that potential in polysulfide melts at 350°C. These results can be taken as good evidence that solid passivating $Na_2S_2$ films form on the carbon surface when reducing $Na_2S_4$. Interestingly, Brennan found the currents on pyrolytic graphite five times larger than on vitreous carbon and interpreted this finding as being caused by a larger surface roughness of pyrolytic graphite. However, it seems that the notion of diffusion limitation of the observed currents is not readily reconciled with drastic effects of surface roughness, unless the diffusion layer thickness is of similar dimensions as the surface roughness. This appears unlikely in the present case.

The anodic oxidation of $Na_2S_{4.8}$ showed reversible and ohmic behavior on a fresh carbon electrode, but increasing effects of passivating film formation on continued cycling. This is shown in Figure 8. Passivation did occur and the curves remained ohmic on cycling if the anodic voltage excursions were limited to within 150 mv versus a reference electrode in the same melt. Once the electrode was passivated, depassivation occurred in the single-phase melt on open circuit. Direct microscopic examination revealed the formation of distinctly yellow spheres near the carbon surface during anodic oxidation of $Na_2S_3$ and $Na_2S_4$ melts. On continued oxidation, sphere growth and occasional coalescence was observed and on opening the circuit, the reverse, namely, the gradual collapse of the spheres was evident.

These results present direct evidence for the formation and passivating effect of sulfur on planar carbon and graphite electrodes under quite well-defined conditions of convection and initially uniform current density.

Carbon Felt/Melt Interface. Conditions in the sulfur compartment of actual sodium-sulfur cells are considerably more complex than those encountered on planar carbon electrodes. The melt is usually contained in a felt of conducting material, most commonly carbon, which is positioned between the beta alumina surface and the surface of a current collector, most often consisting of a metal can. Such a porous electrode structure gives rise to a spatial distribution of reaction rate or current density in a direction perpendicular to the solid electrolyte and current collector surface. The electronic conductivity of the matrix and the ionic conductivity of the melt both have a profound effect on the current distribution in the sulfur compartment. Calculations of Gibson[13] address these questions in detail. Calculations of this type have most recently been

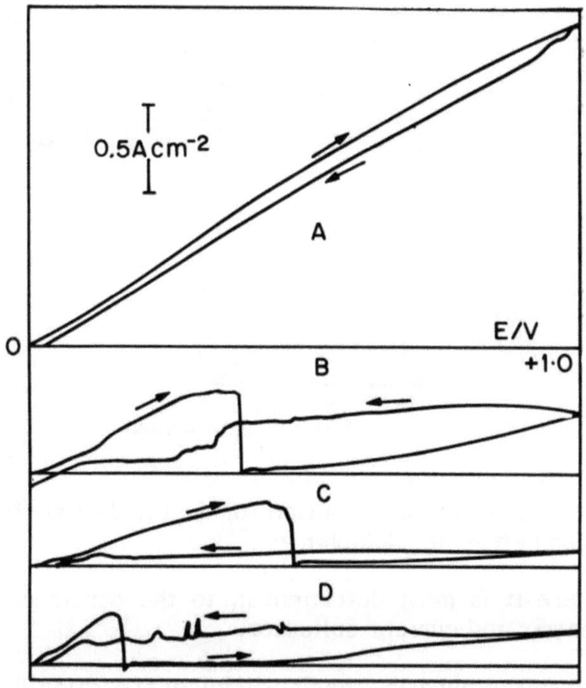


Fig. 8. The current response to anodic linear potential sweeps on a vitreous carbon electrode. A=fresh electrode, B=after 310 cycles, C=after 1492 cycles, D=after 2644 cycles. The melt composition lay within the limits $Na_2S_{4.89}$-$Na_2S_{4.74}$. Sweep rate is 0.032 $Vs^{-1}$.

extended by Breiter and Dunn[14] and applied to provide an interpretation of cycling results of sodium-sulfur cells with different types of carbon matrices[14]. According to Breiter and Dunn, the reaction rate distribution can be changed dramatically by using two carbon matrices with grossly differing electronic resistivities. A thin layer of a high-resistance carbon matrix was positioned next to the beta alumina and a thick layeer of regular, low-resistance carbon matrix next to the current collector. The calculated change in reaction rate distribution for such a dual matrix as compared to the use of only the regular matrix is shown in Figure 9. The reaction rate at the beta alumina interface is predicted to be much lower when using a dual matrix. Hence less insulating sulfur should be formed at the beta alumina interface. Actual cycling of sodium sulfur cells containing a dual matrix indeed showed considerably improved capacity. Such cells could be cycled into the two-phase region whereas a cell with a single matrix could only be cycled within the single-phase region. These results are summarized in Figure 10. An important aspect of this work is that the formation of passivating sulfur films can be largely shifted from the relatively small surface area of the

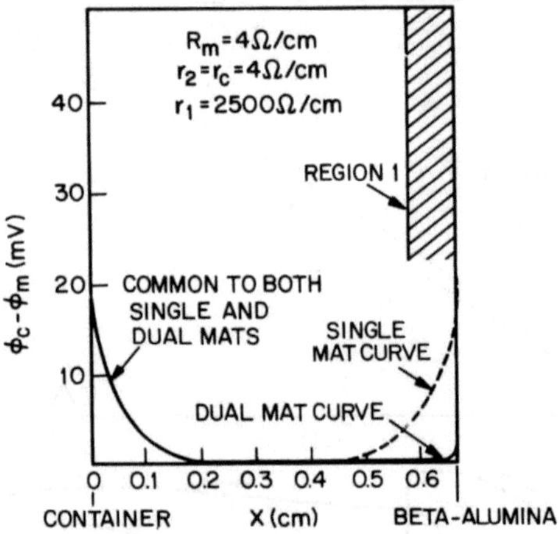


Fig. 9. Reaction rate distribution for dual and single mat geometries ($r_2$=4.0 ohm-cm$^{-1}$).

beta alumina, where it is most detrimental, to the considerably larger surface area of matrix and current collector.

While these results emphasize the significance of reaction distribution, other authors place emphasis on sulfur electrode designs that facilitate the reactant transport within the sulfur compartment[9,15]. Kleinschmager et al.[9] explored two methods aimed at providing facilitated reactant

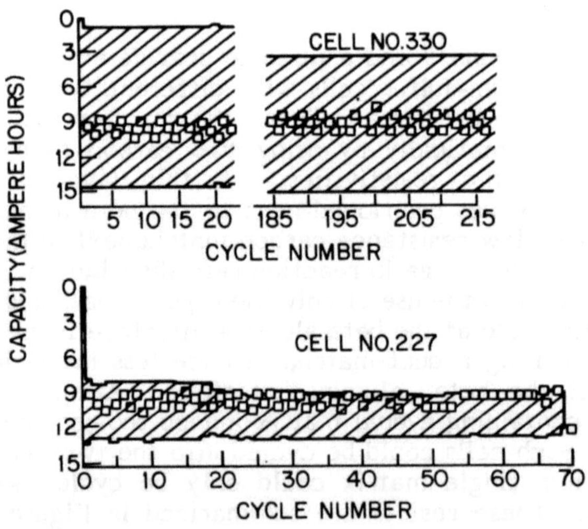


Fig. 10. Capacity performance for cells with and without the graduated resistance sulfur electrode.

transport: (1) the use of voids and vertical channels near the beta alumina interface and (2) the use of additives to decrease the viscosity of the molten sulfur. They obtained considerably improved rechargeability by applying either one of these two methods as is evident from their results shown in Figure 11. Incorporation of a reference electrode into the matrix established the melt/beta alumina interface as the site of the resistance increase toward the end of charging. This was interpreted by the formation of elemental sulfur at that interface. Employment of voids or channels in the carbon matrix next to the interface appeared to confine sulfur formation to limited areas of the beta alumina.

The beneficial effect of additives to the S, such as Se and $B_2S_3$, was ascribed to a lowering of the S vixosity, resulting in facilitated transport of S away from areas where otherwise insulating S-films would form.

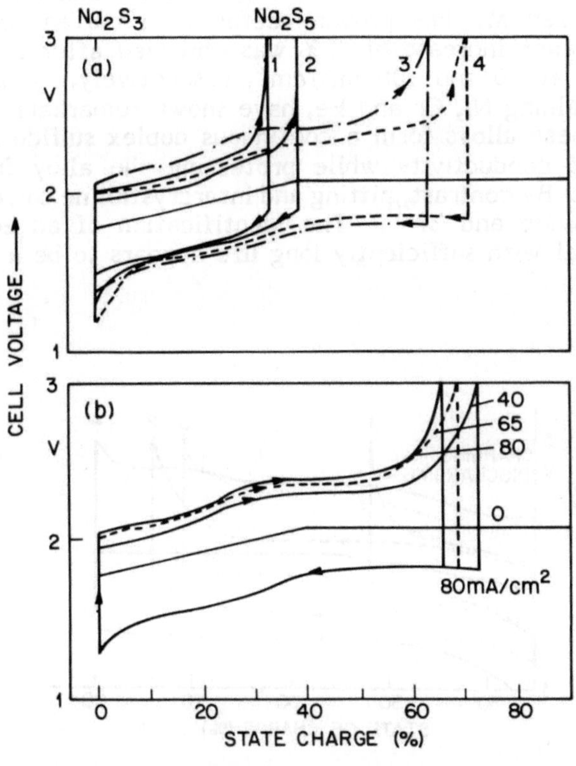


Fig. 11. Charge-discharge behavior of Na-S cells at 310°C. Fig. a: Current density $80 mA/cm^2$ - Curve 1: standard cell with carbon felt, curve 2: standard cell with graphite felt, curve 3: felt-free cavities at the electrolyte tube, curve 4: felt-free vertical channels at the electrolyte tube. Fig. b: Cell homogeneously packed with graphite felt, cathodic reactant S 0.8 atom %Se 0.8 mol %$B_2S_3$.

A quite different additive approach has been described by Fischer et al.[16]. These authors used $C_6N_4$ additivies which dramatically raise the electronic conductivity of sulfur by forming charge transfer complexes. This additive resulted in better rechargeability and simultaneously in lower charge and discharge polarizations. These effects are shown in Figure 12.

Current Collector/Melt Interface. The current collector usually consists of a metal container in contact with the carbon matrix and the sulfur/polysulfide melt. Various metals and metal alloys have been studied with respect to their corrosion and passivation behavior in the melt.

Stainless ssteel was shown to corrode at unacceptably high rates[17]. Aluminum forms passivating layers[18]; Al coated with certain Cr alloys has shown some promise[18], but the large coefficient of thermal expansion of Al is expected to pose problems with regard to the adherence of such coatings. Anodized Mo has proven useful as a current collector[19] although a resistance increase of 20% was observed after 800 charge-discharge cycles at 50 and 100 mA/cm$^2$, respectively. A number of superalloys, containing Ni, Cr and Fe, have shown remarkable corrosion resistance[20]. These alloys form a continuous duplex sulfide layer that retains sufficient conductivity while protecting the alloy from rapid further corrosion. By contrast, pitting and intercrystalline corrosion have been observed in Mo and Zr[20]. The identification of an economical container material with sufficiently long life appears to be a continued challenge.

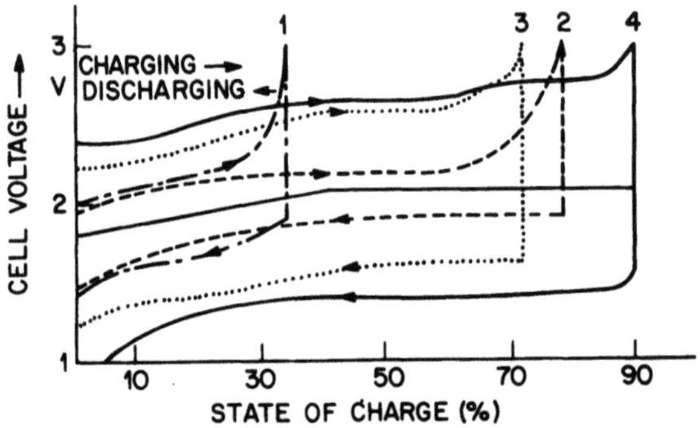


Fig. 12. Voltage of sodium/sulfur cells as a function of state of charge. Current density of cells (1), (2), (3) 80mA/cm$^2$, cell (4) 150mA/cm$^2$. The cathode compartment was filled with: (1) graphite felt and pure sulfur; (2) graphite felt, sulfur and 0.8 mol % $C_6N_4$; (3) graphite felt coated with beta alumina grains in order to create sodium ion concentration cells; (4) same as (3) plus 0.8 mol % $C_6N_4$.

## LITHIUM/ALUMINUM - IRONSULFIDE SYSTEM

This battery is a high-temperature system, which employs solid electrodes in a molten lithium potassium chloride eutectic, typically operated at 430°C. Interfaces of particular interest to cell operation are those between (1) the molten salt electrolyte on the one hand and the active materials and current collectors on the other, (2) the melt and the separator and (3) the active materials and the current collectors.

While the most dominant interfacial phenomenon relates to corrosion, the wetting behavior of the electrodes and the separator by the chloride melt is also of interest. Furthermore, the formation of other than the expected reaction products and their effects on electrode performance will be described briefly in the following section.

### Current Collector/Melt Interface

The current collectors are surrounded by high-porosity active materials which permit the corrosive chloride melt to come into direct contact with the current collectors. A large number of metals and alloys have been studied with respect to their corrosion behavior in this environment[21]. Iron base alloys and 304 stainless steel were rapidly corroded so were nickel, Inconel 625 and Hastelloy B. Low-carbon steel collectors for the FeS active material showed sulfidation followed by dissolution and intergranular attack. Only Mo showed excellent corrosion resistance, possibly resulting from a thin adherent layer of $MoS_2$ which was shown to form. The rate of corrosion layer formation on Mo current collectors used in the positive electrode was found to be less than 4 μ per year.

### Current Collector/Active Material Interface

Interestingly, the corrosion of low-carbon steel collectors in the negative electrode did not result from reaction with the melt but with the active material consisting of Li-Al intermetallic. A brittle corrosion product consisting of $FeAl_2$ was identified by x-ray diffraction. The rate of corrosion layer formation was found to be 35 μ per year.

### Separator/Melt Interface

Boronitride felt (BN) has so far been identified as the best separator, combining the advantages of chemical stability and high porosity. Various oxide separators are being examined[22]. $Y_2O_3$ separators react slowly in contact with $FeS_2$ or FeS to form $Y_2O_2S$. This species has higher molar volume than $Y_2O_3$ and, consequently, leads to a progressive decrease in separator porosity and, hence, increase in cell resistance. Furthermore, formation of $Y_2O_2S$ leads to loss of sulfur from the active positive material, that is, FeS or $FeS_2$.

The wetting behavior of the separator materials by the chloride melt has been studied by measurements of the advancing and receding contact angles[23]. A summary of the results is presented in Figure 13. While $Y_2O_3$ is readily wetted, BN is not. A surprising finding was made in that the application of a thin $LiAlCl_4$ powder coating produced spontaneous wetting of the BN by the chloride melt. The $LiAlCl_4$ melts at the operating temperature of the cell and was found to have a similar effect of promoting wetting on other surfaces such as Fe, FeS, and $FeS_2$. It was suggested that the $LiAlCl_4$ cleans the surface, but the mechanism is clearly not understood as yet.

## Active Material/Melt Interface

The performance of the Li-Al and the FeS and $FeS_2$ electrodes is determined by slow solid state transport processes which control the overall reaction rate.

The charge and discharge of the positive electrode involves the alternate transport of Li out of (charge) and into (discharge) the iron-

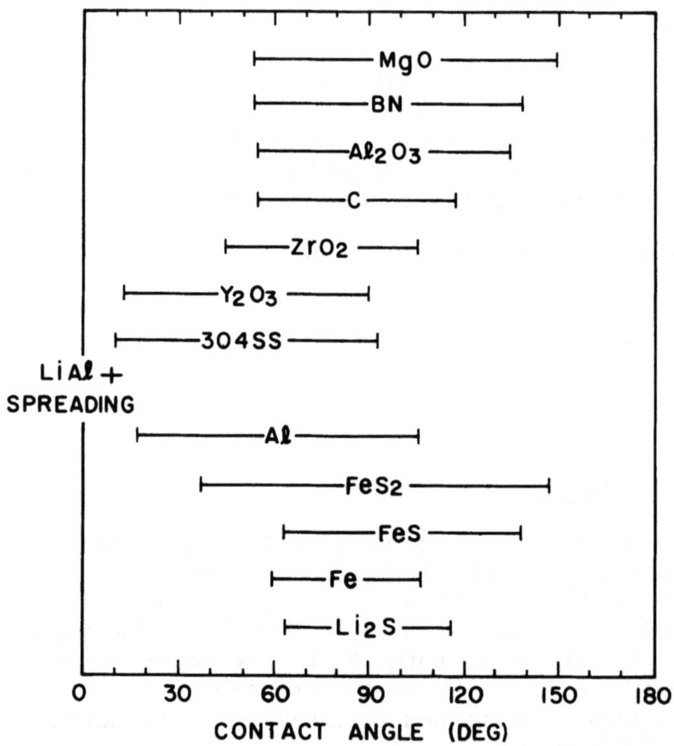


Fig. 13. Contact Angle Ranges for Molten Salt (400°C) on Various Solid Surfaces.

sulfide structure. Figure 14 shows a voltammogram at 0.015 mV/s for the charge (oxidation) and discharge (reduction) of FeS[24]. The two anodic peaks during charge are attributed to the reactions

$$Fe + 2Li_2S = Li_2FeS_2 + 2Li + 2e^-;\ (1.31V) \quad (1)$$

$$Fe + Li_2FeS_2 = 2FeS + 2Li^+ + 2e^-;\ (1.41V) \quad (2)$$

The discharge sequence is considerably more complex, and there is metallographic evidence that the surface reaction is different from the bulk reaction[24] namely

(surface)

$$6Li + 26FeS = LiK_6Fe_{24}S_{26}Cl + 2Fe + 5LiCl;\ (1.35\text{-}1.32V) \quad (3)$$

(interior)

$$2Li + 2FeS = 2Li_2FeS_2 + Fe;\quad (1.32\text{-}1.27V)\ \text{(complex)} \quad (4)$$

(interior)

$$2Li + Li_2FeS_2 = 2Li_2S + Fe \quad (5)$$

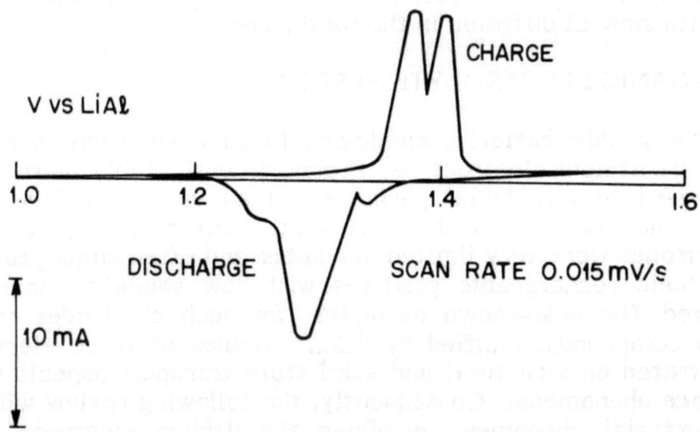


Fig. 14. FeS cyclic voltammogram in LiCl-KCl at 700K. (Working electrode contained 20 mA-hr FeS and 100 mg Fe powder in 5 $cm^2$x0.3 cm molybdenum housing. Reference and counter electrodes were LiAl.)

(surface)

$$46Li + LiK_6Fe_{24}S_{26}Cl + 5LiCl = 26Li_2S + 24Fe + 6KCl; (1.27\text{-}1.22V) \quad (6)$$

The overall reaction, however, is a simple two-electron process:

$$2Li + FeS = Li_2S + Fe \quad (7)$$

The occurrence of the same species, namely, $LiK_6Fe_{24}S_{26}Cl$ in both surface reactions might explain the presence of only one major maximum in the cathodic sweep during discharge. This compound is essentially identical to the mineral djerfisherite. Since it forms on the surface of the FeS particles, it can interfere with the discharge reaction, especially at high current densities. Apparently, the addition of $Cu_2S$ partially nullifies this effect[24].

During charging with high rates and during overcharging, $FeCl_2$ and elemental sulfur are formed. While the latter dissolves in the electrolyte the former is solid at the cell operating temperature. The interfacial effects of such a solid reaction layer have not been studied in detail.

Charge and discharge of the negative Li-Al electrode involve the alternate transport of Li into (charge) and out of (discharge) the intermetallic structure. The process is solid state diffusion - controlled, and Li concentration gradients form from the surface into the interior of the Li-Al particles. The limited rate of Li diffusion is evident from a comparison of voltammograms obtained at different sweep rates[24]. This is shown in Figure 15. Both the anodic (discharge) and the cathodic (charge) maxima shift with increasing scan rate in a manner that can be interpreted with slow Li diffusion in the solid phase.

## LITHIUM-ORGANIC ELECTROLYTE SYSTEM

In rechargeable batteries employing lithium negatives in organic electrolyte, the lithium electrode has received considerably more attention with regard to interfacial phenomena than various positive electrodes. This may be due to the fact that, until recently, reversible positive electrodes were very limited in number and often employed liquid reactants. Solid rechargeable positives with low solubility have only lately emerged. The best-known examples for such electrodes are the intercalation compounds, typified by $TiS_2$. Studies of these electrodes have concentrated on structural and solid state transport aspects rather than on surface phenomena. Consequently, the following review will only consider interfacial phenomena involving the lithium electrode where good progress in understanding has been made in the past few years.

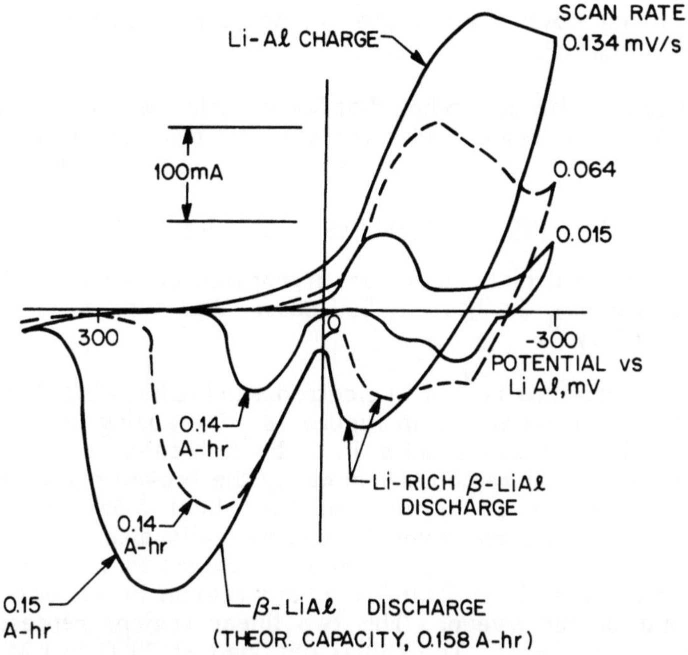


Fig. 15. Cyclic Voltammograms of Aluminum (0.195 g) in LiCl-KCl Electrolyte at 465°C.

It has become apparent that the behavior of the lithium electrode in organic electrolytes is governed by the formation and the properties of surface films or layers. Such important characteristics as the cycling efficiency, cycle life, power density, self discharge and delay time on discharge after open circuit, are all intimately related to film formation and properties.

It has also been recognized that the nature of the organic solvent and the solute play a significant role in film formation and properties.[25-29] In fact, the differences in electrode behavior from one organic electrolyte to another are often so large that few generally applicable statements can be made at this time.

The general features that appear to emerge are that lithium electrodes in organic electrolytes are always covered with a film or layer. For best performance, the lithium electrode should be covered with a continuous film that prevents direct contact between lithium and the solvent molecules while permitting the transport of lithium ions through it. Such an ideal system has obviously not been found, but considerable progress has been made in that direction.

In accordance with the finding that few generic phenomena exist on Li electrodes in different organic electrolytes, the interfacial behavior of lithium will be described separately in selected organic solvents.

## Lithium Interfaces in Propylenecarbonate - Based Electrolytes

The behavior of lithium electrodes in propylenecarbonate (PC)-based electrolytes is probably the best-studied case. This system has recently been reviewed by Dey[28].

Kinetic measurements[30] of Li electrodes in $LiClO_4$ - PC electrolytes showed a dual Tafel behavior as in Figure 16. Increasing anodic polarization of the Li electrode results in a linear region which changes abruptly to another linear region (as shown by the broken curve in Figure 16) of lower overvoltage at current densities about 2-5mA/cm$^2$. Removal of moisture to a 5ppm level did not basically alter the situation. The electrodes appear to be activated by anodization at high current densities of the order of 10 to 20 mA/cm$^2$ and remain in this active state during a second anodic sweep. The two linear regions represent two different exchange currents. The values obtained at 23$^o$C in 1 M $LiClO_4$ in PC were 0.7 mA/cm$^2$ and 1.7 mA/cm$^2$ corresponding to the lower and the upper curves.

The above dual Tafel behavior of the Li electrode was attributed to the formation of a film on the Li electrode in the $LiClO_4$-PC electrolyte. The facts that (a) extensive dehydration[30] and purification[31] of the electrolyte did not cause any change in the Tafel line and (b) the same low

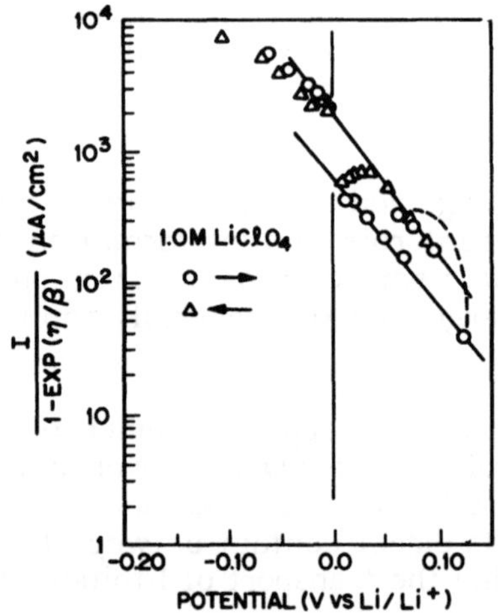


Fig. 16. Tafel plots of the Li electrode polarization data[64]

exchange current (0.7 $mA/cm^2$) was obtained in both cases indicate that the film formation is not due to the reaction of Li with impurities.

It has been established[32] that PC is electrochemically reduced on a graphite substrate at a potential of 0.6 to 1.0V atainst lithium as reference to form propylene and carbonate anions which precipitate as $Li_2CO_3$. Therefore, thermodynamically, the reaction

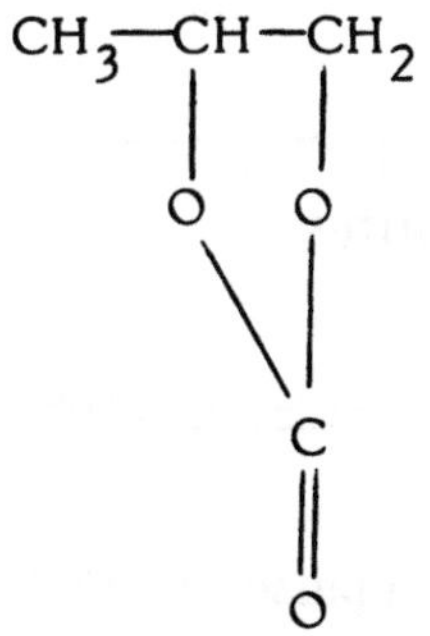

$$+ 2Li = CH_3{-}CH{=}CH_2 + Li_2CO_3 \qquad (8)$$

is favorable. $Li_2CO_3$ is extremely insoluble in PC and precipitates quantitatively. On Li electrodes, reaction (8) results in the formation of a passivating film of $Li_2CO_3$ which prevents any further chemical attack of the Li by PC. This is why Li is stable in an organic solvent such as PC.

A thin film of $Li_2CO_3$ is formed spontaneously when Li is brought into contact with PC. For current densities of the magnitude observed (lower Tafel line in Figure 16) at 23°C, it must be assumed that the film has pin-holes through which $Li^+$ can be transported in the electrolyte phase. When the anodic current density is increased beyond a critical value, the electrode polarization suddenly drops, that is, the electrode becomes "activated".

The phenomenon of activation by anodic pulses has been studied by Dey[28] and the nature of the film characterized metallogrphically. The experiments were performed in IM $LiClO_4$ in PC at -20°C in order to accentuate the effect of the Li film on electrode polarization. The Li electrode was polarized anodically at a constant current density of 3 mA $cm^{-2}$, and the potential was monitored against a Li reference electrode as a function of time. The presence of a passive film on the Li electrode is illustrated in Figure 17a; a pulse of 5 s was applied, and it took at least four individual pulses for the polarization to attain a steady value, indicating gradual removal of the film. The extent of film growth with time of Li exposure to the electrolyte is further demonstrated in Figure 17b; Li electrodes kept in the electrolyte for 1 h and 16 h at room temperature (25°C) and 16 h at -20°C were subjected to 1 min pulses, and the electrode potentials were recorded as a function of time. A more severe initial polarization was observed for the electrode which was kept in the electrolyte for 16 h at 25°C compared with those that were kept for 1 hr at 25°C or 16 h at -20°C. Clearly, the extent of film formation

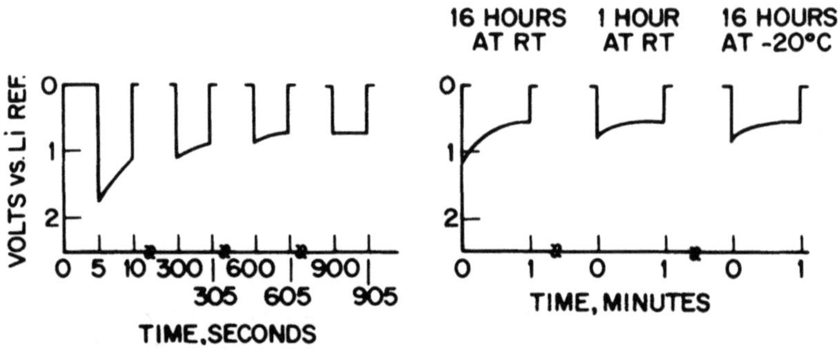


Fig. 17a. The voltage recovery of a Li anode on a 5 s pulse at -20$^o$ in lM $LiCLO_4$-PC.

Fig. 17b. The voltage recovery of a Li anode on a l min pulse at -20$^o$C after storage at 25$^o$C and -20$^o$C.

increased with time and temperature of storage.

The morphology of the film formed on Li in PC was examined[34] using scanning electron microscopy (SEM). The original Li surface and the Li surface after exposure for 16 h at 25$^o$C to PC are shown in Figures 18a and 18b, respectively, at a magnification of 5000X. The film appears to be extremely thin and consists of small granules with an average diameter of approximately 0.1 μm. X-ray diffraction studies by Froment et al[29] identify such layers formed on Li in perchlorate-containing PC solutions as consisting of $Li_2CO_3$, $LiClO_4$ and LiCl.

According to Dey, the activation process of the Li anode by the anodic current does not appear to be electrochemical in nature since the only electrochemical reaction that could occur is anodic Li dissolution. Dissolution must occur through the pinholes and other imperfections in the $Li_2CO_3$ film. The initial voltage drop is due to the low effective surface area of the electrode. The dissolution proceeds through the pinholes and undercuts the $Li_2CO_3$ film which then cracks under mechanical and/or thermal stresses. This process increases the effective area and leads to voltage recovery. Crude double-layer capacitance measurements showed that the effective electrode area corresponding to the upper Tafel line (Figure 16) is twice as much as that corresponding to the lower line. Thus the activation process is postulated to be of mechanical rather than of electrochemical nature and the physical nature of the film (thickness, brittleness, etc.) is expected to be important in the activation process.

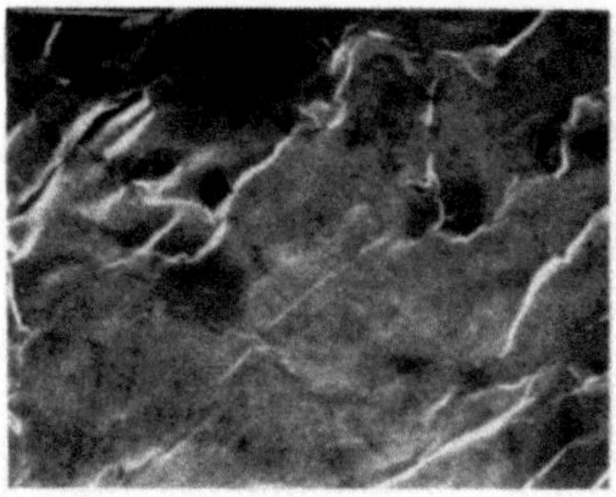

Fig. 18a. An SEM photograph of an untreated lithium surface. (Magnification 5000X).

Fig. 18b. An SEM photograph of a Li surface after 16 h storage in PC at 25°C. (Magnification, 5000X.)

Progressive film formation on Li electrodes when kept on open circuit in PC solution is responsible for a progressive decrease in the amount of Li which is available for anodic discharge. This was clearly shown in the studies of Selim and Bro[35] who demonstrated that in PC solutions containing $LiClO_4$ and 40 ppm of water, essentially no dischargeable Li was present after an open circuit stand of 40 hours at 25°C.

Rauh and Brummer[36] established the rate of capacity loss as initially 30-40 µA/cm$^2$, decreasing after a few hours to 5-10 µA/cm$^2$. A mechanism for the capacity loss was postulated which consists of the gradual encapsulation of Li grains with an insoluble film -- according to Dey and Froment et al $Li_2CO_3$ -- thus resulting in the loss of electrical contact with the Li grains or the current collector underneath. Interestingly, Rauh and Brummer found that certain additives, for example, nitromethane (NM) and $SO_2$ decrease the rate of capacity loss during cycling. This effect was interpreted in terms of the formation of additive-modified films with a certain degree of $Li^+$ ion conductivity. On long open circuit stands, however, the formation of insulating films prevailed and the Li electrode passivated with essentially the same rate as in PC solutions without additives. More studies appear to be needed to establish the probably complex mechanism of dual film formation.

Lithium Interfaces in Other Organic Electrolytes

Considerably less information is available on the behavior of lithium electrode interfaces in organic electrolytes other than PC.

Tetrahydrofuran (THF) has been used extensively in Li batteries. A gel-like thick film has been observed by Dey[28] in $Li/V_2O_5$ and $Li/Ag_2CrO_4$ cells which were stored at 55°C (the cells contained 1 M

$LiClO_4$-THF as the electrolyte). This type of massive film is not formed by direct Li-THF reaction since no such film is observed when Li alone is stored in $LiClO_4$-THF electrolyte at 55°C for a long period of time. The solid cathode materials which are strong oxidizing agents appear to participate in the film formation.

It has been shown[37] that THF can be polymerized anodically on Pt and graphite electrodes using 1 M $LiClO_4$-THF electrolyte at a potential of 3.0 V against a Li reference electrode. In Li batteries, the polymerization is probably initiated at the positive electrode. The polymers then diffuse to the Li anode and can be terminated by Li.

At this point the polymer appears to precipitate on the Li anode, forming a gel-like film. This is an example of an indirectly formed Li film which results in a far more serious voltage delay in Li batteries such as $Li/V_2O_5$ than has been observed in PC-based electrolytes. The flexible gel-like film is virtually impossible to dislodge, and in such cases voltage recovery often does not occur.

Rauh and Brummer[38] have examined methylacetate (MA) as a solvent for use in non-aqueous secondary lithium batteries. The efficiency of cycling lithium on nickel in MA/1 M $LiAlO_4$ containing less than 10 ppm $H_2O$ was less than 10%. Addition of nitromethane (NM), $SO_2$ or small amounts of $H_2O$ improved the efficiency markedly. Compared to propylene carbonate (PC), MA plus $SO_2$ or NM afforded more repeated cycles before failure of the working electrode. Unlike PC, the lithium disappearance rate in MA on open circuit is decreased in the presence of additives, e.g. from about 300 $\mu A/cm^2$ to 70 $\mu A/cm^2$ with > 0.2M nitromethane, and to exceedingly small values with 3 M $SO_2$. The differences in behavior between PC and MA are attributed to the greater solubility of the MA-lithium reaction products, allowing greater opportunity for buildup of additive-induced conductive films on the metallic deposit. These films, formed by the reaction between Li and the additive have low solubility in the organic electrolyte and allow the transport of $Li^+$ ions.

Thus it is seen that different organic solvents and additives lead to the formation of films with grossly different properties resulting in quite different Li electrode performance. The approach of using additives appears to be very promising and is apt to lead to further improvements in Li electrode performance, once a good understanding of the formation mechanism is at hand and the films have been fully characterized.

## ZINC-CHLORIDE HYDRATE SYSTEM

### Systems Description

The zinc-chlorinehydrate ($Zn-Cl_2.8H_2O$) system[39] employs a circulating aqueous solution of zinc chloride and $Cl_2$ storage in a separate reservoir as a solid $Cl_2$ hydrate. The hydrate is kept refrigerated at temperatures below $10^oC$. During discharge, the temperature of the hydrate is raised, by using heat exchange with the warm electrolyte, to allow the release of gaseous $Cl_2$ which is fed into the circulating electrolyte. The $Cl_2$-saturated electrolyte flows through a porous carbon electrode, resulting in its reduction to $Cl^-$ ions. Simultaneously, a Zn layer, covering the surface of a second carbon electrode, is oxidized to $Zn^{++}$ ions. During charge, these processes are reversed and $Cl_2$, formed on the positive carbon electrode, is removed from the circulating electrolyte with a gas pump and sparged at elevated pressure into cold water or dilute $ZnCl_2$ solution. This results in the formation of solid $Cl_2$ hydrate which is filtered from the liquid phase and stored in the refrigerated reservoir.

Types of Interfaces and Interface Phenomena

Important interfaces in a Zn-$Cl_2$ hydrate system include those between solid hydrate and liquid phase, gaseous $Cl_2$ and liquid phase, dissolved $Cl_2$ and solid carbon, Zn and carbon and between Zn and electrolyte.

Some of the interface phenomena occurring in this type of battery are unique to the Zn-$Cl_2$ hydrate system and most of them have not been studied in detail as yet. Discussed below are interface phenomena relating to the hydrate/liquid and $Cl_2$/carbon interface. Zn/electrolyte interface phenomena will be discussed in the section addressing the Zn-$Br_2$ system.

Chlorinehydrate Interfaces

$Cl_2$ hydrate forms when $Cl_2$ gas is dispersed in cold water or aqueous salt solutions. Heat and mass transfer across the hydrate/solution interface determine the rate of hydrate formation. The temperature of formation depends upon the pressure and the concentration of the salt solution. For conditions of thermodynamic equilibrium, the relationship between $Cl_2$ vapor pressure and temperature is shown in Figure 19[39] for $Cl_2$ hydrate in water and in $ZnCl_2$ solutions of three different concentrations. The straight lines in the semi-logarithmic plot signify the stability ranges of $Cl_2$ hydrate in these four environments. Above a given straight line, $Cl_2$ hydrate forms and below the line, it decomposes into $Cl_2$ and $H_2O$. Increasing salt concentration decreases the stability of $Cl_2$ hydrate, that is, lower temperatures or higher pressures are required to form $Cl_2$ hydrate in salt solutions than in pure water.

In practice it is observed that some supercooling is required to initiate hydrate formation and that the amount of necessary supercooling depends upon whether the solution is cooled at constant pressure or pressurized at constant temperature. This is evidenced by Figure 20[39] for the case of $Cl_2$ hydrate formation in water. The broken curve refers to pressurization at constant temperature and shows a relatively small deviation

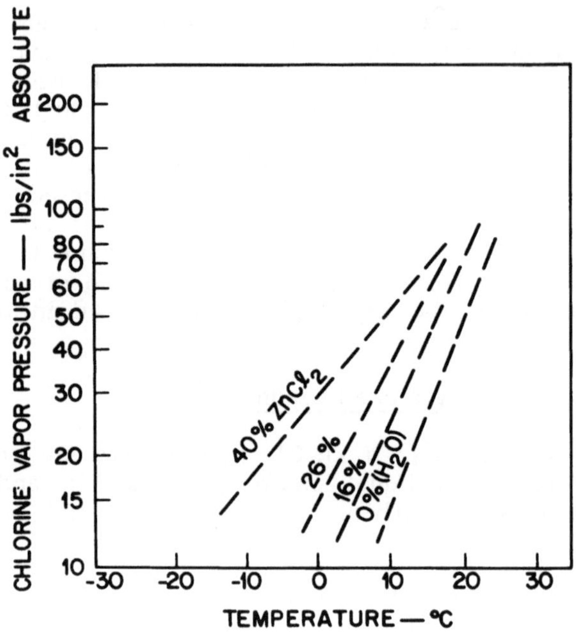


Fig. 19. Vapor pressure of chlorine over chlorine hydrate.

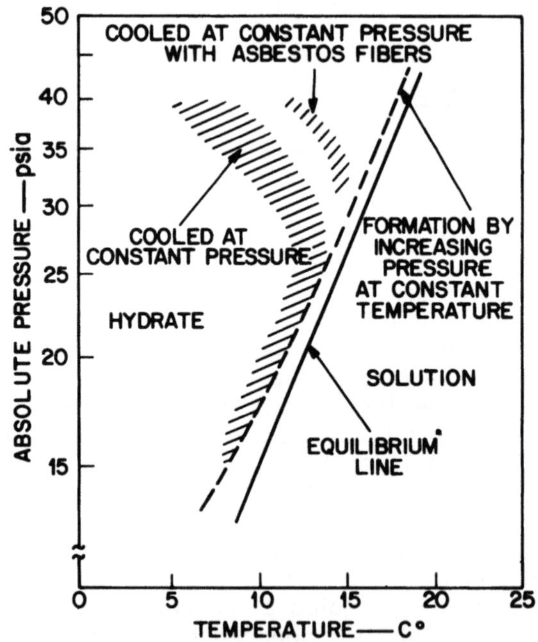


Fig. 20. Supercooling requirement to initiate formation of chlorine hydrate.

from the equilibrium line. Considerably higher supercooling is required when cooling at constant pressure, but the introduction of nucleation agents reduces the supercooling requirements. Small hydrate crystals and certain foreign materials, such as asbestos fibers, act as nucleation agents.

Proper control of the location and rate of hydrate formation is essential for proper operation of the Zn-$Cl_2$ hydrate system. Formation of $Cl_2$ hydrate ideally should be confined to the bulk of the hydrate forming pump and the hydrate reservoir. Formation of hydrate films or accumulation of particulate hydrate in the heat exchanger, pipes and pump walls constricts the flow of solution and has to be avoided. Phenomena at the interfaces between nucleation sites and $Cl_2$, between hydrate and $Cl_2$ and between $Cl_2$ and liquid are significant in this respect.

It has been found[39] that the local rate of heat removal often controls the formation rate. Hydrate formation requires heat removal to the extent of 18.6 kcal/mole. For the desired formation of small hydrate particles to occur in the solution, high surface areas between $Cl_2$ and solution are necessary for rapid heat transfer from the crystallization site to the bulk of the solution. In addition, the morphology and chemical nature of the surfaces, the thermal conductivity of the substrates and the hydrate itself and, finally, the hydrodynamic conditions all affect the rate of hydrate formation.

While the effects of these parameters are known empirically, no clear understanding of these effects and their interrelationship exists as yet.

## Chlorine Electrode Interface

Porous uncatalyzed carbon electrodes have replaced ruthenized titanium electrodes for use as chlorine electrodes. $Cl_2$-containing electrolyte flows through the porous carbon electrodes during discharge, resulting in the cathodic reduction of $Cl_2$. When charging the battery, $ZnCl_2$ solution is circulated past the porous carbon electrode and gaseous $Cl_2$ is formed.

The interface reactions that are significant for $Cl_2$ electrode performance are the dissociation of $Cl_2$ molecules, the charge transfer between carbon and Cl atoms or $Cl^-$ ions and the recombination of Cl atoms during charging. Important for the long-term performance is the resistance of the carbon to corrosion, particularly under conditions of anodic polarization, that is, charging. The evolution of oxygen, simultaneous to the formation of chlorine, has to be kept at a minimum for good electrode life. This can be effected by adjusting and controlling the solution pH to low values within the limits of reasonable hydrogen evolution rate from the Zn electrode.

Certain activation procedures have been employed to increase the electrocatalytic activity of the porous carbon electrode for $Cl_2$ reduction and formation. It has been found[39] that either heat treatment of carbon in nitric acid for several weeks or anodic polarization in dilute aqueous solutions, such as $ZnCl_2$, leads to enhanced electrode activity. Figure 21 shows that both the anodic and cathodic performance of carbon electrodes are substantially improved by electrolytic activation. The improvement in performance increases as the amount of charge during the anodic activation increases.[39] The improvement was found[39] to be related to an increase in BET surface area by a factor of 3 to 4 and the creation of a significant number of small pores in the surface layers of the porous carbon. The observed formation of $CO_2$ during electrode activation makes it apparent that the activation process consists of the controlled oxidation and corrosion of the carbon during anodization in dilute $ZnCl_2$ solution. In actual battery operation, these processes are minimized by acidifying the concentrated $ZnCl_2$ solution.

## ZINC-BROMINE SYSTEM

Like the Zn-$Cl_2$ hydrate system, the Zn-$Br_2$ employs circulating electrolyte and storage of the reactants separate from the electrochemical cells. The system operates at ambient temperature. Since the solubility of $Br_2$ in bromide ion-containing electrolytes is significantly higher than the solubility of $Cl_2$, special provisions must be made to prevent rapid self-discharge due to direct chemical reaction of $Br_2$ with Zn. Two major approaches have been taken: In one, a cation exchange

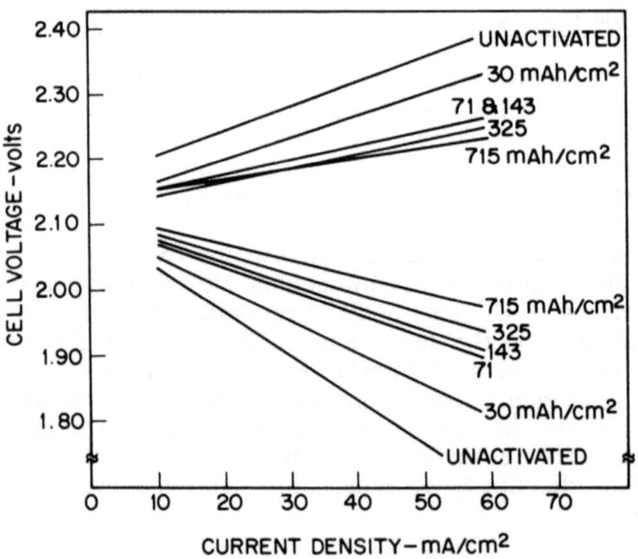


Fig. 21. Chlorine-electrode voltaic performance as a function of electrochemical activation level (espressed in $mAh/cm^2$.

membrane is used[40-42] to keep the reactants separate; in the other, an organic $Br_2$ complex is employed[43-45] to lower the solubility of the $Br_2$ in the aqueous phase.

Important interfaces exist between the oil-like organic complex and the aqueous phase, aqueous phase and ion exchange membrane, aqueous $Br_2$-containing phase and positive electrode and between the aqueous phase and the negative Zn electrode. Inasmuch as the Zn electrode/solution interface is common to both the Zn-$Cl_2$ hydrate and Zn/$Br_2$ systems, emphasis in the following discussion will be placed on that interface.

<u>Zinc Electrode/Electrolyte Interface</u>

The dominant interface phenomena relating to the Zn/solution interface are the tendency of the Zn deposits to develop dendritic growth forms and non-uniform thickness across the electrode surface, in particular, in the direction of the gravitational field.

Extensive literature exists on the initiation and propagation of dendrites in general[46,47] and of Zn dendrites deposited from alkaline solutions in particular[48]. Once nucleation has occurred, preferably on locations of the surface that have microscopic or macroscopic inhomogeneities, such as crystal imperfections or geometric protrusions, dendrite propagation proceeds under mass transfer-controlled conditions.[49] Thus, all factors which affect the rate of mass transfer also affect dendrite growth rate and habit. Very significant in this respect are the hydrodynamic conditions and electrolyte composition. Electrolyte additives can be selected in favorable cases that decrease the rate of charge transfer reaction to such an extent that the reaction is no longer under mass transfer control. Dendrite growth can then be prevented as long as some critical current density is not exceeded. A number of dendrite inhibitors have been found to be effective for a few successive Zn plating events. However, in most cases the dendrite inhibitor undergoes chemical modification by reduction at the Zn electrode and becomes ineffective within a few cycles.[50]

Another problem relates to the gradual change in the Zn distribution during repeated cycling, phenomena that have been called "shape change" and "slumping". These phenomena are much more pronounced on the alkaline than on the acid Zn electrode. This is due to the much lower solubility of the Zn discharge product in alkaline solutions than in acid halide solutions. A preferred mode of operation of Zn-$Br_2$ and Zn-$Cl_2$ hydrate cells consists of completely removing the Zn deposit from the inert carbon or graphite substrate once every few cycles by an extended discharge at low rates or by shorting the cell. This procedure prevents uneven Zn thicknesses from building up over a large number of cycles.

Electrolyte circulation also greatly helps to decrease the rate of shape change and slumping while at the same time increasing the critical current density at which dendrite formation occurs. Figures 22 and 23 show the results of depositing Zn from $ZnBr_2$ solution onto vitreous carbon foam without (Fig. 22) and with (Fig. 23) electrolyte circulation.[51] In both cases, plating was carried out at 25 $mA/cm^2$ for 4 hours. Without electrolyte circulation, profuse Zn dendrite growth occurs. With electrolyte circulation, the crystal habit is completely different and the growth form changes from dendritic to nodular.

Inspite of electrolyte circulation, however, the Zn deposit has very low density and forms predominantly on the surface of the vitreous carbon substrate. The addition of certain dendrite inhibitors not only prevents dendrite growth, but, at the same time produces Zn deposits of considerably higher density and with quite even distribution within the porous substrate rather than predominantly on its surface. This is shown in Figures 24 and 25 for plating of Zn at 25 $mA/cm^2$ for 4 hours without (Fig. 24) and for 8 hours with electrolyte circulation (Fig. 25).[51] In both cases, the electrolyte contained 1% by weight of a fluorocarbon surfactant[51] that acted as a very effective and stable dendrite inhibitor.

In contrast to the large amount of information available on the alkaline Zn electrode, very few fundamental studies have addressed the acid Zn electrode. Clearly such studies are needed to bring about a better understanding of the similarities and differences between the alkaline and acid Zn electrode.

Fig. 22. Optical micrograph of Zn deposit at 6X after plating from 50% Zn $Br_2$ solution at 25 $mA/cm^2$ for 4 hours <u>without</u> electrolyte circulation.

Fig. 23. Scanning electron micrograph at 14X under similar conditions as in Fig. 22 but with electrolyte circualtion at $N_{RE}$=2500.

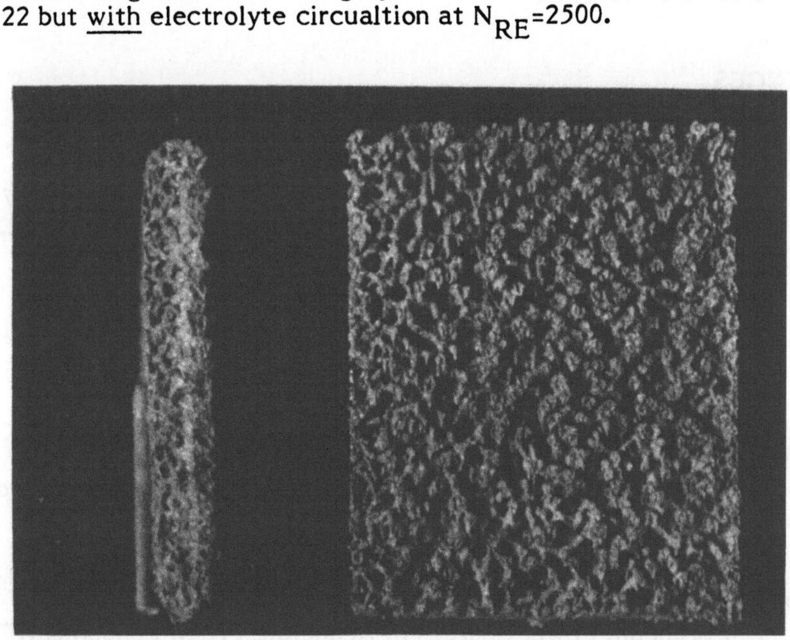

Fig. 24. Optical micrograph at 6X of Zn deposit; 50% $ZnBr_2$ solution plus 1% fluorosurfactant additives; plating for 4 hours at 25 mA/$cm^2$ without electrolyte circulation.

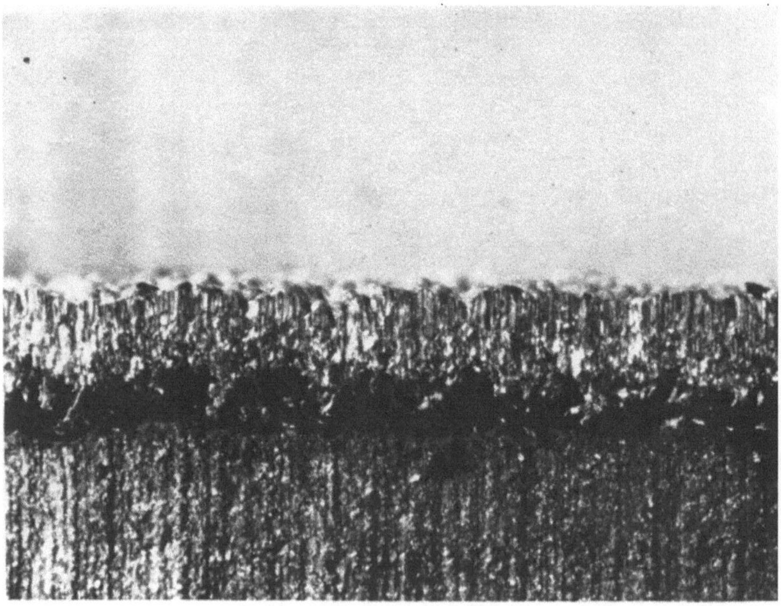

Fig. 25. Optical micrograph of 1 mm thick 100 pore RVC foam; plating for 8 hrs at 25 $mA/cm^2$; 30% $ZnBr_2$ solution plus 1% fluorosurfactants with electrolyte circulation.

## REFERENCES


1. R. D. Armstrong, T. Dickinson and J. Turner, Electroanal. Chem. 44, 157 (1973).
2. A. Gibson, Power Sources 6, Academic Press, London (1977) p. 673.
3. M. W. Breiter, B. Dunn and R. W. Powers, Electrochem. Soc. Mtg., Los Angeles, Oct. 1979; to be published.
4. F. G. Will, J. Electrochem. Soc. 123, 834 (1976).
5. D. S. Demott and P. Hancock, Proc. Brit. Ceram. Soc. 19, (1969).
6. L. J. Miles and I. Wynn Jones, ibid. 19, 179 (1969).
7. S. A. Weiner, Proceedings Energy Storage and Conversion Symposium, Electrochemical Society (1976), p. 141.
8. S. P. Mitoff, M. W. Breiter and D. Chatterji, Proceedings 12th IECEC, Vol. 1 (1977), p. 359.
9. H. Kleinschmager, W. Haar, G. Weddigen and W. Fischer, Power Sources 6, Academic Press, London (1977) p. 712.
10. F. G. R. Zobel, J. Power Sources 3, 29 (1978).
11. R. D. Armstrong, T. Dickinson and M. Reid, Electrochim. Acta 20, 709 (1975).
12. M. P. J. Brennan, Power Sources 6, Academic Press, London (1977), p. 693.
13. J. G. Gibson, J. Appl. Electrochem. 4, 125 (1974); Intern. Symp. Molten Electrolytes and High Temperature Batteries, Brighton, Sept. 1977.

14. M. W. Breiter and B. Dunn, J. Appl. Electrochem. 9, 291 (1979).
15. S. A. Weiner, NSF Contract Report No. C805 (1975, 1976, 1977).
16. W. Fischer, W. Haar, B. Hartmann, H. Meinhold and G. Weddigen, J. Power Sources 3, 299 (1978).
17. R. J. Bones, R. J. Brook and T. L. Markin, Power Sources 5, Academic Press, London (1975) p. 539.
18. B. Hartmann, ibid 3, 227 (1978).
19. T. L. Markin, A. R. Junkison, R. J. Bones and D. A. Teagle, Power Sources 7, Academic Press, London (1978), p. 757.
20. R. Bauer, W. Haar, H. Kleinschmager, G. Weddigen and W. Fischer, J. Power Sources 1, 109 (1976/77).
21. J. E. Battles, Argonne National Laboratory Report ANL-78-94, Nov. 1978, p. 138.
22. R. B. Swaroop. G. Bandyopadhyay, J. T. Dusek and T. M. Galvin, ibid, p. 147.
23. J. G. Eberhart, ibid., p.157.
24. R. K. Steunenberg and M. F. Roche, Symposium on Electrode Materials and Processes, ECS Vol. PV 77-6, (1977) p. 869.
25. A. N. Dey and C. R. Schlaiker, Proc. 26th Power Sources Symp. Atlantic City 1974.
26. J. O. Besenhard and G. Eichinger, J. Electroanalyt. Chem. 68, 1 (1976).
27. G. Eichinger, ibid. 74, 183 (1976).
28. A. N. Dey, Thin Solid Films, 43, 131 (1977).
29. M. Froment, M. Garreau, J. Thevenin and D. Warin, J. Microsc. Spectrosc. Electron., 4, 111 (1979).
30. R. F. Scarr, J. Electrochem. Soc. 117, 295 (1970).
31. S. G. Meibur, ibid, 117, 56 (1970).
32. A. N. Dey and B. P. Sullivan, ibid., 117, 222 (1970).
33. C. C. Liang, "Solid State Batteries", Appl. Solid State Sci. 4, 95 (1974).
34. A. N. Dey, U.S. Patent No. 3,947,289 (1976).
35. R. Selim and P. Bro, J. Electrochimica Acta 22, 75 (1977).
36. R. D. Rauh and S. B. Brummer, Electrochimica Acta 22, 75 (1977).
37. A. N. Dey and E. J. Rudd, J. Electrochem. Soc. 121, 1294 (1974).
38. R. D. Rauh and S. B. Brummer, Electrochimica Acta 22, 85 (1977).
39. P. Symons et. al., "Development of the Zinc-Chlorine Battery for Utility Applications", EPRI Report No. EM-1051, Parts 1-4, Apr. 1979.
40. F. G. Will, Proc. 12th IECEC, p. 250 (1977).
41. H. S. Lim, A. M. Lackner and R. C. Knechtli, J. Electrochem. Soc. 124, 1154 (1977).
42. F. G. Will, Power Sources 7, Academic Press, London, p. 313 (1979).
43. G. Clerici, M. DeRossi and M. Marchetto, Power Sources 5, Academic Press, London, p. 565, (1975).
44. M. A. Walsh, F. Walsh and D. Crouse, Proc. 10th IECEC, p. 1144 (1975).
45. R. A. Putt et. al., "Assessment of Technical and Economic Feasibility of Zinc/Bromine Batteries for Utility Load-Leveling", EPRI Final Report No. EM-1059, May 1979.
46. J. L. Barton and J. O'M. Bockris, Proc. Roy. Soc. 268, 485 (1961).

47. S. Tajima and M. Ogata, Electrochim. Acta 13, 1845 (1968).
48. J. W. Diggle, A. R. Despic and J. O'M. Bockris, J. Electrochem. Soc. 116, 1503 (1969).
49. R. V. Moshtev and P. Zlatilova, J. Appl. Electrochem. 8, 213 (1978).
50. H. H. Tombrink, Proc. 8th Congress Int. Union of Electrodeposition, Basel, Sept. 5-9 (1972).
51. F. G. Will, C. D. Iacovangelo, J. S. Jackowski and F. W. Secor, "Assessment of the Zinc-Bromine Battery for Utility Load Leveling", DOE Final Report No. COO/2950-1, March 1978.

# SECTION II

# SHORT COMMUNICATIONS

LEAD-ACID BATTERY OVERVIEW

S.M. Caulder and A.C. Simon

International Lead-Zinc Research Organization
Naval Research Laboratory, Washington, DC 20375

## INTRODUCTION

The lead-acid battery has become so dependable in its usual applications of automobile starting, emergency lighting, and telecommunications, that the belief has arisen erroneously that no further investigation of this battery is necessary or desirable. While there has been a slow but continuous improvement in lead-acid battery performance and dependability over the long period of its existance (20wH/Kg in 1920 to 42wH/Kg presently), these improvements have been mainly limited to better engineering of parts, such as through the cell wall connectors and better current collector designs as well as the use of improved construction materials such as plastics for cell cases. These improvements have lead to less unreactive materials being incorporated within the battery thus giving higher energy densities per unit weight. Considering the long time that this battery has existed, there has been relatively little effort to understand the complex chemistry and electrochemistry upon which the successful operation of the battery system is based. This lack of research arises partly from the fact that this is a highly competitive and cost conscious industry which does not encourage research expenditure, and partly from the fact that it has been possible in the past to produce a product, without extensive research, that has been adequate for most prior applications.

The use of the lead-acid battery for other applications such as large scale energy storage and limited mission electric vehicle propulsion will place much greater demands upon its performance, and it has become increasingly evident that the required increase in performance cannot be expected from engineering changes alone.

Clearly, more emphasis must be placed upon understanding the fundamental battery processes if the lead-acid battery is to be successful in these new and critical applications.

## FACTORS AFFECTING THE COMPETITIVE POSITION OF THE LEAD-ACID BATTERY

In determining whether the lead-acid battery can compete as an effective means of energy storage against present and proposed alternative systems, several factors must be considered:

The first is energy density, watt-hrs/kilogram. The lead-acid battery has the unenviable position of possessing the lowest energy density of any commercial battery. Fortunately, in batteries for stationary applications, such as energy storage, this is not as an important a consideration as would be the case for vehicle propulsion batteries, where a decided penalty is paid for low energy density. However, it is evident that any improvement in energy density would improve the competitive position of the lead-acid battery and make it more desirable for applications of the future.

A second factor concerns life and reliability. The lead-acid battery shows well in this comparison, as such batteries can be made with proven life in excess of 20 years or a capability for thousands of 50 percent charge-discharge cycles. Although these characteristics can be obtained, it is not without penalty. The increased thickness of plates, the thicker separators and the necessary fiberglass mats or other wrappings, all add up to give a reduced energy density for the battery.

Another factor that must be considered is how well the battery retains an applied charge during periods of no use, and also what portion of the original capacity can be retained by the battery after a period of charge-discharge cycling. It is an unfortunate fact concerning all types of batteries that the capacity to store electricity and convert it subsequently to useful work is continuously reduced as the battery is cycled. Also, the charge already within any type of battery at any given time of its life is also slowly lost while the battery stands unused. The amount of energy lost by self-discharge varies with the different electrochemical systems.

Batteries are also compared on the basis of cost per kilowatt-hr. This cost factor depends not only upon the service life of the battery but also upon its original cost to manufacture. While the figures presented for some of the proposed batteries are only projections, the figures for the lead-acid battery are based on actual data and have proven considerably lower than for any other battery now in production.

While the above list is not complete, it represents the principal points to be considered in comparing battery systems and also serves to indicate the various ways that batteries are imperfect and where improvement is needed. While some of these factors may appear unrelated, investigation has shown that they are merely different manifestations of the same basic imperfection in battery chemistry.

## FACTORS AFFECTING THE IMPROVEMENT OF THE LEAD-ACID BATTERY

In considering the problem of energy density, the lead-acid battery materials are themselves very heavy. The lead compounds comprising the active material and the lead alloys required to give the current collector the necessary corrosion resistance have very high densities. Although the actual weight percentages of the various constituents may vary with the type of construction used (as required for different applications), generally the electrolyte constitutes about 20% of the total battery weight, the active material about 40%, the grids about 27%, and the container, separators, internal cell connectors and terminals make up the remaining 13%.

Very little can be done about the weight of the electrolyte. Almost all of it is required for the cell reaction and any remainder is required to furnish conductivity between the plates when they are in a discharged condition. A large number of experiments have demonstrated that the presently used specific gravities of acid are the best compromise of factors such as viscosity, conductivity, heat transfer, diffusion, grid corrosion, and resistance to freezing.

There have been numerous attempts to reduce the weight of the grids, since these represent about a fourth of the total weight of the battery and yet contribute nothing to the cell reactions that furnish energy. Such grids are unfortunately required, both as a support for the active material and as internal current collectors. In recent years, computer studies have produced grid designs that appear to offer the best approach to uniform current distribution throughout the plate, but there is very little reduction in grid weight, because the conductivity of the active material is so low in the discharged condition that numerous grid wires, of rather close spacing, are required to insure that current is carried to and from all parts of the active material. The inability to provide a corrosion proof material for this function also requires the presence of a greater amount of lead than is required just for strength and conductivity. Although many grid alloys have been tried, most have proven unsuccessful for any of a number of reasons and all investigators have been forced to employ lead as a major constituent of whatever alloys they tested.

There exists the possibility that greater corrosion resistance could be obtained by a protective overcoat of some material such as titanium, or even that a better conductor such as aluminum could be used as a grid, if properly protected. At the present time, none of these ideas appear practical, especially if the low cost and materials availability of the lead-acid battery is to be preserved.

Although the active material makes up but 40% of the battery weight, it represents 100% of the useful energy that can be either stored in or removed from the system. Since it is a well-known fact that less than 50% of the active material is presently utilized, it is obvious that any increase in this utilization would result in an increase in the energy density. As a theoretical possibility, if the utilization could be made high enough, the energy density could be doubled.

As for the remaining 13% (comprising the separators, internal connectors, terminals and container), recent improved plastics and separators appear to have resulted in about as much improvement in these materials as can reasonably be expected (at least as far as weight is concerned) and because of the corrosion problem, internal connectors and terminals will probably remain at their present weights.

## AREAS OF RESEARCH IN THE LEAD-ACID BATTERY

The above summary indicates that the only areas of research offering real possibilities of major improvement in this battery system are those that concern the active material and the grid. The most productive effort that could be made would be in determining the mechanisms of active material failure and grid corrosion so that greater utilization and longer life could be imparted to the active material and more corrosion resistant lighter weight grids could be developed.

## QUESTIONS TO BE ANSWERED THROUGH RESEARCH:

1) Why does the $PbO_2$ electrode loose capacity without softening and shedding of the active material (i.e., loose electrochemical activity)?
2) Why can't we increase and prolong the % utilization of active material above 50%.
3) Why does the $PbO_2$ active material soften and shed?

AREAS OF RESEARCH THAT MAY ANSWER SOME OF THE QUESTIONS

1) What is the effect of changes in $PbO_2$ atomic structure on softening, shedding, and electrochemical activity?
2) What role does $H_2O$, $H^+$, and $OH^-$ play on electrochemical activity and electrode microstructure?
3) What is the relationship between the various crystal structures of $PbO_2$ and plate microstructure?
4) How do changes in $PbO_2$ atomic structure effect plate microstructure and cycle life?

In order to provide answers to some of the above questions, use of the more modern experimental techniques such as Auger and X-ray photoelectron spectroscopy, nuclear magnetic resonance, neutron and X-ray diffraction, and scanning electromicroscopy will have to be utilized.

DESIGN AND DEVELOPMENT OF MICRO-REFERENCE ELECTRODES

FOR THE LITHIUM/METAL-SULFIDE CELL SYSTEM

Laszlo Redey
Chemical Engineering Faculty
Technical University of Budapest
Egry J.U. 20, Hungary

Donald R. Vissers
Argonne National Laboratory
Chemical Engineering Division
Argonne, Illinois 60439

Reference electrodes are being designed and developed to investigate the polarization characteristics of electrodes in Li/MS engineering cells and to carry out other electrochemical studies in these cell systems.[1] The Li/MS cells, which operate at 400-450°C, utilize negative electrodes of Li-Al alloy, positive electrodes of either FeS or $FeS_2$, and a molten LiCl-KCl electrolyte (mp, 352°C). The reference electrodes must possess a well-defined potential, be electrochemically stable for long periods (months), and be rugged and adaptable to miniaturization. The high lithium and sulfur activities present in this cell system place stringent limitations on the materials which are suitable for the reference electrode.

Five cell couples were investigated as potential reference electrode systems: $Ag/AgCl/Cl^-$, $Ag/Ag_2S/S^=$, $Ni/Ni_3S_2/S^=$, Fe/Li-Al $(\alpha + \beta)/Li^+$, and Al/Li-Al$(\alpha + \beta)/Li^+$. (Some of these couples had been investigated in earlier studies.[2-5]) The stabilities of these cell couples were evaluated against a Li-Al electrode with a capacity of 3 to 5 A-hr; this Li-Al electrode has a very stable potential in LiCl-KCl melts, but because of its large size, it is impractical as a reference electrode in engineering cells. Consequently, all emf values presented in this paper are given *vs* the two-phase $(\alpha + \beta)$ Li-Al alloy electrode.

The engineering cell requires the use of very small reference electrodes because of the limited space available. To minimize the size of the reference electrode, we selected a beryllium oxide tube, 17.8 cm in length and 3.2 mm in diameter, as the insulator housing; a $Y_2O_3$ plug on top of a short section of $Al_2O_3$ tube forms the diffusion barrier, which develops a resistance of 7 kΩ/cm. A drawing of one of the reference electrode couples is shown in Fig. 1. All the electrochemical couples evaluated in this paper utilized this electrode design; and, with the exception of the Fe/Li-Al(α + β)/$Li^+$ couple (fabricated using particulate Li-Al), were formed electrochemically by pulse charging.

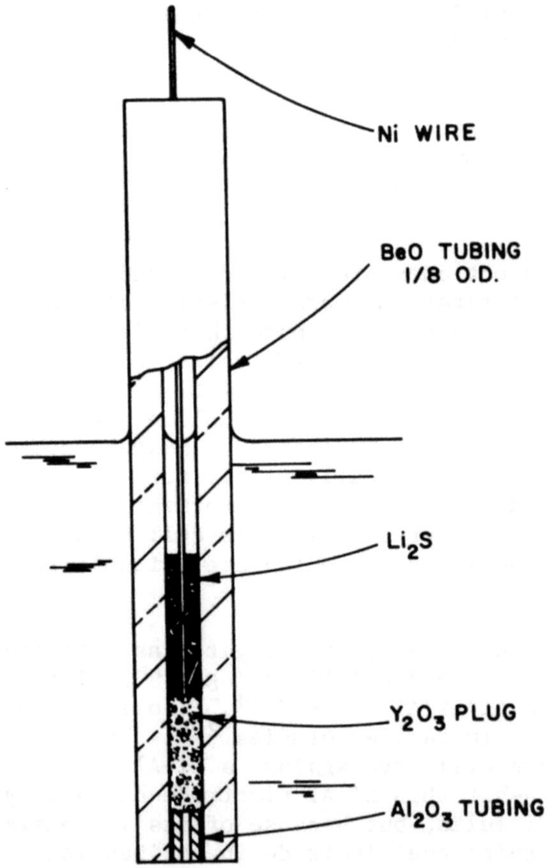


Fig. 1. Reference Electrode Design

The stabilities of the five electrode couples are presented in Table I. The data indicate that the $Ni/Ni_3S_2/S^=$ couple possesses the best long-term stability and also that the $Al/Li\text{-}Al(\alpha + \beta)/Li^+$ electrode couple has excellent short-term (∿30 min) stability. Therefore, the $Ni/Ni_3S_2/S^=$ couple is the most suitable reference electrode, and the $Al/Li\text{-}Al(\alpha + \beta)/Li^+$ has potential as an *in situ* reference electrode calibration standard for the $Ni/Ni_3S_2/S^=$ electrode. For the latter purpose, an aluminum wire is intermittently charged with lithium, thereby resulting in a well-defined potential. One can characterize this charging technique by the potential recorded during the procedure (see Fig. 2). The constant potential value indicated by the consecutive steady-state, open-circuit values correspond to the well-defined potential of the two-phase ($\alpha + \beta$) Li-Al alloy. This series of potential values can be obtained with a reproducibility of ± 1 mV on new aluminum wires or on the same wire with the consecutive charging procedures. After a period of time, depending on the quantity of lithium deposited, the steady-state potential of this alloy electrode shifts up toward positive values, but a repeat of the intermittent charging technique can again achieve the well defined value. After a few repetitions of this charging technique, the same wire can no longer be used; but replacement of this monitoring electrode is very simple.

Table I. Summary of Data on Reference Electrode Couples

| Couple | Observed Potential, V *vs* Li-Al(α+β)* | Observed Stability *vs* Li-Al(α+β)* |
|---|---|---|
| $Fe/Li\text{-}Al(\alpha+\beta)^*/Li^+$ | 0.001 ± 0.001 | ±0.0015 V/day** |
| $Al/Li\text{-}Al(\alpha+\beta)/Li^+$ | 0.001 ± 0.001 | ±0.0005 V/hr |
| $Ag/Ag_2S/S^=$ | 1.610 ± 0.003 | ±0.010 V/day |
| $Ag/AgCl/Cl^-$ | 2.3 ± 0.1 | ±0.100 V/day |
| $Ni/Ni_3S_2/S^=$ | 1.361 ± 0.0015 | ±0.0005 V/day |

*43 at. % Li

**The potential of this couple tended to drift upward with time.

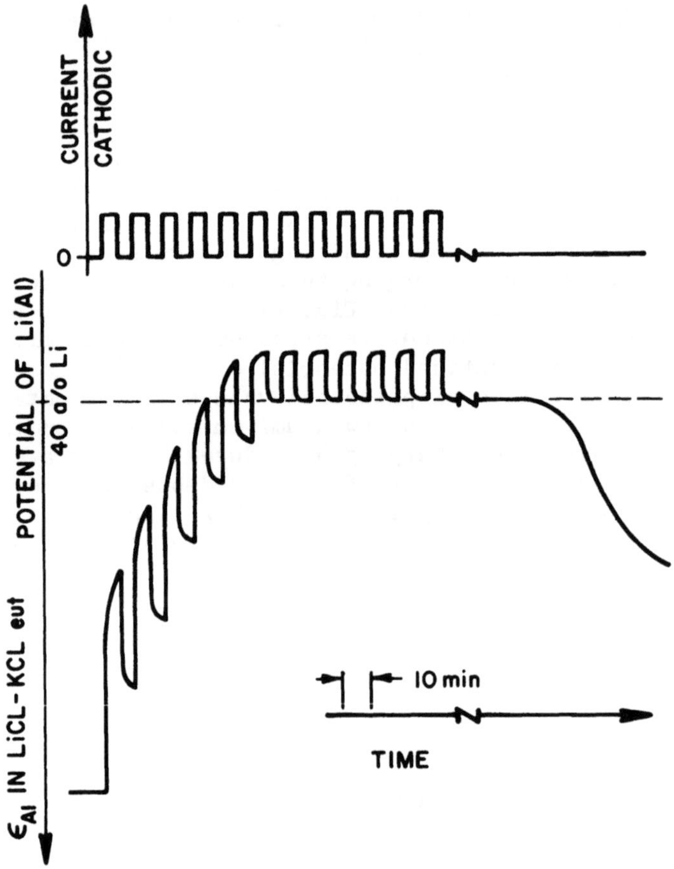


Fig. 2. Formation of Al/Li-Al(α+β)/Li$^+$ Reference Electrode (Primary Standard)

A schematic diagram of the $Ni/Ni_3S_2/S^=$ reference electrode inserted in an engineering cell is shown in Fig. 3. As shown in the figure, the reference electrode is encased in a stainless steel tubing. A saw-toothed loop at the bottom of the nickel wire holds the BeO tubing firmly in place, thereby avoiding the need for a rigid connection between the reference electrode and the stainless steel tubing. The polarization characteristics of the electrodes in 100 A-hr Li-Al/FeS bicells are currently being evaluated using the $Ni/Ni_3S_2/S^=$ reference electrode described in this paper.

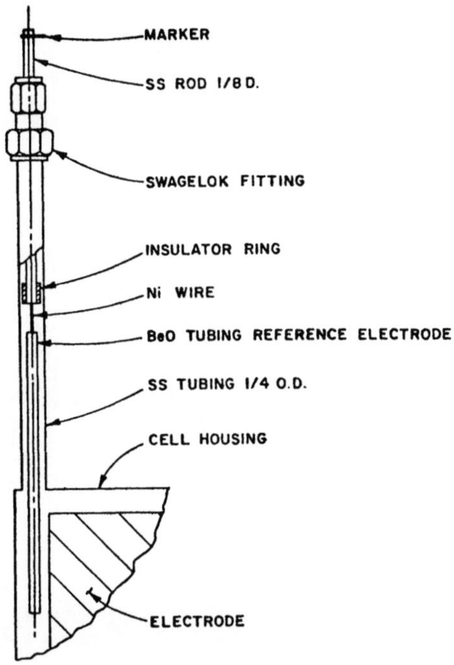


Fig. 3 Reference Electrode Enclosure

ACKNOWLEDGMENT

This work was performed under the auspices of the U.S. Department of Energy.

REFERENCES

1. P. A. Nelson *et al.*, High Performance Batteries for Electric-Vehicle Propulsion and Stationary Energy Storage: Progress Report for Period October 1978-March 1979, Argonne National Laboratory Report ANL-79-39 (1979).
2. N. P. Yao *et al.*, Electrochem. Soc., 118, 1059 (1971).
3. C. H. Liu, A. J. Zielen, and D. M. Gruen, J. Electrochem. Soc. 120, 67 (1973).
4. M. L. Saboungi, J. Marr, and M. Blander, J. Chem. Phys. 68, 1375 (1978).
5. M. L. Saboungi, J. Marr, and M. Blander *et al.*, 125, 1567 (1978).

MOLTEN SALT ELECTROCHEMICAL STUDIES AND

HIGH ENERGY DENSITY CELL DEVELOPMENT

A. Bélanger, F. Morin, and M. Gauthier

Institut de recherche de l'Hydro-Québec (I.R.E.Q.)
Varennes, Québec J0L 2P0, Canada

and

W.A. Adams and A.R. Dubois

Defence Research Establishment Ottawa (D.R.E.O.)
Department of National Defence (D.N.D.)
Ottawa, Ontario K1A 0Z4, Canada

SUMMARY

Rechargeable cells of the type LiAl/LiCl-KCl/FeS having 75Ah capacity have been constructed and cycled. Utilization of active materials is 70% in a cell containing 15 mole percent $Cu_2S$ as an additive to the FeS electrode and over 120 cycles have been achieved in this cell. Coulombic efficiency is 99 percent with a selfdischarge of 1 percent per day at 450°C. Temperature, stand-time, and rate of discharge-charge studies are described. Details of the mechanical design of the cell and cell assembly are discussed.

Electrochemical measurements using cyclic voltammetry in eutectic LiCl-KCl and in mixed halide eutectic lithium melts have been made on $TiS_2$, $TiSe_2$ and $VS_x$. These results are discussed in terms of the possible use of these systems to develop high energy density batteries.

INTRODUCTION

The need for development of advanced secondary batteries was recognized at DND in Canada in connection with the under-ice propulsion of small submersibles prior to the current molten salt battery program (1). The desirable characteristics of molten salt batteries are a result of the features of the molten salt electrolyte (2). High power is possible in principle because of current densities approaching 1 $Acm^{-2}$ whereas, since the conductivity of aqueous systems are an order of magnitude lower, the available power on these batteries is less. Organic and solid state electrolytes have conductivities that are lower in general by several orders of magnitude. The molten salt electrolyte cell requires high temperature for operation. This generally has a beneficial effect on the electrode kinetics, so that reaction rates are fast resulting in lower polarization losses than in aqueous or organic systems and therefore a higher energy storage capacity, i.e. higher voltage for a given electrode combination or higher power output for a given voltage. Finally, highly reactive electrode materials can be used due to the high decomposition potential of the molten salt which results in a higher theoretical specific energy.

Although there are molten salt primary batteries currently being manufactured and used in certain specialized military applications where, besides the high specific power and energy, the unlimited shelf-life and safety of the inert battery at ambient temperature is attractive, it si prospective secondary storage batteries based on $Li/FeS_x$ which are of interest for many applications. There are no molten salt storage battery systems currently being manufactured on a commercial scale.

It is the intention of this program to develop a Canadian capability to construct the Argonne National Laboratory (ANL) type of engineering scale $LiAl/FeS_x$ type cells (3) in order to master the relevant technologies. The experience gained doing this work should help us in developing novel high energy density cells based on a wide variety of possible materials for electrodes and other cell components. In evaluating electrode materials, electrodes containing about 20 mg of active material on a variety of substrates were built and tested by cyclic voltammetry at I.R.E.Q. (4,5) in a glove box containing argon with $O_2$ < 5ppm & $H_2O$ < 10ppm (Vacuum Atmospheres Corp.). The reference electrode contained 30 atomic % Li in the Li-Al alloy powder and the auxiliary electrode was 50% Li-Al (Foote Mineral Corp.).

## CELL DEVELOPMENT

In the earlier studies small pre-prototype $Li-Al/LiCl-KCl/FeS_x$ cells (5Ah, 25$cm^2$) were made and cyclic voltammetric measurements conducted in an argon atmosphere. These types of cells are still being used to investigate new electrode or electrolyte combinations.

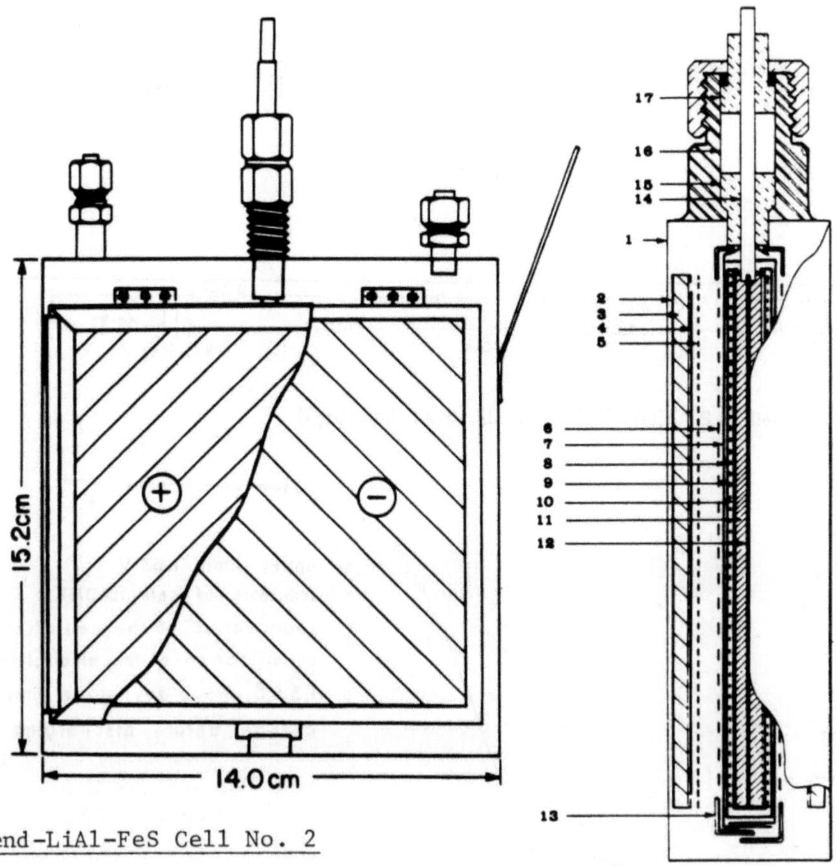


Legend-LiAl-FeS Cell No. 2

1.Stainless steel enclosure,1.27mm

Negative electrode

2.Stainless steel plate,o.127mm
3. Ni foam,3.81mm=LiAl powder

4.$ZrO_2$ cloth
5.Stainless steel screen,325 mesh

Positive electrode

6.Stainless steel screen,60 mesh

7.BN separator
8.$ZrO_2$ cloth

9.Stainless steel screen-325 mesh

10.$ZrO_2$ cloth

11.Ni foam,0.381mm+FeS powder
12.Stainless steel,1.27mm support plate
13.Stainless steel,1.27mm frame spot-welded to screen (6)

Conex feedthrough

14.Stainless steel rod,3.175mm diameter
15.Lower sleeve of BN
16.BN powder,≃325 mesh
17.Upper sleeve of alumine

Fig. 1 Li-Al/LiCl-KCl/FeS non-optimized engineering in prototype cell

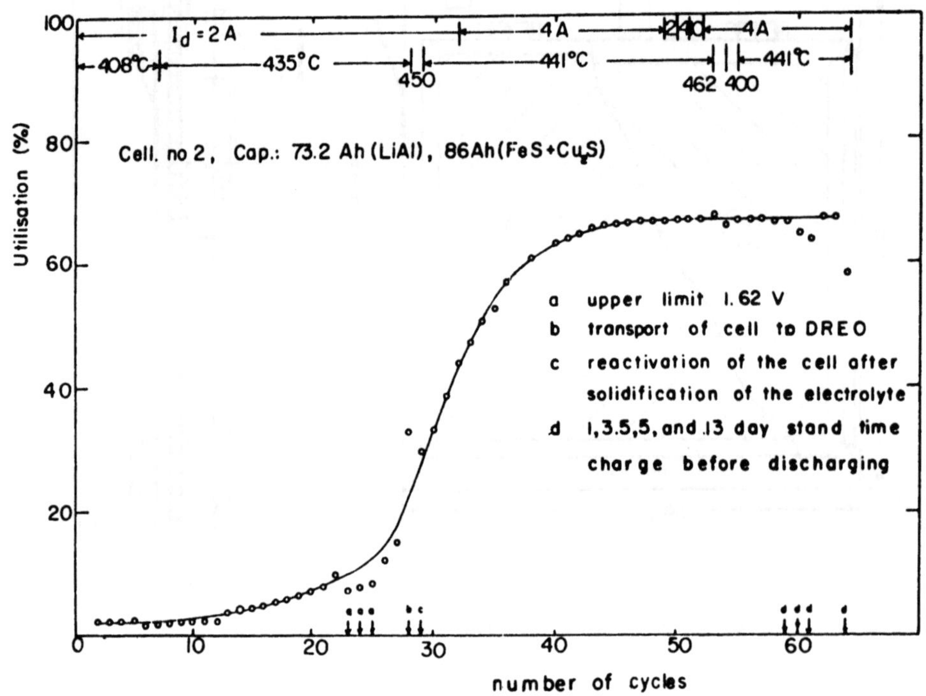


Fig. 2 Utilization of reactants vs. number of cycles for cell IREQ-2

The first of two full size engineering prototype cells was constructed in 1978 at I.R.E.Q. and tested at DREO in early 1979. They were not optimized for weight and volume. The first cell capacity was 72.2Ah (LiAl) and 134.3Ah (FeS), weight was about 3kg, electrode surface was $330cm^2$, and cell resistance was 16m$\Omega$ at 440°C. A schematic of the cell is shown in Figure 1. The cell components are indicated. The cell is a bi-cell configuration consisting of an FeS positive electrode sandwiched between two Li-Al negative electrodes and operated with LiCl-KCl eutectic electrolyte (mp, 352°C). The positive electrode, assembled in air, consists of FeS or FeS+$Cu_2S$ powder imbedded in a porous nickel structure then covered with layers of zirconia ceramic cloth and the BN separator, both held in place by stainless steel screens. The negative electrodes, assembled under argon, were also made by impregnating nickel foam and preventing the loss or movement of the fine powder with zirconia cloth and screens. We have used the boron nitride (BN) woven fabric for our first engineering cell builds rather than felt which has proven less reliable during tests at ANL. Due to the production cost of BN woven fabric however, commercial batteries will have to be built with the BN felt if a less expensive replacement cannot be found. Assembly precedures have been described (4,5). These cells were built with a thick-walled stainless steel rigid case so that heavy clamping plates were not necessary to prevent swelling and distortion of the cell during cycling as is the case for the optimized prototype cells that we (D.R.E.O.) have obtained for evaluations from Gould Manufacturing of Canada (case wall thickness, 0.58mm) under a separate contract.

As has been found with the small 5Ah electrodes (4), the maximum utilization for these cells (65-75%) occurs only after several formation cycles presumably since the Li-Al powder is finely dispersed and may be partially oxidized. Difficulty was experienced with the heating system of the first cell and after three successive solidifications of the electrolyte a short circuit developed inside the cell. The temperature rose rapidly and the cell was completely discharged (Li was absent from the Li-Al electrode and the aluminum melted indicating $T \geqslant 660°C$).

A second cell is still (Sept. 1979) undergoing cycling tests ($>$ 120 cycles). A summary of the cycling tests is given in Figure 2. The reagent utilization began to increase after cycle 10 and stabilized after about 40 cycles at 67%. The internal cell resistance was about 12m$\Omega$. The utilization is defined in terms of the negative electrode, which was limiting (73.2Ah) compared to the positive electrode (86Ah) consisting of FeS + 15% mole $Cu_2S$. Coulombic efficiency was 99% $\left(\frac{\text{discharge}}{\text{charge}}\right)$.

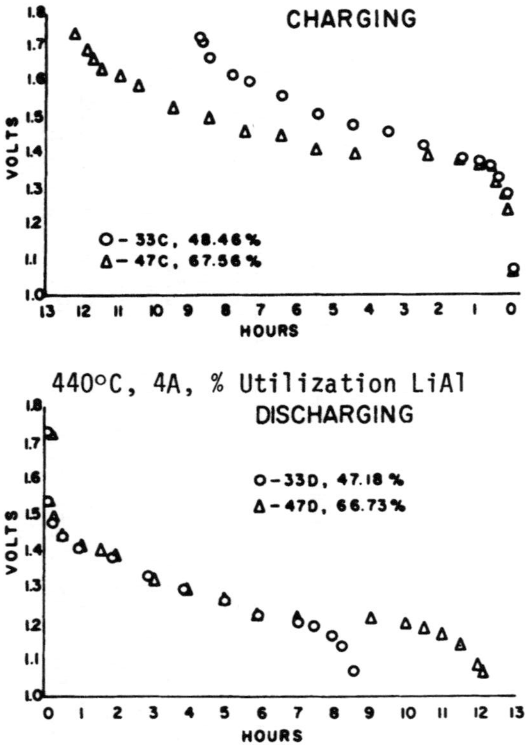


Fig. 3 Charge and discharge curves for cell IREQ-2 for cycle 33 and cycle 47 (partial and ultimate % utilization)

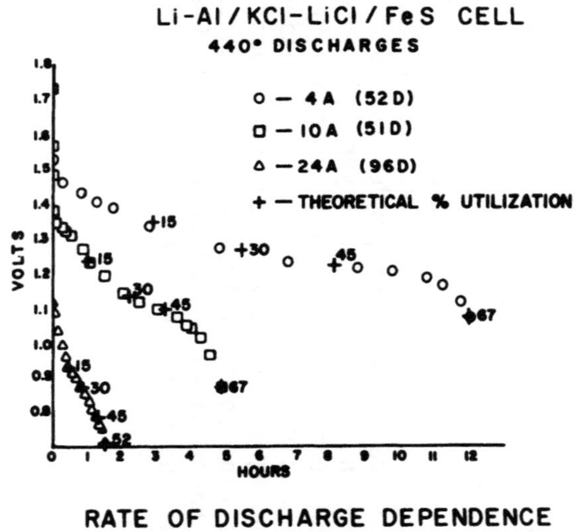


Fig. 4 Cell IREQ-2 discharge curves at three currents

The shape of the discharge and charging curves at 440°C and 4A for the partial and ultimate percent utilizations are shown in Figure 3. Cycling voltage limits were 1.1 V to 1.7 V.

The effect of discharge rate is shown for current densities between 12 and 72 $mAcm^{-2}$ in Figure 4. Utilizations were found to be constant, 67%, between 6 and 31 $mAcm^{-2}$, but fell to 52% at 72 $mAcm^{-2}$. However, the utilization was again 67% for a subsequent 4A cycle.

Figure 5 indicates that between 400°C and 462°C the 4A discharges were identical and the utilization remained constant, 67%. Additional standtime tests were conducted during which the cell was left at 440°C in a fully charged state. The longest period was 13 days during which there was a 14% capacity loss, suggesting about a 1% capacity loss per day at 440°C. The mechanism of the loss is related to the slight solubility of $Cu_2S$ and lithium in the electrolyte. Furthermore, this solubility will also lead to the deposition of metallic copper in the BN separator giving rise to additional self-discharge through electronic conductivity.

The initial shape of the discharge curves in Figure 5 were due to the reduction of $Cu_2S$. The addition of the electronically conductive $Cu_2S$ to the FeS was made to enhance the utilization of materials in the positive electrode (6) especially at higher current densities (> 75mA $cm^{-2}$). The plateau at mid-discharge is related to the formation of an intermediate called X phase. Activation of these two prototype cells was very slow, taking some 30 cycles before achieving a 50% reagent utilization. The origin of this slow activation has been traced to the grain size of the Li-Al in the negative electrodes which results in poor electrical contact between the Li-Al and the current collector, in this case the cell body.

## ELECTROCHEMICAL STUDIES

Cyclic voltammetry on FeS and $FeS_2$ has been conducted on a variety of substrates and our results are in good agreement with the literature (7). In general, the cyclic voltammetry results support our decision to choose FeS for engineering scale prototype cell development since this system, although complicated by the J phase, is more reversible and does not suffer capacity loss during cycling.

Titanium disulfide obtained from a variety of sources has been found to differ in stoichiometry from $Ti_{1.00}S_2$. Stoichiometrically pure $TiS_2$ was produced by reaction of commercial product (Alpha Products) with sulphur vapour (8). Figure 6 was obtained using this material. Cycling tests on this treated product, on

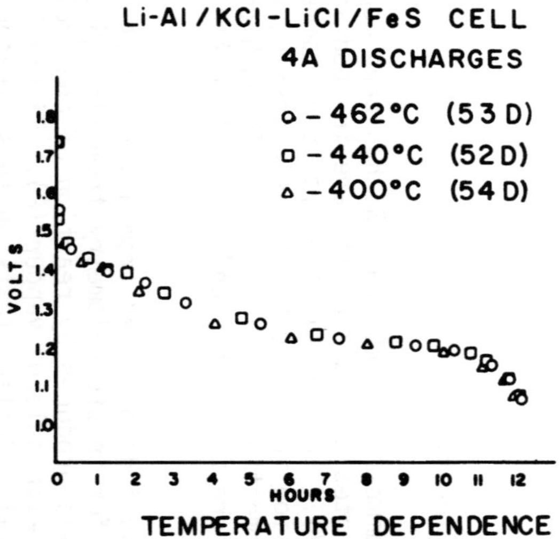


Fig. 5 Cell IREQ-2 discharge curves at three temperatures

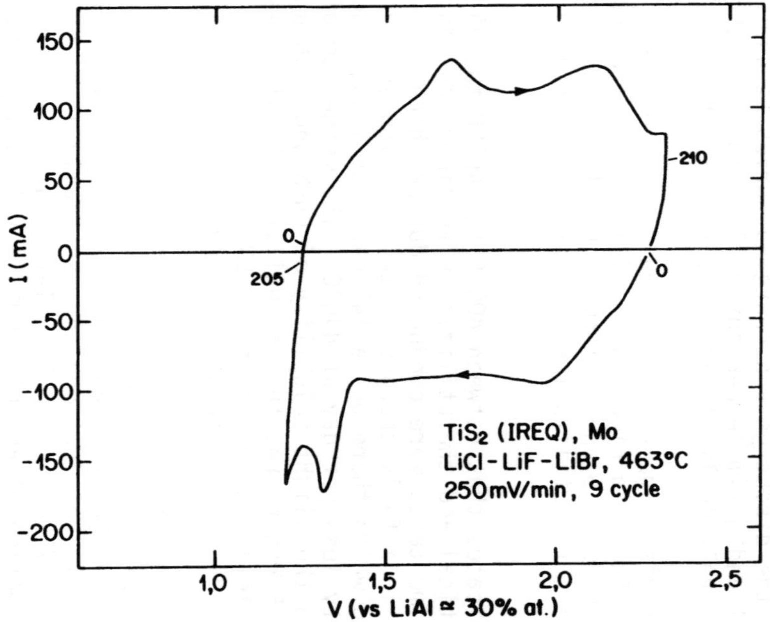


Fig. 6 Cyclic voltammogram of $TiS_2$ in LiCl-LiF-LiBr at 463°C on molybdenum

material provided by Dr. M. Armand of Grenoble, France, on CERAC $TiS_2$ and on electrochemically prepared $TiS_2$ (from $Ti_2S$ + Ti), gave essentially identical voltammograms such as that shown in Figure 6 which is the "whale" shape characteristic of intercalation reactions at high temperatures. A variety of electrolytes have been used to verify that the intercalation reaction is independent of the electrolyte compositions. The capacity for these electrodes has been found to be about 85% of theoretical and to remain constant through several hundred cycles. The fully discharges state appears to be approximately $Li_1TiS_2$.

Cyclic voltammetry on $VS_2$ and $TiSe_2$ has been done in LiCl-KCl eutectic, $VS_2$ (at 447°C) and $TiSe_2$ (at 408°C). An intercalation plateau between 2.0V and 1.3V in $TiSe_2$ corresponds to $LiTiSe_2$ stoichiometry and was found to have a shape independent of scane rate, $(\frac{dE}{dt})$, between 1000 and 25 mV/min.

However, as can be seen in Figure 7, the current is directly proportional to the scan rate. In the case of $VS_2$, it is seen that the current is proportionalto the square root of the scan rate. This system differs from $TiSe_2$ or $TiS_2$ in that the rate of reaction shows similar behaviour to a diffusion controlled reaction.

## PROSPECTS FOR HIGH TEMPERATURE INTERCALATION ELECTRODES

As has been illustrated in the cyclic voltammogram shown, nonstoichiometric intercalation compounds such as $TiS_2$ or $TiSe_2$ show broad plateaus over several hundred millivolts whose heights depend on scan rate. By comparison, materials undergoing displacement reactions and involving several phases, e.g. FeS or $FeS_2$, at low scan rates, have narrow peaks and are often diffusion limited.

Results obtained by Whittingham (9) on $TiS_2$ at 25°C have been compared to the molten salt results. Figure 8 indicates that the open circuit voltage of the electrode relative to $Li^\circ/Li^+$ decreases much more rapidly at high temperature with degree of intercalation than at room temperature. This is interpreted in terms of a model developed by Armand (10) in which the high temperature produces an increase in the intercalated metal ion-metal ion interaction energy. Such a rapid increase with temperature in the slope of the discharge plateau is characteristic of $TiS_2$ intercalation and may outweight advantages of the rapid kinetics of the lithium-cathode intercalation reaction in molten salt systems.

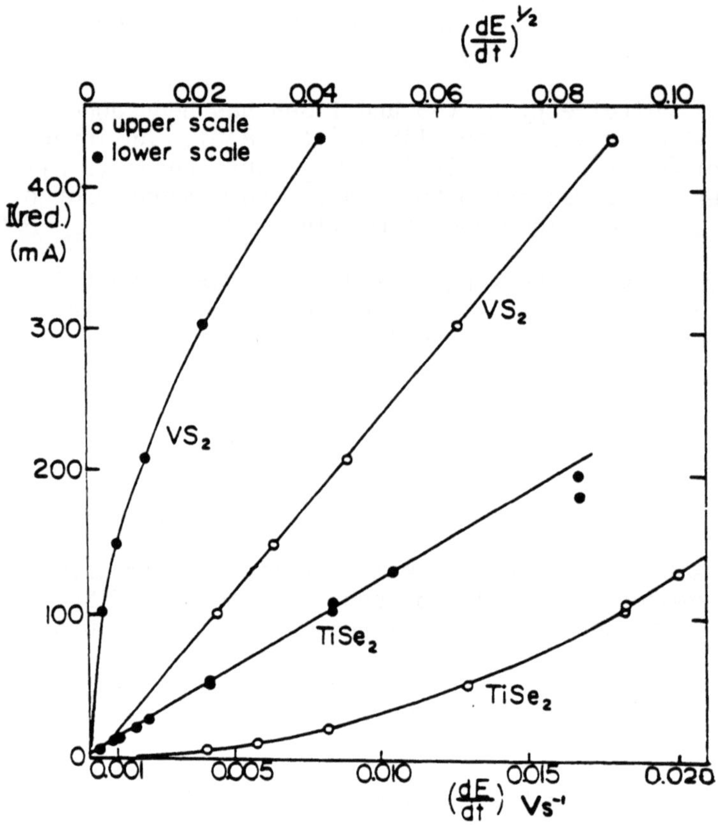


Fig. 7 Dependence of I upon the sweep rate for $TiSe_2$ and $VS_2$

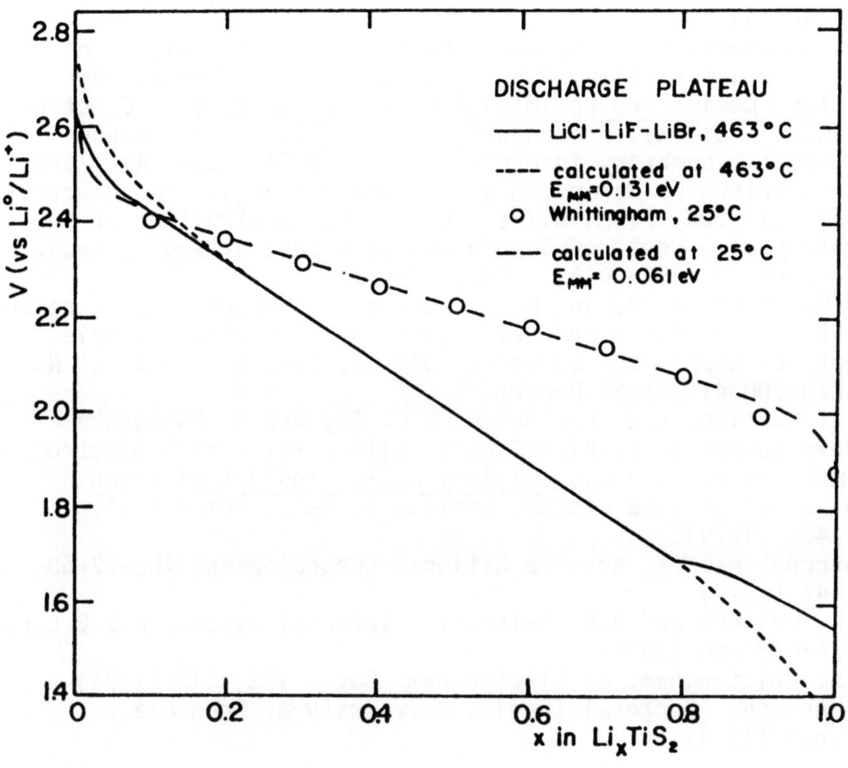


Fig. 8 Calculated and observed discharge curves of $TiS_2$ at 25°C and high temperature

REFERENCES

1. T.E. King and W.J. Moroz. Survey of Continuous Sources of Electric Power for Under-Ice Propulsion of Small Submersibles. Part II: New High Energy Density Systems. Dept. of National Defence, Ottawa, DREO Report No. 757 (1977).
2. D.A.J. Swinkels in Ad. in Molten Salt Chemistry, Vol. 1, ed. J. Braunstein, G. Mamantov and G.P. Smith, Chap. 4, p. 165 (1977).
3. H. Shimotake, E.C. Gay and P.A. Nelson. "The Design and Optimization of Li-Al/$FeS_x$ cells", Symposium Proceedings, Battery Design and Optimization, ed. S. Gross (The Electrochemical Society, Princeton), p. 408 (1979) and Argonne National Laboratory Internal Reports: 8038. 8057, ANL-76-9.
4. M. Gauthier, F. Morin and R. Bellemare (IREQ). "An electrochemical study of Li-Al/$FeS_x$ cells and construction of a prototype". Dept. of Supply and Services, Ottawa, Canada. Contract No. 2SR-00162. Final Report.
5. M. Gauthier, F. Morin, R. Bellemare, G. Vassort and A. Bélanger (IREQ). "Research and Development on Li-Al/FeS Batteries". Dept. of Supply and Services, Ottawa, Canada. Contract No. 2SD78-00034. Final Report.
6. F.J. Martino, L.G. Bartholme, E.C. Gay and H. Shimotake. "Development of Li-Al/FeS cells with Li-Cl - rich electrolyte". Symposium Proceedings, Battery Design and Optimization, ed. S. Gross (The Electrochemical Society, Princeton), p. 425 (1979).
7. Internal Report, Argonne National Laboratories, ANL-77-68, p. 47 (1977).
8. G.L. Holleck and J.R. Driscoll. Internal Report, E.I.C.Corp., AD/A-023 496 (1976).
9. M.S. Whittingham. J. Electrochem. Soc., 123, 315 (1976).
10. M. Armand. Doctoral Thesis, University of Grenoble, France (1978).

## SELENIUM(IV) CATHODE IN MOLTEN CHLOROALUMINATES

Roberto Marassi[+], Daniele Calasanzio[+], and Gleb Mamantov[*]

[+]Istituto Chimico
Universita di Camerino
Camerino, Italy

[*]Department of Chemistry
University of Tennessee
Knoxville, TN 37916

### Introduction

During the last few years we have been concerned with the electrochemical characterization of several potential cathodes for batteries using molten sodium tetrachloroaluminates ($AlCl_3$-NaCl mixtures) as solvents and aluminum or sodium anodes[1-5]. The most promising cathode system found was that involving sulfur cations with formal oxidation states ranging from +4(S(IV)) to 1/8($S_{16}^{2+}$).

It is now well established[6-8] that the solute chemistry in chloroaluminates critically depends on the melt acidity which is determined by the $AlCl_3$/NaCl molar ratio through the following equilibria

$$2AlCl_3 = Al_2Cl_6$$

$$AlCl_4^- + AlCl_3 = Al_2Cl_7^-$$

$$2AlCl_4^- = Al_2Cl_7^- + Cl^-$$

and is usually described by $pCl^-$. In the case of sulfur the acidity of the medium plays a fundamental role in determining the stability of the different oxidation states. In basic (NaCl saturated melts [9,10] only S(I), probably present as $S_2Cl_2$, is stable. In acidic melts ($AlCl_3$/NaCl ratio > 1)[10,11] the electrochemical oxidation of sulfur, which dissolves[12] as $S_8$, proceeds in steps involving the reversible or quasi-reversible couples $S_{16}^{2+}(S_8^{2+})/S_8, S_2^{2+}/S_8^{2+}$ and $S(IV)/S_2^{2+}$. Either in basic or in acidic melts sulfur is reduced to sulfide which is involved in acid-base equilibria with the solvent anions.

S(IV), present in the melt[13] as $SCl_3^+$, has been shown to be a suitable cathode for a rechargeable molten salt battery when coupled either with an aluminum or a sodium/β-alumina anode. However, because of the stability of $SCl_3^+$, the mole fraction of $AlCl_3$ must be greater than one throughout the discharge process resulting in the formation of sulfur.

This paper describes some of the work performed on another possible cathode which uses Se(IV) instead of S(IV) as the active component. The aim was to attempt to increase the cathode capacity by using a cation which appears to be stable over a wider melt acidity range.

## Electrochemistry of Selenium in Molten Chloroaluminates

The electrochemistry of dilute selenium solutions in molten chloroaluminates is quite similar to the electrochemistry of sulfur. It has been described in a preliminary communication by Marassi et al.[14]; a more detailed study has been published by Robinson and Osteryoung[15].

Selenium, like sulfur, may be either reduced or oxidized in chloroaluminates. The reduction process leads to the selenide ion which exists in the melt as either AlSeCl or $AlSeCl_2^-$ depending on the acidity. The oxidation of selenium to Se(IV) proceeds directly through a quasi-reversible four electron step in acidic melts. In the basic melt selenium is first oxidized to Se(II) and then to Se(IV) in a further two electron step. At all melt acidities the reduction of Se(IV) to Se occurs in a single four electron step. While the electrochemical results by Robinson and Osteryoung[15] indicate the presence of equilibria such as

$$SeCl_5^- + Cl^- = SeCl_6^{2-}$$

our Raman results[16], obtained either on the solid $SeCl_3AlCl_4$ (prepared from selenium, $AlCl_3$ and chlorine as the corresponding sulfur compound)[13] or on its solution, indicate that the predominant Se(IV) species in the melt is $SeCl_3^+$ in both acidic and basic chloroaluminates.

Potentiometric results in 63/37 $AlCl_3$-NaCl melts indicate a formal potential of about 2.41 V for the couple Se(IV)/Se and of about 1.53 V for the couple $Se/Se^{2-}$ at 250°C vs. an Al(III)/Al reference electrode in NaCl saturated melts. These values may be taken as the OCV's of batteries obtained by coupling each redox system with an aluminum anode operating in NaCl saturated melts. If a Na/β-alumina electrode is used, the corresponding values are 4.01 V for the Na/Se(IV) cell and 3.13 V for the Na/Se system. OCV values about 0.6V lower may be predicted for cells in which the cathode operates in NaCl saturated melts.

It should be pointed out that these electrochemical results refer to dilute solutions. Potentiometric and spectrophotometric results obtained by Bjerrum et al.[17] in more concentrated Se-$SeCl_4$ solutions in acidic chloroaluminates indicate the presence of a variety of polymeric species with formal oxidation states ranging from 1/2 to 1/8 which may also be formed during normal battery operation.

Discharge Experiments

Several cells have been run to test the cathode operation. Na-β-alumina electrodes have been used as anodes. The cell consisted of a β-alumina tube, sealed to glass, surrounded by an outer glass compartment containing a solution of Se or $SeCl_3AlCl_4$. The cathode current collector was tungsten spiral wrapped around the ceramic tube. The cells were equipped with an Al(III)/Al reference electrode separated from the main solution by a thin glass membrane. Both sodium and the cathode mixture were loaded into the cell, after proper treatment, inside a nitrogen filled dry-box and then sealed under vacuum.

The results obtained so far have demonstrated that a cell may be started either by oxidizing a selenium solution in molten chloroaluminates of proper composition (the charging process causes an increase of melt acidity in the cathode compartment) or by loading directly a mixture of $SeCl_3AlCl_4$ and the $AlCl_3$-NaCl. In both cases the OCV of a charged cell was in the range of 4 to 4.1 V; the potential of the W spiral and of the sodium electrode vs. the Al(III)/Al reference were about 2.4 V and -1.6 V, respectively.

The best results have been obtained with a cell of 1.6 Ahr capacity which has been charged and discharged more than 40 times at current densities of 5 to 10 $mA/cm^2$ (separator area). The initial solvent composition was adjusted to obtain a basic melt at the end of the discharge. This was generally limited to the first process (reduction of Se(IV) to Se) because the second process (reduction of Se to $Se^{2-}$) occurs at a lower potential and the second plateau is always shorter than the expected value because of the low solubility of Se in the melt.

Depending on the current density and temperature the discharge occurs at a potential in the range 2.35 to 2.25 V vs. the external reference. The battery CCV is somewhat lower because of the high resistance of the ceramic; the cracking of the ceramic was the main cause of cell failure. Both coulombic and energy efficiencies were in the range of 80-90% for the cathode.

At the end of the discharge, when the melt composition turns basic, some gaseous products escape to the region above the melt. This, however, does not seem to interfere with battery operation because no degradation or cathode capacity loss has been observed during cell life.

Acknowledgement

The support of this work by the Italian CNR (Rome) is gratefully acknowledged.

References

1. G. Mamantov, R. Marassi, and J. Q. Chambers, "High Energy Cathodes for Fused Salt Batteries," Technical Report ECOM-00600F, April 1974.

2. G. Mamantov, R. Marassi, and J. Q. Chambers, "Molten Salt Electrochemical Systems for Battery Applications," U.S. Patent 3,966,391, June 29, 1976.

3. G. Mamantov and R. Marassi, "A Cathode for Molten Salt Batteries," U.S. Patent 4,063,005, December 13, 1977.

4. G. Mamantov, R. Marassi, J. P. Wiaux, S. E. Springer, and E. J. Frazer, "S(IV) Cathode in Molten Chloroaluminates," in Proceedings of the Symposium on Load Leveling, N. P. Yao and J. R. Selman, eds., The Electrochemical Society Symposium Series, Vol. 77-4, Princeton, N.J., (1977), pp. 379-383.

5. G. Mamantov, R. Marassi, M. Matsunaga, Y. Ogata, J. P. Wiaux, and E. J. Frazer, unpublished results.

6. G. Torsi and G. Mamantov, "Potentiometric Study of the Dissociation of the Tetrachloroaluminate Ion in Molten Sodium Chloroaluminates," Inorg. Chem., 10:1900 (1971).

7. L. G. Boxall, H. L. Jones, and R. A. Osteryoung, "Solvent Equilibria in Aluminum Chloride-Sodium Chloride Melts," J. Electrochem. Soc., 120:223 (1971).

8. G. Mamantov and R. A. Osteryoung, "Acid-Base Dependent Redox Chemistry in Molten Chloroaluminates," in Characterization of

Solutes in Non-Aqueous Solvents, G. Mamantov, ed., Plenum Press, 1978, pp. 251-271.

9. R. Marassi, G. Mamantov, and J. Q. Chambers, "Electrochemical Behavior of Sulfur and Sulfide in Molten Sodium Tetrachloroaluminate Saturated with Sodium Chloride," J. Electrochem. Soc. 123:1128 (1976).

10. K. A. Paulsen and R. A. Osteryoung, "Electrochemical Studies on Sulfur and Sulfides in $AlCl_3$-NaCl Melts," J. Am. Chem. Soc., 98:6866 (1976).

11. R. Marassi, G. Mamantov, M. Matsunaga, S. E. Springer, and J. P. Wiaux, "Electrooxidation of Sulfur in Molten $AlCl_3$-NaCl (63-37 mole percent)," J. Electrochem. Soc., 126:231 (1979).

12. R. Huglen, F. W. Poulsen, G. Mamantov, R. Marassi, and G. M. Begun, "Raman Spectral Studies of Elemental Sulfur in $Al_2Cl_6$ and Chloroaluminate Melts," Inorg. Nucl. Chem. Letters, 14:167 (1978).

13. G. Mamantov, R. Marassi, F. W. Poulsen, S. E. Springer, J. P. Wiaux R. Huglen, and N. R. Smyrl, "$SCl_3AlCl_4$: Improved Synthesis and Characterization," J. Inorg. Nucl. Chem., 41:260 (1979).

14. R. Marassi, G. Mamantov, and J. Q. Chambers, "Electrochemical Behavior of Iodine, Sulfur, and Selenium in Aluminum Chloride-Sodium Chloride Melts," Inorg. Nucl. Chem. Letters, 11:245 (1975).

15. J. Robinson and R. A. Osteryoung, "The Electrochemical Behavior of Selenium and Selenium Compounds in Sodium Tetrachloroaluminat Melts," J. Electrochem. Soc., 125:1454 (1978).

16. G. Mamantov, R. Huglen, and R. Marassi, unpublished results.

17. R. Fehrmann, N. J. Bjerrum, and H. A. Andreasen, "Lower Oxidation States of Selenium. I. Spectrophotometric Study of the Selenium-Selenium Tetrachloride System in a Molten NaCl-$AlCl_3$ Eutectic Mixture at 150°C," Inorg. Chem., 14:2259 (1975).

BEHAVIOUR OF HARD AND SOFT IONS IN SOLID ELECTROLYTES

Finn Willy Poulsen

Metallurgy Department
Risø National Laboratory
DK 4000 Roskilde
Denmark

## INTRODUCTION

The concept of polarizability is frequently referred to in qualitative descriptions of superionic conductors. A high polarizability of the diffusing ion, as for the soft silver ion, is considered essential. This paper treats some of the problems involved in defining a consistent scale of polarizabilities and the possible use of such a tool.

A structure independent theory for solid electrolytes, neglecting the influence of the specific crystal structures in question, would be of great value in prospecting new solid electrolytes for use in advanced batteries, fuel cells and electrolyzers. At least two indications point towards the possible existence of such a theory:

i) Pearson's principle of hard and soft ions predicts the chemistry in liquid, mainly aquous electrolytes. The principle tells that hard ions react preferentially with (tend to be surrounded by) hard ions, and similarly do soft ions prefer the environment of other soft ions.[1] The principle predicts stability and solubility of solids. The two classical examples used [4] to illustrate this rule are identical with two classical solid electrolytes AgI and $CaF_2$. They constitute typical cases of soft-soft and hard-hard ion combinations respectively.

ii) A correlation between the conductivity of an ion, $\sigma_{ion}$, and the viscosity of the surrounding media, $\eta_{medium}$, exists through the well known rule of Walden:

$$\sigma_{ion} \cdot \eta_{medium} = \text{const.} \qquad (1)$$

This relation holds approximately for inorganic and organic liquids, and for glassy systems, at temperatures where viscosity can still be measured.

## POLARIZABILITY OF IONS

The polarizability of an electronic system is the tensor which relates the induced dipolmoment to the electric field causing the charge redistribution. This parameter is clearly of utmost importance in the transfer of an ion from one site to another in a solid, since the ion will experience a (probably strongly) varying electrical field strength along its diffusion path.

Polarizabilities have been evaluated and applied for numerous purposes in solution- and crystal chemistry. Goldschmidt regarded the "opposite" property of ions, the polarizing power, to be more important, and assumed a hard sphere behaviour (zero polarizability) of many ions. Ahrland [2] and Klopman [3] have developed the most extensive scales of softness for ions. Their tabulations are of no use in solid state problems, since the ions are assumed to be hydrated in their standard state. Furthermore, many ions present in important solid electrolytes have no existence in aquous media, e.g. $O^{--}$.

For the present purpose the obvious approach will be to derive polarizabilities from solid state data. Such a procedure is furnished by the concept of the molar refractivity, defined by (Lorenz-Lorentz formular)

$$R = 0.392 \frac{n^2 - 1}{n^2 + 2} \frac{M}{d} = \frac{4}{3} \pi N \cdot \alpha \qquad (2)$$

where R is molar refractivity ($Å^3$), n is refractive index, M is molar mass, d is density, N is Avogadros number, and $\alpha$ is the molar polarizability. R will be used as polarizability or softness parameter. Four crude assumptions are needed when molar refractivities are to be split into individual refractivities of ions and complex units. It is primarily assumed that a set of selfconsistent ionic refractivities can be obtained, which obeys an additivity law. Average refractive indices have to be used for anisotropic solids. Refractive indices show frequency dispersion and are quite arbitrarily compared at a single frequency, normally the sodium D-line. A reference ion has to be gived an arbitrary reference refractivity or polarizability. This is because the resulting system of equations in molar and individual refractivities are linearly related (fixed oxidation numbers of the ions).

We use the values compiled by Jørgensen [4], which are based on the assumption that $Al^{3+}$ ions are nonpolarizable, R = 0.

Molar refractivities for individual ions, deduced this way, range from 0 Å$^3$ ($Be^{2+}$, $Al^{3+}$) to 6.3 Å$^3$ for $I^-$. The assumption of additivity generally holds within 0.1 - 0.2 Å$^3$ for metal-, Halide- and complex ions. A large scatter of the refractivity of oxide ions is revealed when this set of consistent (primarily metal ion) refractivities are used to compute $R(O^{--})$ from refractive index data of oxides. $R(O^{--})$ varies from 1.25 in BeO to 2.05 Å$^3$ in $Li_2O$. This may be caused by various degrees of covalency in oxidematerials. $R(O^{--})$ in $SiO_2$, $Al_2O_3$ and $ZrO_2$ amounts to o.30, 1.35, and 1.36 Å$^3$, respectively. A value of 1.35 will be used below.

Ion vacancies and free protons can formally be attributed zero polarizabilities. Protons in condensed phases are, however, always covalently bonded, in which case the concept of individual polarizabilities is inapplicable.

## HARD- SOFT ION CORRELATIONS

We look for correlations between the magnitude of the ionic conductivity of a system and the polarizabilities of the ions composing the system. Two ways have been explored.

The conductivity, at constant T, of a series of ion-conductors, e.g. all silver ion conductors, are plotted against the polarizabilities of the immobile ions in the respective compounds. It will then be revealed that conductivity increases steply with increasing polarizability of the immobile lattice; qualitatively as expected from Walden's rule. No convincing correlation will be observed for oxide conductors, where assignment of individual polarizabilities is dubious. The general trend can be formulated:

The conductivity of a certain ion in various solids increases with increasing polarizability of the ions in the immobile lattice.

A second way of revealing hard-soft ion correlations in known solid electrolytes is to map the compounds using cation refractivities as abscissa and anion refractivities as ordinate, see figure 1. Oxoanion containing Li-, Na- and K-conductors are in this representation shown by single points, e.g. Li, O. This is due to the fact that individual polarizabilities of the oxide ions composing the bottlenecks in aluminate-, silicate-, etc. anionlattices have not yet been worked out.

Good anion conductors are located in the region where cation softness is higher then anion softness, and vice versa. The line R(cation) = R(anion) has been drawn to guide the eye. The earth alkaline halides do not follow this general trend.

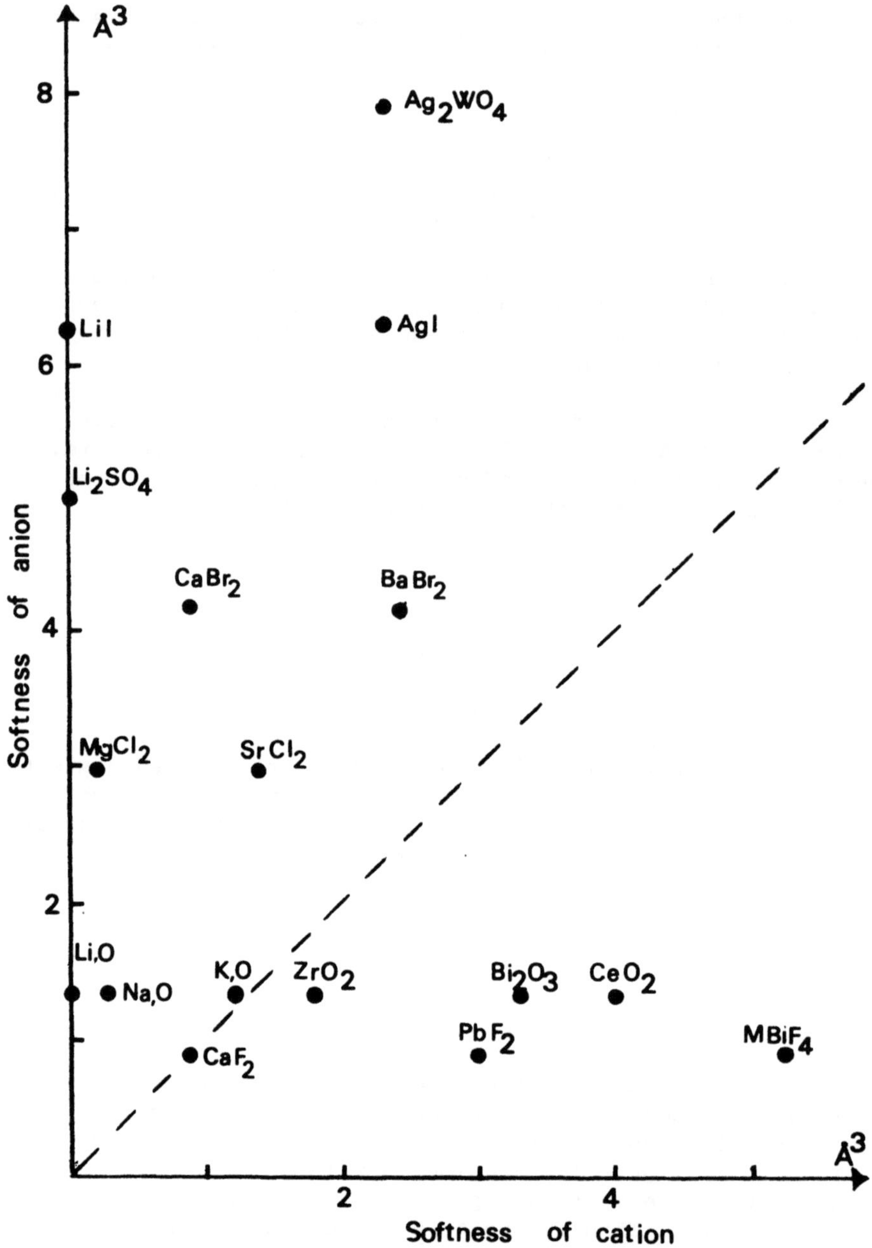


FIGURE 1. Solid electrolytes mapped according to molar refractivities (Å$^3$) of the constituent ions.

They may be anion conductors by virtue of speciel structures and the trivial fact that low charged ions potentially have much higher mobilities than high charged ions.

The general trend revealed from figure 1 can be postulated as:

Cation conduction in solids is favoured, when the cation is harder than the immobile lattice. Anion conduction is favoured, when the anion is harder than the immobile lattice.

The usefulness and validity of these postulates are presently under examination. They are not able to account for phenomena such as mixed cation-anion conduction and preferred ion diffusion in mixed cation systems, viz. (Li/Na) beta-alumina and $RbAg_4I_5$.

REFERENCES

1. R. G. Pearson, J. Amer. Soc. 85, 3533 (1963)
2. S. Ahrland, Structure and Bonding 5, 118 (1968)
3. G. Klopman, J. Amer. Soc. 90. 223 (1968)
4. C. K. Jørgensen, Topics in Current Chemistry, 56, 1, (1975)

# USE OF THE PROTON CONDUCTOR HYDROGEN URANYL PHOSPHATE TETRAHYDRATE AS THE SOLID ELECTROLYTE IN HYDRIDE-AIR BATTERIES AND HYDROGEN-OXYGEN FUEL CELLS

Peter E. Childs[†] and Arthur T. Howe

Department of Inorganic and Structural Chemistry
University of Leeds
Leeds LS2 9JT, England

## INTRODUCTION

Recent investigations[1-9] of the good proton conductor $HUO_2PO_4.4H_2O$ (HUP) have opened up the prospects for its use as a solid electrolyte in a range of electrochemical cells involving transport of $H^+$ ions. Hydrogen concentration cells of the type $MH_x|HUP|M'H_y$[2] and electrochromic cells of the type $H_xWO_3|HUP|H_yWO_3$[9] have been previously described. Perhaps the most unique application of HUP is to hydride-air cells of the type $MH_x|HUP|O_2,H_2O$, which we describe in this paper, and compare them to the more conventional hydrogen-type fuel cell of the kind $H_2|HUP|O_2$.

In the hydride-air cells hydrogen gas is replaced by a metallic hydride ($MH_x$) such as $PdH_x$, $LaNi_5H_x$, $TiFeH_x$ or $MgNiH_x$ which afford a compact and convenient store of hydrogen. The cell thus acts as a fuel cell in the discharge mode, and, provided water vapour is present, will act as an electrolysis cell in the charging mode. Such high energy density cells would have potential as secondary batteries, which may be able to be prepared in thin film form for incorporation on component boards as emergency power supplies. On a layer scale, cells of this sort may be valuable as rechargeable fuel cells for vehicular use.

Such hydride-air cells have not previously been possible because of the lack of a suitable electrolyte. The properties of HUP have

† Present address: Thomond College, Limerick, Ireland.

been previously compared[2,4] with those of other proton-conducting solids and solutions. HUP will neither dissolve in excess moisture, as phosphomolybdic/tungstic acids will, nor will it dry out in air, as acid solutions on the plastic membrane Nafion will. The stability of HUP as the discreet tetrahydrate phase over a wide range of water vapour pressures[5] renders it unique amongst proton conductors for this application.

## EXPERIMENTAL

Construction of the hydride cell was done in the following stages.

1. Pressing of a HUP disc. HUP was prepared as described elsewhere[4,9] by adding 25 g uranyl nitrate dissolved in 50 ml water to 22 ml 2.3 M phosphoric acid at room temperature. After about 30 min the precipitate was filtered and allowed to dry until it contained only about 10% excess water, whereupon it was placed between the plungers in the perspex die, as in fig. 1(i), and pressed hydraulically at 20 MN $m^{-2}$ (2 ton $cm^{-2}$) for several hours to produce a clear disc[8] about 1 mm thick and 1.1 cm diameter. Clear discs have a density of 99.5 ± 0.5%[4] that of the theoretical value, and are essentially impermeable to gaseous hydrogen at an over pressure of 5 cm Hg of hydrogen.

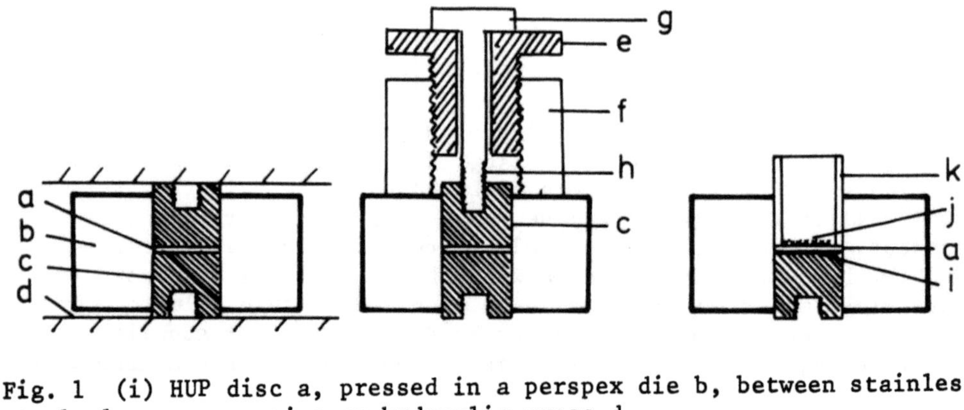


Fig. 1 (i) HUP disc a, pressed in a perspex die b, between stainless steel plungers c, using an hydraulic press d.
(ii) Device used to extract plungers. Bolt e is unscrewed against nut f to raise the slip-screw g, which is screwed into the plunger c by a reverse thread h.
(iii) Final cell with Pd foil i, HUP disc a, air electrode j and hollow current collector k.

2. Placement of the Pd electrode. One of the plungers was extracted using the extraction screw arrangement shown in fig. 1(ii) A Pd foil, 25 μm thick, which had been previously electrochemically Pd-blacked, was inserted, and the plunger was reinserted and light pressure applied to impress the foil onto the HUP disc. Araldite was then placed around the external rim of the plunger to provide a seal.

3. Placement of the air electrode. The other plunger was now removed, and the air electrode, consisting of teflon-bonded Pt-black on a Pt-mesh support,[10] was inserted and impressed onto the disc. The plunger was then removed again and replaced by a stainless steel tube which contacted the electrode around its circumference and served as the current collector, as shown in fig. 1(iii).

The $H_2$-$O_2$ fuel cell was constructed differently by using a HUP disc as a separator between two acidic solutions, in a cell similar to that already described.[3] The gases were bubbled across the blacked Pt mesh electrodes, which were positioned about 1 mm from each disc face.

For both the hydride and gas cells, conductivity occurred from face to face of the HUP discs, that is in a direction perpendicular to the lie of the constituent platelets, giving an ac conductivity of approximately 0.1 $ohm^{-1}$ $m^{-1}$ at 290 K, which is about one quarter of that exhibited by both discs[4] and single crystals[7] when measured parallel to the structural layers of the material.

## RESULTS AND DISCUSSION

The cells were charged in an atmosphere of moist air at 1.5 V. In the fully charged state typically 13 C had been passed, sufficient to hydride the palladium to $PdH_{0.5}$. Upon discharge in air at constant current the voltage showed an initial plateau at between 0.9 and 1.0 V which persisted during a large fraction of the discharge. Currents of typically 5 A $m^{-2}$ could be drawn under short circuiting conditions. Repeated charging and discharging of the cells resulted in a loss of capacity, but not of voltage, of about 70% each time.

The results show that the cells achieved a voltage and capacity close to the theoretical values expected. The voltage plateaus upon discharging correspond to the co-existence of the α phase (Pd-$PdH_{0.03}$) and the β phase ($PdH_{0.57}$-$PdH_{0.69}$), which corresponds to a voltage of 50 mV lower than the RHE.[11] For comparison with a cell run at 1 atm hydrogen we determined the voltages of the $H_2$-$O_2$ fuel cells using HUP, which varied between 0.95 and 1.0 V, and a typical voltage-current plot is shown in fig. 2. The voltage is slightly lower than the theoretical value of

1.23 V, probably due to over-voltage effects. The voltage of 0.9-1.0 V achieved for the hydride cells is thus a satisfactory voltage.

Both the hydrogen gas and metal hydride cells showed that continuous currents could be drawn through the HUP electrolyte. However, in both cases the calculated dc conductivity of HUP was only about 50% of the measured ac values. For the hydride cells this may be caused by the limiting rate of hydrogen absorption into Pd, which is known[11] to show limiting effects above 10 A $m^{-2}$

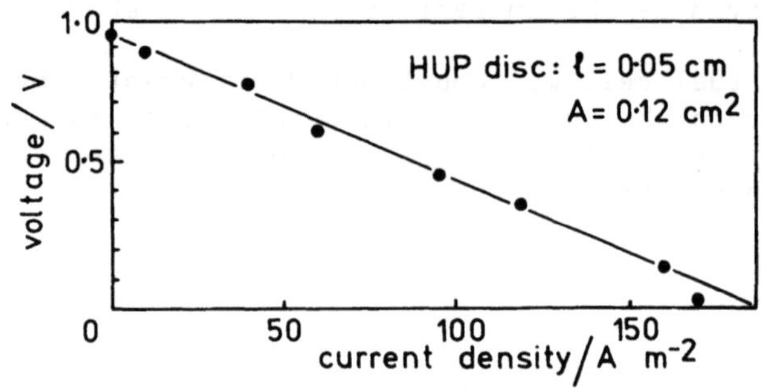


Fig. 2 Voltage-current plot for the fuel cell $H_2$,acid|HUP|$O_2$,acid. The straight line reveals the absence of polarisation.

The observed drop in capacity with successive cycling is a feature which is presently being investigated. Possible causes might be a developing irreversibility of part of the palladium, perhaps caused by physical isolation during the expected[11] 11% expansion and contraction upon hydriding and dehydriding. The capacity loss was not caused by incorporation of $Pd^{2+}$ into HUP,[9] since there was no drop in the ac conductivity values just after the tests, and over a year later.

We have demonstrated the potential of a solid state hydride-air cell as a compact high energy density secondary cell. Assuming a continuous discharge voltage of 1 V, the theoretical energy densities, in Wh $kg^{-1}$, for various hydride-air system are, for magnesium-nickel (assumed $MgNiH_2$) 576, for $FeTiH_2$, 520, for $LaNi_5H_6$, 367 and for $PdH_{0.7}$, 177. The comparatively light weight of HUP and the plastic container materials may reduce these figures perhaps by a factor of 50%, leaving values which compare favourable with those of the present commercial silver/zinc and mercury/zinc button cells of 100-150 Wh $kg^{-1}$.

Larger current densities would be expected using thinner HUP discs. Films of HUP, 6 cm in diameter and 100 μm thick have been prepared,[9] which would give a current density of 50 A $m^{-2}$ in a hydride-air cell. The material cost of such films is approximately 0.1 p $cm^{-2}$.

The operational temperature range of pure HUP is 273 to about 310 K in air. However, we have shown that mixed $NH_4^+/H^+$ compounds are stable in air up to 340 K and the conductivity is only reduced by a factor of about 50% by the substitution of $NH_4^+$. The low temperature limit can be lowered to about 250 K by a lowering of the antiferroelectric transition,[6] by using a different crystallographic modification of HUP,[4] which has essentially the same conductivity, and which is supplied by Alfa Chemicals.

P.E.C. wishes to thank N.R.D.C. for a Post-doctoral Fellowship.

REFERENCES

1. M.G. Shilton and A.T. Howe, Rapid $H^+$ Conductivity in HUP - a Solid $H^+$ Electrolyte, Mat. Res. Bull. 12:701 (1977); A.T. Howe and M.G. Shilton, U.K. Pat. Appl. 47470 (1976).
2. P.E. Childs, A.T. Howe and M.G. Shilton, Battery and Other Applications of a New Proton Conductor: HUP, J. Power Sources 3:105 (1978).
3. P.E. Childs, T.K. Halstead, A.T. Howe and M.G. Shilton, N.M.R. Study of Hydrogen Motion in HUP and HUAs, Mat. Res. Bull. 13:609 (1978).
4. A.T. Howe and M.G. Shilton, Studies of Layered Uranium (VI) Compounds. Part I. High Proton Conductivity in Polycrystalline HUP, J. Solid State Chem. 28:345 (1979).
5. M.G. Shilton and A.T. Howe, idem Part II. Thermal Stability of HUP and HUAs, 33 (1980) in press.
6. ibid Part III. Structural Investigations of HUP and HUAs below the Respective Transition Temperatures of 274 and 301 K, 33 (1980) in press; idem J. Chem. Soc. Chem. Comms. 194 (1979).
7. A.T. Howe and M.G. Shilton, Part IV. Proton Conductivity in Single Crystal HUP and in Polycrystalline HUAs, J. Solid State Chem. 33 (1980) in press.
8. P.E. Childs, A.T. Howe and M.G. Shilton, idem Part V. Mechanisms of Densification of HUP, idem, submitted for publication.
9. A.T. Howe, S.H. Sheffield, P.E. Childs and M.G. Shilton, Fabrication of Films of HUP and their Use as Solid Electrolytes in Electrochomic Displays, accepted for publication in Thin Solids Films.
10. A.C.C. Tseung and P.R. Vassie, A Study of Gas Evolution in Teflon Bonded Porous Electrodes. Part III. Performance of Teflon Bonded Pt black Electrodes for $H_2$ evolution,

Electrochim. Acta 21:315 (1976).
11. F.A. Lewis, The Palladium-Hydrogen System, Academic Press, N.Y. (1967).

# BEHAVIOUR OF THE PASSIVATING FILM IN $LI/SOCL_2$ CELLS UNDER VARIOUS CONDITIONS

G. Eichinger

Anorganisch-chem. Institut der Techn. Univ.
München; Lichtenbergstr. 4
8046 Garching (F.R.G.)

## INTRODUCTION

The good storability of $Li/SOCl_2$ cells is due to the formation of a protective LiCl film at the surface of the lithium electrode[1,2]. The protective film, however, often causes a voltage delay if the cell is discharged. The time of full voltage recovery varies from seconds to several hours[2]. This voltage delay may be unfavourable in some applications.

The intention of this work was to observe the behaviour of the protective film and the voltage delay associated with that film under various conditions.

For that purpose commercially available $Li/SOCl_2$ D-size cells (Tadiran) with an overall capacity of 10 Ah were tested. The cell contains a lithium foil anode with an area of about 54 $cm^2$ that is swaged against the inner wall of the cylindrical can. Impedance measurements were applied to obtain information about the cell resistances and accordingly the film growth, without destroying the cells. The measuring equipment is described elsewhere[3].

## RESULTS AND DISCUSSION

During the experiments with several cells it was observed, that some of the test cells had a very high internal resistance ($> 10^7 \Omega$). After a relatively long time of short circuiting such a cell (about 30-60 min.), this resistance broke down and the usual current drains could be applied for discharge. If the applied load was only short, these cells exhibited after a few days storage again a very high overall-resistance.

One of these cells was opened and SEM pictures of the electrodes were taken. Pictures of a cross section of the lithium electrode showed a very thick film (more than 700 μ) at the surface (about three quarters of the total electrode thickness).

It is somewhat surprising that in spite of the relatively dense structure of the film its growth is obviously not limited. It follows that transport of lithium ions from underneath the film to the surface or of chloride ions from the surface into the interior of the electrode occurs. Ionic transport of lithium ions in LiCl, however, is surely very low at room temperature (much lower than in lithium iodide which has an ionic conductivity of about $10^{-7}\ \Omega^{-1}\ cm^{-1}$ at room temperature[4]). From this consideration it follows that transport of ions should overwhelmingly occur through cracks and pores in the film.

A forced discharge of a cell with an initially very high internal resistance under galvanostatic conditions (50 mA) showed that the cell capacity had decreased from 10 Ah to only 2 Ah (using a cut-off voltage of 1V). This effect is obviously due to the consumption of lithium by excessive film formation.

In fig. 1 impedance curves of a $Li/SOCl_2$ cell with an internal cell resistance of about 105 Ω are shown. The measuring points were recorded at frequencies of 35, 20, 7,5, 3, 1 kHz and 400, 150, 50, 15, 10, 5, 2, 1, 0,5 Hz.

From plot B it may be readily seen that there are two somewhat superposed semicircles representing the impedance of the system in the complex plane. It may be assumed that the high frequency intercept with the real axis represents the $R_e$-value (due to the electrolyte resistance).

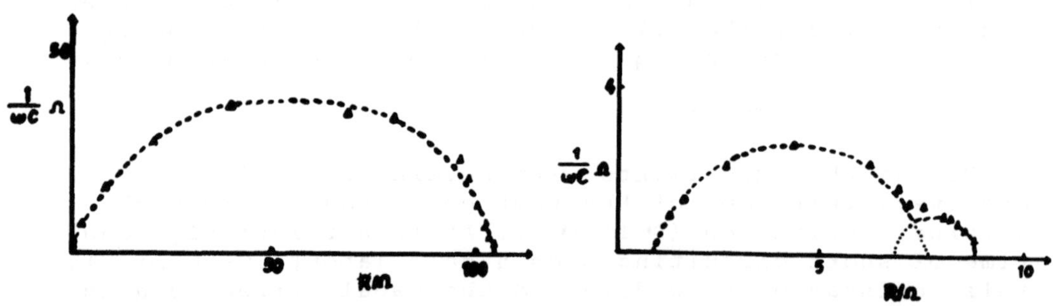


Fig. 1. Impedance plots of a $Li/SOCl_2$ cell

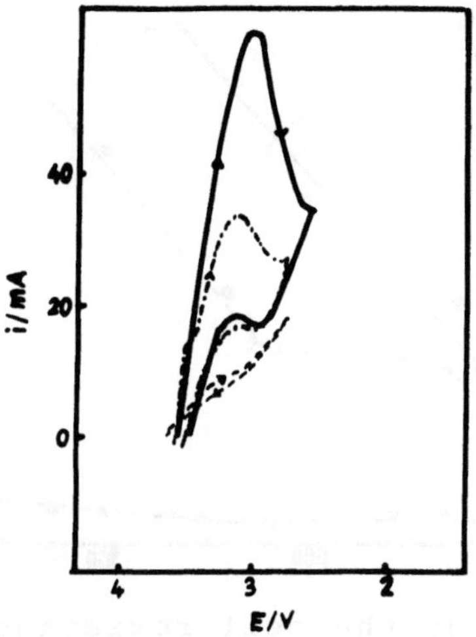


Fig. 2. Cyclic voltammograms of the lithium electrode in a Li/SOCl$_2$ cell

The two following semicircles are obviously due to the resistances and capacities associated with the electrodes. Because of the strong variation of the high frequency semicircle with different loads and storage times of the cell, this is attributed to the lithium electrode.

Fig. 1 shows the strong influence of a load on the lithium electrode. Whereas curve A was recorded after an undisturbed storage of the cell, curve B was recorded after measuring cyclic voltammograms with this cell (fig.2). These voltammograms show directly, how the chloride film at the lithium electrode surface breaks down with an increasing number of cycles.

These measurements were also done at complete Li/SOCl$_2$ cells without applying a reference electrode. In that case, the potential of the counter electrode was used as a reference point. The experiments proved that this remains sufficiently stable during cycling. In fig. 2, the first, fourth and seventh cycle of the lithium electrode at scan rates of 10 $mV \cdot s^{-1}$ are plotted.

As already mentioned, the cell resistance after such a load decreases considerably (see fig. 1B). It was interesting to observe the reformation of the lithium chloride film or the closing of the pores in the film during storage (fig. 3). In that figure the resistance behaviour of the cell used for the earlier experiments (fig. 1 und 2) is shown.

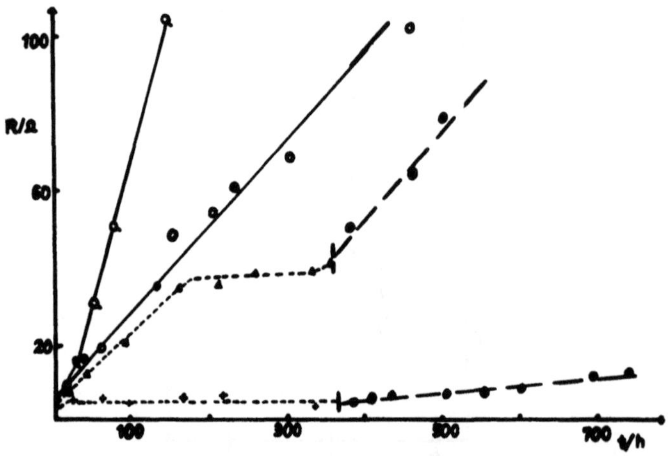


Fig. 3. Variation of the cell resistance of $Li/SOCl_2$ cells with storage time under various conditions

The curve marked by the symbol ● was obtained after recording of the cyclic voltammograms shown partly in fig. 2. After a storage time of about 100 h, the cell resistance is again very high. Nevertheless, if this cell is discharged again (50 mA, 2 min.), the voltage delay is very small. Storing the cell results in a fast increase of the resistance due to a reformation of the film (symbol Q ). With this cell additional tests on the behaviour of the film were done. For that purpose, continuous discharge currents of 100 μA (symbol ▵ ) and 500 μA (sym bol +) were applied (after a galvanostatic discharge of 50 mA for 5 min. in any case, to break up the rebuilt film The results (fig. 3) prove that there is a significant influence of the discharge current on the film growth. Even with the very low current drain of 100 μA (for the complete D size cell) a stagnation of the cell resistance is obtained. After stoping this discharge, again a linear increase of the resistance due to film growth is observed (symbol ◑ ). With 500 μA continuous discharge current the effect is even more pronounced and the cell resistance remains very low. In that case also a linear increase of the cell resistance is obtained (symbol ○ ) after stoping the discharge.

As known from the work of Dey[2], film growth is also sensitive to temperature. In these experiments, two cells were stored at 0°C (fig. 4). In that case it was observed, that the cell resistance after a load of 50 mA, 11 min., did not linearely increase, as is the case at room temperature, but remained fairly constant at values of about 20Ω (symbols + and ◑ ). At room temperature, how-

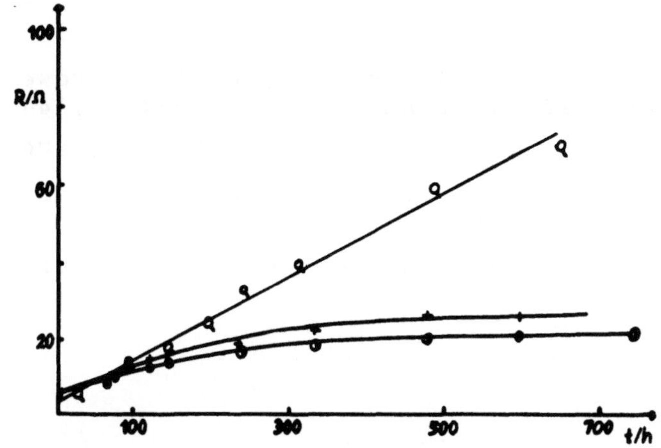


Fig. 4. Variation of the cell resistance of Li/SOCl$_2$ cells at various temperatures

ever, a linear increase is observed (symbol Q ). As is to be expected from the relatively low cell resistance, no voltage delay occurs if this cell is discharged at 50 mA.

CONCLUSIONS

An important fact is that at room temperature or at higher temperatures the growth of the LiCl film in Li/SOCl$_2$ cells is not limited and thus considerable amounts of the lithium anode may be consumed. By this effect, the internal resistance of the Li/SOCl$_2$ cells becomes very high( $> 10^7 \Omega$ ) and the total charge capacity is considerably diminished. If the cells are stored, however, at 0$^{\circ}$C or lower temperatures, film growth is obviously minimized. In that case, the cell resistance remains relatively constant. Even after a prolonged storage time of about 1000 h, no voltage delay is observed in these cells if they were subjected to a load prior to the storage. Film growth may be also minimized if the Li/SOCl$_2$ cells are kept under a low, permanent discharge current. In the case of the cells investigated here, 500 μA are sufficient to keep the overall cell resistance low. The same effect may be achieved by a pulsed discharge (e.g. current drain of 50 mA, 2 min. every 25 h). In that case, the LiCl film at the lithium electrode remains also open. From the results it may be concluded that there are some possibilities to keep the overall resistance of Li/SOCl$_2$ cells low and thus to avoid the voltage delay that is generally observed during the initial discharge.

References

1. A. N. Dey and C.R. Schlaikjer, Proc. Power Sources Symp. 26th, Atlantic City, 1974, p. 47.

2. A. N. Dey, Electrochim. Acta 21:377 (1976).

3. G. Eichinger, G. Deublein and W. Foag, Electrochim. Acta, in press.

4. C. C. Liang, J. Electrochem. Soc. 120:1289 (1973)

# NOVEL SOLID ELECTROLYTES IN THE SYSTEM $Li_4GeO_4$-$Li_2ZnGeO_4$-$Li_3PO_4$

J. Schoonman, J.G. Kamphorst, E.E. Hellstrom

Department of Solid State Chemistry
Physics Laboratory
State University of Utrecht
P.O. Box 80.000, 3508 TA Utrecht, The Netherlands

## INTRODUCTION

It has been established[1,2] that fast alkali ion transport can occur in materials which adopt a rigid three-dimensional, cation-anion network structure with interconnected interstitial sites that are partially occupied by alkali ions. These crystallographic concepts have led to the discovery of a novel class of solid electrolytes of composition $Li_{16-2x}D_x(TO_4)_4$, D = Zn, Mg and T = Si, Ge, $0 < x < 4$.[2] Within this class, compositions with D = Zn, T = Ge and x = 1 or 2 exhibit very high ionic conductivities at 300 and 400 °C. Of these, $Li_{14}Zn(GeO_4)_4$ (lisicon) has been shown[2] to conduct $Li^+$ ions not only better than $Li_{12}Zn_2(GeO_4)_4$, but also better than the well-known fast ionic conductors Li-β-alumina[3] and $Li_{3.5}Si_{0.5}P_{0.5}O_4$.[4,5]

The structure of lisicon has a rigid three-dimensional network of $Li_{11}Zn(GeO_4)_4$, with three mobile $Li^+$ ions in interstitial sites.[2] In $Li_3PO_4$ all the $Li^+$ ions form part of a network. Consequently, the ionic conductivity is low. In solid solutions of $Li_3PO_4$ and $Li_4SiO_4$ interstitial $Li^+$ ions compensate for the substitutionally incorporated $Si^{4+}$ ions, and lead to high values of ionic conductivity. It has been suggested[2] that the mobility of these interstitials can be even enhanced by enlarging the apertures between the intersitial sites. In fact, attempts to enlarge the apertures in $Li_{16-2x}D_x(TO_4)_4$ by replacement of $Si^{4+}$ (0.26 Å) by $Ge^{4+}$ (0.40 Å), and $Mg^{2+}$ (0.57 Å) by $Zn^{2+}$ (0.60 Å) have led to the discovery of lisicon.

Although the ionic conductivity of $Li_4GeO_4$ has been reported to be low[6,7] pseudobinary solid solutions of $Li_3PO_4$ with $Li_4GeO_4$ seem attractive to compare with their counterparts in the system $Li_3PO_4$-$Li_4SiO_4$, of which the terminal phases also are poor ionic conductors.[4,5] As the high-temperature structure of $Li_3PO_4$ and lisicon are isostructural ($\gamma_{II}$-$Li_3PO_4$), incorporation of $Li_3PO_4$ into lisicon can be expected to occur, and this may lead to novel $Li^+$ conducting solid electrolytes. Therefore, we have studied selected phases in the $Li_4GeO_4$-$Li_2ZnGeO_4$-$Li_3PO_4$ system. The present study includes lisicon since Von Alpen et al.[8] observed conductivity values for lisicon that are about an order of magnitude lower than those observed by Hong.[2]

## EXPERIMENTAL

Samples with compositions as shown in Table I were prepared from $Li_2CO_3$, ZnO, $GeO_2$, and $(NH_4)_2HPO_4$, following Hong's preparation technique.[2] Detailed conditions will be reported elsewhere.[9] X-ray diffraction patterns were obtained using Cu-Kα radiation with KCl added as internal standard for lattice parameter determination. Lattice parameters (Table I) were fitted by starting with the orthorhombic lisicon indexing, followed by the use of a refining and 4-parameter least-squared-error fit program.[9] AC conductivity measurements[10] were performed on pellets with sputtered Pt electrodes in nitrogen as ambient. Admittance parameters were recorded in the temperature region 25-450 °C over the frequency range $10^{-3}$ - $5 \times 10^4$ Hz with a signal of 50-100 mV rms.

## RESULTS

An example of the AC conductivity response plotted in the complex admittance plane is presented in Fig.1, along with the equivalent circuit used to fit the high-frequency semicircle. $C_G$ is the geometric capaticance, $R_B$ and $R_{GB}$ are the ionic bulk and ionic grain boundary transfer resistances, respectively. $Z_{GB}$ is of the form $A(i\omega)^{-\alpha}$, where $0 < \alpha < 1$, and is thought to be related with current inhomogeneities in the polycrystalline material.

The conductivity $G = (R_B + R_{GB})^{-1}$ could be determined reliably for the solid solutions up to about 450 °C. In several instances true bulk conductivities $(R_B)^{-1}$ could be determined at low temperatures. Figure 2 presents conductivity data ($(R_B + R_{GB})^{-1}$) for lisicon, and two bulk conductivity values. A conductivity isotherm at 200 °C is given in Fig.3. Included are data for lisicon of Von Alpen et al.,[8] and data for the $Li_4SiO_4$-$Li_3PO_4$ systems of Hu et al.[4] Conductivities reported here were measured during the initial heating. The conductivity of lisicon (Fig.2) decreased with temperature for $T > 320$ °C, and had irreversibly decreased on temperature cycling. This was found in all the phases containing $Li_2ZnGeO_4$ except #10 upon standing above about 320 °C for several

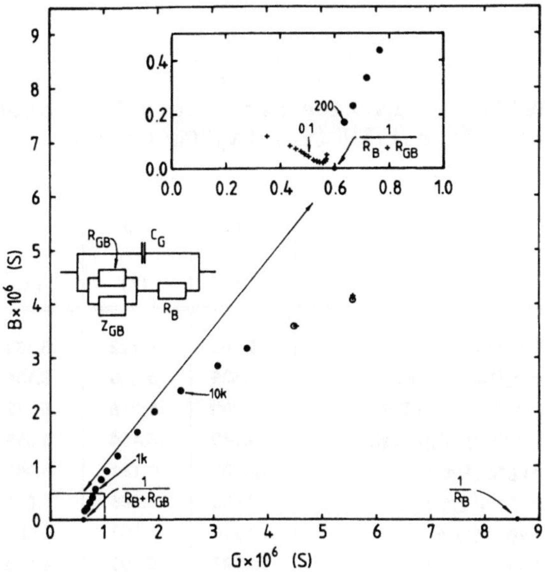


Fig.1. Experimental data (+) and equivalent circuit fitting (0) for solid solution 9 at 50 °C. $R_B = 1.16 \times 10^5\ \Omega$, $R_{GB} = 1.55 \times 10^6\ \Omega$, $C_g = 4.9$ pF, $Z_{GB} = 9.45 \times 10^8\ (i\omega)^{-0.74}$.

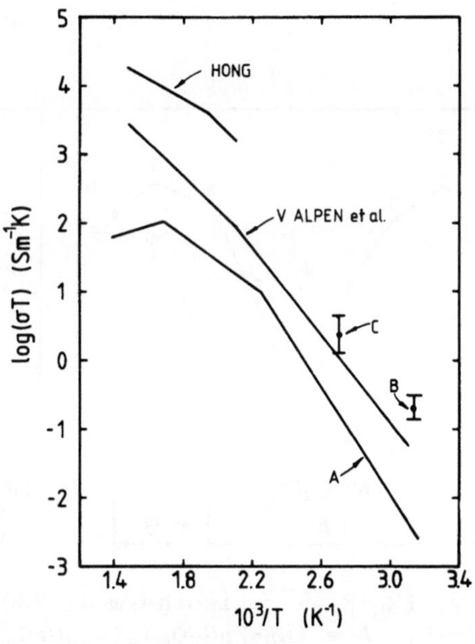


Fig.2. A: Conductivity, $(R_B+R_{GB})^{-1}$, during the initial heating of lisicon. B,C: bulk ionic conductivities. Data of Hong[2] and Von Alpen et al.[8] are shown for comparison.

Table 1. Lattice parameters for solid solutions 1-10 ($\gamma_{II}$-$Li_3PO_4$ type).

| Solid solution | $a_0$ (Å) | $b_0$ (Å) | $c_0$ (Å) | $\beta$ |
|---|---|---|---|---|
| 1. $Li_{14}Zn(GeO_4)_4$ | 10.868 | 6.222 | 5.168 | 90.23 |
| 2. $Li_{14}Zn_{0.75}(Ge_{7/8}P_{1/8}O_4)_4$ | 10.835 | 6.226 | 5.138 | 90.09 |
| 3. $Li_{14}Zn_{0.5}(Ge_{3/4}P_{1/4}O_4)_4$ | 10.790 | 6.215 | 5.115 | 90.11 |
| 4. $Li_{14}Zn_{0.25}(Ge_{5/8}P_{3/8}O_4)_4$ | 10.740 | 6.194 | 5.093 | 90.08 |
| 5. $Li_{14}(Ge_{0.5}P_{0.5}O_4)_4$ | 10.700 | 6.181 | 5.060 | 90.14 |
| 6. $Li_{13.33}(Ge_{1/3}P_{2/3}O_4)_4$ | 10.650 | 6.166 | 5.027 | 90.10 |
| 7. $Li_{14.4}(Ge_{0.6}P_{0.4})_4$ | 10.716 | 6.193 | 5.087 | 90.11 |
| 8. $Li_{15}(Ge_{3/4}P_{1/4}O_4)_4$ | 10.783 | 6.199 | 5.119 | 90.06 |
| 9. $Li_{14.5}Zn_{0.5}(Ge_{7/8}P_{1/8}O_4)_4$ | 10.833 | 6.219 | 5.139 | 90.21 |
| 10. $Li_{12}Zn_2(GeO_4)_4$ | 10.885 | 6.252 | 5.155 | 90.08 |

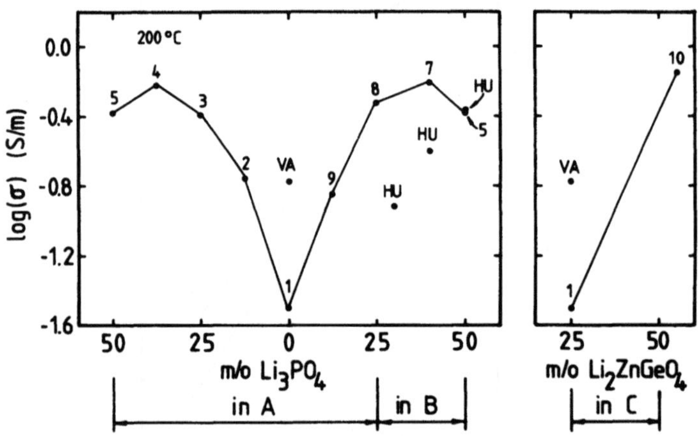


Fig.3. Conductivity, $(R_B+R_{GB})^{-1}$, isotherm at 200 °C for solid solutions 1-10. A = $(Li_2ZnGeO_4)_x(Li_4GeO_4)_y(Li_3PO_4)_{1-x-y}$, B = $(Li_4GeO_4)_x(Li_3PO_4)_{1-x}$, C = $(Li_2ZnGeO_4)_x(Li_4GeO_4)_{1-x}$.

hours (> 6). X-ray analysis of these samples after the conductivity runs revealed the presence of orthorhombic $Li_4GeO_4$.[9]

DISCUSSION

In the system $Li_4GeO_4$-$Li_3PO_4$ stable solid solutions are formed as in the analogous $Li_4SiO_4$-$Li_3PO_4$ system. The conductivity and X-ray pattern of solid solution 10 did not change on heating, thus suggesting that composition 10 is within a one phase region, while lisicon and the other solid solutions containing $Li_2ZnGeO_4$ are metastable phases at low temperatures, that decompose at a measurable rate for T > 320 °C.

The high-temperature data for lisicon show a negative deviation from Arrhenius-behavior, of which the origin seems to be related with electrode effects.[8] The present structural information and $(R_B+R_{GB})^{-1}$ data indicate that at high temperatures segregation of low-conducting $Li_4GeO_4$ causes $R_{GB}$ to increase, while above about 320 °C $R_{GB}$ starts to govern the conductivity behavior in the present samples. Fig.2 reveals that analysis of complex impedance/admittance data lead to fairly consistent bulk conductivity values for lisicon. This bulk conductivity of lisicon is lower than that of $Li_{3.5}Si_{0.5}$-$P_{0.5}O_4$ (Fig.3), while the conductivity activation enthalpies are comparable: 0.56 eV[8] and 0.52 eV,[4] respectively. More important, however, is the recognition that most of the solid solutions studied here have larger conductivities at 200 °C than lisicon, and the solid solutions in the system $Li_4SiO_4$-$Li_3PO_4$. The differences between the bulk conductivities will be even larger, since for the materials studied here $(R_B+R_{GB})^{-1}$ data are present.

Hong[2] generally found the conductivity in $Li_{16-2x}D_x(TO_4)_4$ to increase with increasing number of $Li^+$ interstitials, while for constant stoichiometry the conductivity is usually larger with larger size of the D and T cations, assuming for all materials a rigid network of $Li_{11}D(TO_4)_4$. The present data for solid solutions 1-4 (*constant total Li*) reveal increasing conductivities with decreasing number of $Li^+$ interstitials. In this sequence the $PO_4^{3-}$ content in the network increases while the $Li^+$ to $Zn^{2+}$ ratio on the 8d sites[2] increases. The conductivity data for solid solutions 1,9,8 (*constant mobile Li*) reveal a very similar behavior. This strongly indicates that the conductivity changes are primarily due to small structural changes within the network, which increase the mobility of the $Li^+$ interstitials significantly. Their number seems to play a minor role. In view of the ionic radii - $Zn^{2+}$ (0.60 Å), $Li^+$ (0.59 Å), $Ge^{4+}$ (0.40 Å), $P^{5+}$ (0.17 Å) - the structural changes in the network can be ascribed largely to the incorporated phosphate. A similar situation is encountered in the $Li_4SiO_4$-$Li_3PO_4$ system below 50 mole per cent phosphate.[4] The role of the D cations present in the network needs further study, since it is found that for the pair Zn-Ge the composition with x = 2 (#10) has a better conductivity than lisicon contrary to Hong's observations,[2] but in line with the conductivity data for the Zn-Si, and Mg-Si pairs.[2]

Although an increase in size of the D and T cations leads to fast ionic conductivity in $Li_{16-2x}D_x(TO_4)_4$, further optimization of the conductivity is achieved by partly replacing $Ge^{4+}$ by $P^{5+}$ in the network. In the system $Li_4TO_4$-$Li_3PO_4$ replacement of $Si^{4+}$ by $Ge^{4+}$ indeed favors fast ionic conductivity.

As lisicon actually stands for *li*thium super*i*onic *con*ductor, one can argue that all the materials discussed here are good lisicons, though falicons would be a better general name for the *fa*st *li*thium *con*ductors. The present data indicate that it is still a challenge to design the LISICON among the falicons.

REFERENCES


1. H. Y-P. Hong, Mat. Res. Bull. 11:173 (1976).
2. H. Y-P. Hong, Mat. Res. Bull. 13:117 (1978).
3. M. S. Whittingham and R. A. Huggins, Solid State Chem.; R. S. Roth, and S. J. Schneider, NBS Spec. Publ. 364:139 (1972).
4. Y-W. Hu, I. D. Raistrick and R. A. Huggins, J. Electrochem. Soc. 124:1240 (1977).
5. R. D. Shannon, B. E. Taylor, A. D. English and T. Berzins, Electrochim. Acta 22:783 (1977).
6. W. Grätzer, H. Bittner, H. Nowotny and K. Seifert, Z. Krist. 133:260 (1971).
7. B. E. Liebert and R. A. Huggins, Mat. Res. Bull. 11:533 (1976).
8. U. Von Alpen, M. F. Bell, W. Wichelhaus, K. Y. Cheung and G. J. Dudley, Electrochim. Acta 23:1395 (1978).
9. J. G. Kamphorst and E. E. Hellstrom, J. Solid State Ionics, submitted.
10. R. W. Bonne and J. Schoonman, J. Electroanal. Chem. 89:289 (1978)

# THE ANALYSIS OF SURFACE INSULATING FILMS ON LITHIUM NITRIDE CRYSTALS#

A. Hooper*, C.J. Sofield+, J.F. Turner+ and B.C. Tofield*

*Materials Development Division and +Nuclear Physics Division

AERE Harwell, Oxfordshire OX11 ORA, U.K.

Single crystals of lithium nitride[1] have been studied as part of a programme to determine the effects of dopants on the ionic conductivity, and to assess the compatibility of lithium nitride with electrode materials in high energy density rechargeable cells. The ionic conductivity has been measured using complex impedance techniques and, in the course of this work, it has become clear that surface films on the lithium nitride crystals contribute strongly to the observed response. In particular, examination of complex impedance and related plots of electrical response versus frequency[2] reveal the presence of two resistive and capacitative components arising from the response of the crystals themselves.

One resistive component scales linearly with the crystal thickness and is ascribed to the bulk conduction process. Typical values of room temperature conductivity and activation energy, perpendicular to the crystal c-axis, are 2-3 x $10^{-3}$ $ohm^{-1}cm^{-1}$ and 0.23-0.25 eV, respectively. These values are in accord with those previously described[3,4]. The other component is dependent more on the crystal history, in particular the degree of exposure to the atmosphere, than on crystal thickness. From the associated capacitative component, a thickness of about 20 nm to 1000-2000 nm is derived, assuming a dielectric constant of 10. This implies a

# This work was partially supported by CEC contract 316-78-1 EE UK and the Department of Industry, U.K.

room temperature resistivity of about $10^9$ ohm cm. A higher activation energy than for the bulk crystal, in the range 0.6–0.7 eV is observed. Even with such a thin insulating layer present on crystals of several millimetres thickness, a careful examination of the plots of electrical response against frequency is necessary to avoid interpreting the response of this layer as characteristic of the bulk material; for the more heavily exposed crystals the layer resistance is the dominant series component. The conductivity work is described in more detail elsewhere[5].

The atmosphere-dependent properties of the surface layer, and its electrical response indicate it to be LiOH. This assignment is supported by infra-red measurements which reveal both the $OH^-$ stretching vibration characteristic of LiOH at 3680 $cm^{-1}$, with the intensity dependent on the degree of exposure to the atmosphere, and also, on more heavily exposed samples, the 1430 $cm^{-1}$ band characteristic of $Li_2CO_3$ which is presumably formed as a secondary product.

It is difficult to calibrate the IR intensity to give an accurate, independent assessment of the layer thickness or homogeneity. We show that nuclear reactions involving incident deuterons, protons and alpha-particles are particularly useful for providing quantitative information both on the layer composition and on its depth-profile. The incident particle beams used have been 880 keV deuterons and 3 MeV protons from the 6 MV Van der Graaf generator at Harwell and 4 MeV alpha-particles from the Harwell Tandem Accelerator. Energy analysis of protons and alpha-particles arising from (d,p) (d, $\alpha$ ) and (p,$\alpha$) reactions gives information on the depth concentration of C, O, N and Li, while energy analysis of protons arising from alpha-particle-hydrogen collisions gives hydrogen depth-concentration analysis. The techniques have been described previously for other applications[6,7]. $CaCO_3$, $Si_3N_4$ and $LiNbO_3$ were used as standards and further experimental details are given in Ref. 5.

The results of the (d,p) and (p,$\alpha$) reactions show that the Li:O atomic concentration is approximately unity in the surface layer. The thicknesses found agree within a factor of two or three with those estimated from the electrical measurements. At the surface layer-bulk crystal interface a fairly sharp transition to a lithium-nitride-like composition with Li:N approximately 3:1 is observed. The absolute concentrations of Li and O in the layer indicate considerable porosity which is consistent with the lack of protection against further hydrolysis provided by the lithium hydroxide layer. The oxygen content of the bulk crystal is below the detectable limit (about 5 atom % of the nitrogen atomic content). A carbon content is observed at the exterior of the

LiOH layer which may be attributed to the presence of $Li_2CO_3$. This suggests that the initial surface reaction is hydrolysis, followed by slow conversion to lithium carbonate at the air-layer interface.

The hydrogen profile, on the other hand, indicates that the effect of moisture attack is not precisely confined to a surface layer of LiOH. Although, for example, for a surface LiOH-layer approximately 2000 nm thick, the hydrogen concentration agrees with that of Li and O over this depth, so completing the chemical analysis of LiOH, it does not drop rapidly to a lower background level at greater depths, but decreases fairly smoothly to a depth of at least 5000 nm which is near the penetration limit in this type of experiment. At this point the concentration is roughly 10% of that near the surface. An investigation of this feature of the surface corrosion is continuing.

In conclusion, we have shown that surface corrosion films can strongly modify the electrical response of lithium nitride single crystals, and that, for crystals which have been exposed to the atmosphere even for short times, because of the high resistivity of the surface layer and the high conductivity of the bulk crystals, some care must be exercised in the interpretation of data if the conductivity and activation energy of the films are not to be confused with bulk crystal response. We imagine that this distinction may become more difficult to make in the case of polycrystalline samples, where similar activation energies ($\sim 0.7$ eV) are reported for intercrystalline conductivity[8]. Although the thickness and chemical composition of the corrosion layer can be qualitatively identified by electrical and IR-spectroscopic measurements respectively, particle beam probes are particularly useful, in combination with single crystal samples, in providing quantitative analysis of the film composition and porosity, and of variations in these quantities as a function of depth. In the case of lithium nitride crystals in particular, because of the several mass coincidences possible (for example, Li + Li = N) the nuclear probes have been more useful in the analysis of the surface films than alternative techniques such as secondary ion mass spectrometry.

REFERENCES

1. This work is part of an Anglo-Danish project of research on materials for advanced batteries partially funded by the European Community Research and Development Programme in Energy Conservation. The crystals were grown by T. Lapp, S. Skaarup and K. Fl. Nielsen, Physics Laboratory III, Technical University of Denmark, Lyngby.

2. Data were collected using a Solartron 1191 computer-controlled, frequency response measurement system between 1 Hz and 1 MHz, from ambient temperature to 200°C.
3. U.v. Alpen, A. Rabenau and G.H. Talat, Appl. Phys. Letts., 30, 621, (1977).
4. J. Wahl, Solid State Commun., 29, 485, (1979).
5. A. Hooper, T. Lapp and S. Skaarup, to be published.
6. N.E.W. Hartley in "Ion Implantation, Sputtering and their Applications", Eds. P.D. Townsend, J.C. Kelly and N.E.W. Hartley, p. 210, Academic Press, (1976).
7. B. Terreault et al and C. Brassard et al, Advances in Chemistry, 153, 295, (1976).
8. B.A. Boukamp and R.A. Huggins, Mat. Res. Bull., 13, 23, (1978).

# KINETIC AND MORPHOLOGICAL ASPECTS OF LITHIUM ELECTRODES AT AMBIENT TEMPERATURE IN PROPYLENE CARBONATE

Israël Epelboin, Michel Froment, Michel Garreau
Jacques Thevenin and Dominique Warin

"Physique des Liquides et Electrochimie"
Groupe de Recherche n°4 du C.N.R.S. associé à
l'Université Pierre et Marie Curie
4, place Jussieu 75230 PARIS Cedex 05 (FRANCE)

Advance in the realization of high energy secondary batteries using lithium as negative electrode is hindered by the impossibility of obtaining a great number of cycles with lithium electrodeposits strippable with high coulombic efficiencies. The difficulties encountered are linked to the reactivity of Lithium which, with all the electrolytes known, leads to the formation of passivating layers whose influence on the lithium electrochemical behavior has not been elucidated yet.

In this paper, we present an attempt to reduce the difficulties provoked by the chemical passivation, by turning to account the favourable effects linked to the cathodic incorporation of the lithium in a suitable structure such as an alloy [1,2]. We will summerize a systematic comparison of the behavior of the two types of electrodes : solid lithium substrate and lithium incorporated into an aluminium substrate in a molar solution of lithium perchlorate in propylene carbonate.

## EXPERIMENTAL

The experiments were carried out, at 25°C, in a glove box filled with argon whose residual water content was reduced to 5 ppm. The propylene carbonate (Merck) was stored over molecular sieves 3A. The lithium perchlorate (Smith Chem.) dried by heating in vacuum at a temperature close to its melting point. The residual water content of the solution was about 10 ppm.

The working surfaces of the electrodes were cross-sections of 5 mm high cylinders tightly fit into 5 mm inner diameter glass tubes for lithium (Alfa Ventron ; 99.9 %), and into teflon tubes for aluminium (Johnson-Matthey ; 99.99%). The potential measured between the working electrode and a lithium reference electrode can be obtained free from ohmic drop,using the interrupted current technique and free from concentration gradients,using a rotating disc electrode.

The electrode impedance was measured in a wide frequency range ($10^{-3}$ - $10^5$ Hz) with a transfer function analyser (Schlumberger , Solartron 1174) in a sinusoidal mode,working with an ohmic compensator device. The electrode morphology was observed with a scanning electron microscope (SEM) and a special transfer apparatus. The layers were analysed with the transmission electron diffraction technique and the X ray photoelectron spectroscopy (ESCA-XPS).

## RESULTS AND DISCUSSION

### Lithium substrate

#### Anodic preparation

The SEM observations showed that a cleaning of the lithium surface is obtained by submitting the electrode to a preliminary anodic dissolution with a free ohmic drop overpotential lower than 200 mV. The cristallographic structure of the underlaying metal was best observed after an anodic dissolution carried out with an overpotential of about 50 mV and a charge density exceeding 10 $C/cm^2$.

#### Electrode morphology

The SEM observations showed that the lithium deposit began to grow principally on some crystalline defects of the metal such as the grain boundaries. This deposit can be entirely removed by a subsequent appropriate anodic stripping as long as the charge density does not exceed 1$C/cm^2$. In the other conditions, the presence of a passivating layer led to the formation of dendrites which were not removed during the further anodic stripping.

#### Layer analysis

The observation by electron transmission microscopy of thin films obtained from the electrode surface showed that the electrode was always covered with a plastic and porous layer. The electron microdiffraction patterns showed the presence of very small cristallites (often 50 Å or less). The salts so identified were : $Li_2CO_3$, LiCl and $LiClO_4$. The analysis of the electrode surface by ESCA-XPS spectroscopy confirmed the presence of the ions $CO_3^{=}$, $ClO_4^-$ and $Cl^-$.

The presence of the lithium carbonate agrees with the chemical decomposition of the PC in presence of lithium with a formation of propylene which would polymerise to give the polymeric membrane [3].

### Kinetic parameters

Polarization curves with a cleaned rotating disc electrode were obtained reversibly provided than the anodic and cathodic overpotentials did not exceed 50 mV. A classical treatment of the curve led to define an exchange current density $J_o$ of 3.3 mA/cm$^2$ and a transfer coefficient $\alpha$ equal to 0.5 for the charge transfer reaction $Li \rightleftarrows Li^{+} + e^{-}$.

### Cycling operations

The previous results imply limiting values for the current and charge densities available for cycling. A quite reversible behavior was only obtained with current and charge densities values equal to 1 mA/cm$^2$ and 1C/cm$^2$ with a motionless electrode. Thus, it became possible to achieve more than 500 galvanostatic cycles without any important evolution of the overpotentials [4].

## Lithium-Aluminium substrate

### Cathodic preparation

The cathodic reduction of $Li^+$ ions carried out galvanostatically on a pure thick aluminium substrate led to the insertion of lithium into aluminium. The presence of the inserted lithium deposit led to the formation of the Li-Al compound which was detected by X ray diffraction analysis.

### Electrode morphology layer analysis

The SEM observations confirm that the cathodic insertion of lithium avoided the formation of dendrites below a threshold $Q_d$ whose value depends on the current density. For example,
$Q_d \neq\neq$ 50C/cm$^2$ with J = 10 mA/cm$^2$ and $Q_d \neq\neq$ 100C/cm$^2$ with J = 1 mA/cm$^2$
The incorporation of Lithium into Aluminium does not permit however to entirely avoid the PC decomposition. The plastic porous film is formed in the very beginning of the electrolysis. Its composition is similar to those observed on lithium substrate but its thickness is always low.

### Electrode rest-potential

The Al-Li electrodes present two distinct rest open-circuit potentials. After a deposition process, the electrodes reach a rest potential $V_c$ = 335 mV versus Li/Li$^+$. After dissolution of about 1/10 of the initial deposit, the electrode takes another rest potential $V_a$ = 385 mV. The latter value corresponds to the free enthalpie of

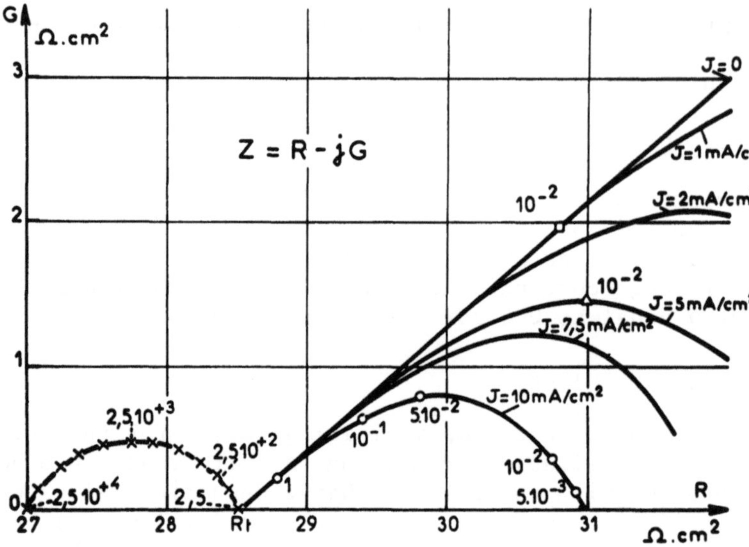


FIGURE : Electrochemical impedance of the aluminium-lithium electrode measured for various cathodic current densities. Rotating disc electrode Ω = 2000 rpm.

formation for the intermetallic compound LiAl ($\Delta G$ = 8.9 kcal/mole). The difference between $V_c$ and $V_a$ is probably due to the fact that a fraction of the lithium incorporated into the Al is not engaged into the coumpound LiAl.

Kinetic parameters

Anodic and cathodic polarization curves with a rotating disc electrode had been drawn using as overpotential $\eta$ the difference between the free ohmic drop potential measured while the current flows, and the instantaneous potential $V_a$ obtained after the circuit was interrupted. The analysis of this overpotential-current density curve led to the determination of an exchange current density much higher than those calculated with the solid lithium electrode ($J_o$ ≠≠ 15 mA/cm$^2$)

The previous rough evaluation of $J_o$ has been confirmed with greater accuracy thanks to the electrode impedance measurements (see figure). The half circle corresponding to high frequencies is characteristic of the charge transfer reaction, with an usual double layer capacity value of the order of 40 μF/cm$^2$ and a transfer resistance nearing 1.5 Ω.cm$^2$ which leads to a $J_o$ = 17 mA/cm$^2$ value. The second half circle corresponding to low frequencies seems to be due probably both to the evolution process of the instantaneous potential $V_o$ versus time and to the transport process inside the passivating layer covering the electrode.

### Cycling operations

An improvement of the cycling was obtained with the lithium-aluminium substrate. For example, with an initial charge density of Q = 50C/cm$^2$ of lithium into an aluminium motionless electrode, it was possible to achieve more than 1500 galvanostatic cycles with current and charge densities respectively 2 mA/cm$^2$ and 1C/cm$^2$ and an average coulombic efficiency of about 97%. However, as with the lithium substrate, the number of cycles decreases when using high charge and current densities. For example, only 100 cycles were obtained with a current density of 2 mA/cm$^2$, a charge density of 5C/cm$^2$ and an average cycling efficiency of 95%.

1. A.N. Dey "Electrochemical alloying of lithium in organic electrolytes", J. of Electrochem. Soc. 118 : 1547 (1971).
2. J.R. Van Beck and P.J. Rommers "Behavior of the secondary electrode on alloying substrates in propylene carbonate based electrolytes" 11th International Power Sources Symposium Brighton (1978) paper 37
3. M. Froment, M. Garreau, J. Thevenin and D. Warin "Sur la nature des couches passivantes formées sur l'électrode de lithium en présence de carbonate de propylène et de perchlorate de lithium" J. Microsc. Spectrosc. Electron. 4 : 111 (1979).
4. M. Garreau, J. Thevenin and D. Warin "Cycling behavior of lithium in propylene carbonate-$LiClO_4$ electrolyte" Progress in Batteries and Solar cells 2 : 54 (1979).

INVESTIGATION OF THE Li-$Bi_2O_3$ BATTERY SYSTEM

USING THE A.C. IMPEDANCE TECHNIQUE

U. von Alpen, M.F. Bell, and F.J. Kruger

Max-Planck-Institut für Festkörperforschung
D-7000 Stuttgart 80
VARTA Batterie AG, FEZ, D-6233 Kelkheim

## INTRODUCTION

The a.c. impedance technique has been increasingly used in the elucidation of the mechanism and kinetics of complex electrochemical reactions. However until recently this technique has suffered from being either slow and cumbersome or very complicated to use and this has restricted the use of this technique in industrial situations. However with the development of automated equipment, such as those developed on the basis of the Solartron 1170 Frequency Response Analyser, [1,2] this technique could become a useful tool in the study of problems associated with many sections of industry.

It is the purpose of this article to further demonstrate the usefulness of this technique in the field of battery research and development. As, in recent years, there has been an increasing interest in the development of high energy battery systems based on a lithium anode and an electrolyte containing a lithium salt dissolved in an organic solvent, it was decided to study the lithium-bismuth trioxide ($Bi_2O_3$) system. The active cathode material, $Bi_2O_3$, has a high specific capacity (ca. 0.7 A-hr/g) and the density of this oxide is 8.9 g/$cm^3$ which would lead to high volumetric capacities and energy densities. Fiordiponti et al. [3] have studied the discharge characteristics of this cathode in a number of electrolytes and conclude that it is an attractive material for such applications.

## EXPERIMENTAL

The Li-$Bi_2O_3$ cells used in this work were built with the same design as the conventional, alkaline $Ag_2O$-Zn cells with a diameter of 11.6 mm and a height of 3.6 mm. The anode was prepared from lithium foil supplied by Foote Mineral. To prepare the cathode, bismuth oxide was thoroughly mixed with around 2 % polyvinylalcohol (PVA) and pressed into a disc. The disc is pressed into the stainless stell can. This was then heated to 600°C to remove the PVA. The electrolyte (64 % propylene carbonate, 27.5 % 1,2 Dimethoxyethane and 8.5 % lithium perchlorate) was prepared in a glovebox ($\leq$ 20 ppm $H_2O$) from the purified materials. The solvent was purified by distillation and dried over $Al_2O_3$. The lithium perchlorate ($LiClO_4$, Alfa anhydrous) was dried by heating in vacuum at 200°C.

The behaviour of the lithium electrode was investigated using a three-compartment glass cell of standard design in which the reference electrode, a small strip of lithium, was separated from the main compartment by use of a Luggin capillary. The working electrode was prepared by pressing lithium into the end of a high pressure polypropylene cylinder. The counter electrode was a large platinum sheet separated from the main cell compartment by use of a frit to avoid contamination by the products occurring via reaction at this electrode. A.c. impedance measurements were carried at the rest potential, the potential being controlled by a Jaissle potentiostat. All measurements were carried out at room temperature in a glovebox in which the argon atmosphere contained less that 5 ppm of $N_2$, $O_2$ and $H_2O$ as examined by a quantitative gas chromatograph.

The automatic impedance measurement equipment has been described previously[2] and is based on the Solartron 1174 Frequency Response Analyser linked via an IEC-Bus to the H.P. 9825 A calculator. For measurement of the battery, a special circuit had to be designed to prevent discharge of the cell through the sine wave generator (30 Ω output impedance). This is shown in Figure (1). Thus a large capacitance, C, is placed in series with the cell to prevent discharge. However it should be noted that the input cnannels (X and Y; input impedance 1 MΩ) are connected directly across the cell and current measuring resistance and so no effect of the capacitance is observed in the resulting impedance diagram.

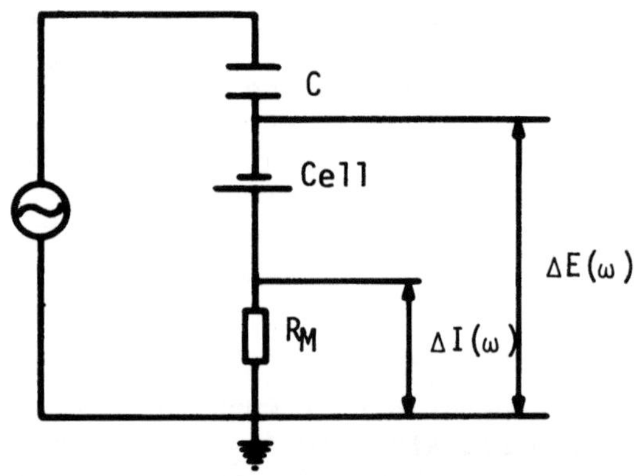


Figure 1. Circuit for measuring the impedance of Li-$Bi_2O_3$ cells

## RESULTS AND DISCUSSION

A representative impedance spectrum obtained under open-circuit conditions is shown in Figure (2). The simplest equivalent circuit which would account for the behaviour is also given, and, for clarity, the various electrical components will be discussed separately.

The high frequency intercept gives a value of $R_E$, which, although designated as the electrolyte resistance, is in fact the sum of the resistances of the electrolyte, the electrodes and the contact resistances between the electrodes and current collectors. In this case, the resistance of the bismuth trioxide electrode is not negligible if one considers that the conductivity ($\sim 10^{-3}\ \Omega^{-1}cm^{-1}$) calculated from the average value of the high frequency intercept (90 ± 11 Ω) is rather low. Measurement of the electrolyte conductivity between two symmetrical lithium electrodes immersed in the same electroylte gave a value of $3.8 \cdot 10^{-2}\ \Omega^{-1}cm^{-1}$.

The rest of the spectrum (figure (2)) represents the superposition of two shapes and it is reasonable to assign each of these shapes to the separate behaviour of each electrode. Thus the appearance of a semicircle at high frequency is considered to be associated with a parallel combination of a resistance ($R_{Li}$) and capacitance ($C_{Li}$) due to the behaviour of the lithium electrode, whereas the low frequency portion is assumed to arise from the capacitive behaviour of the $Bi_2O_3$ electrode. To verify this conclusion, the behaviour of a lithium electrode in this electrolyte was investigated with a potentiostatic,

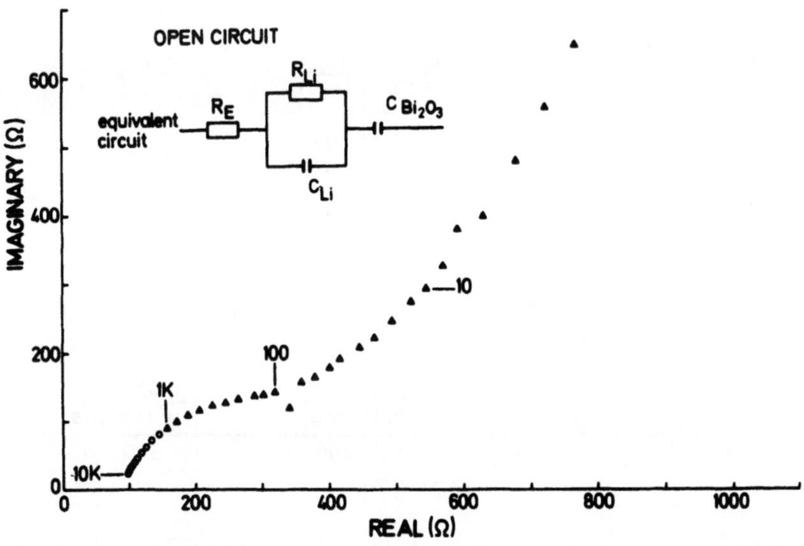


Figure 2. A typical impedance spectrum for a $Li\text{-}Bi_2O_3$ cell under open circuit conditions

three-electrode technique described previously. The impedance spectra obtained at the rest potential as a function of time are shown in Figure (3). On initial immersion (t = 0) the impedance contains two semicircles. The second, low frequency semicircle grows with time until it completely dominates the impedance spectrum (t = 66 hrs.). This behaviour would be consistent with the formation of a passivating film on the electrode surface in which case the low frequency semicircle would then reflect the resistance and capacitance of the film. After several days immersion, the resistance of this film becomes constant suggesting that full and uniform coverage of the electrode has been reached and further growth is prevented. It should be noted that passivation of lithium electrodes has been postulated earlier [4,5] and, in the case of propylene carbonate, the postulated film is thought to be composed of mainly lithium carbonate.[5] Therefore it would seem that the semicircle which was assigned to the lithium electrode in the battery is primarily connected with a film on this electrode surface. By comparison of the resistance value at full coverage (195 $\Omega cm^2$) with the value obtained in the full cell (390 ± 30 Ω) it is possible to calculate the surface area of the lithium electrode which is in contact with the electrolyte in the cell. Our calculations suggest this to be approximately 0.7 $cm^2$ corresponding to a round sheet of 1 cm diameter, a value to be expected for this size of cell.

Although the error in the measurement was high, the value of the capacitance at the bismuth oxide electrode showed a considerable variation between the first (batteries 1 - 10) and the second (11 - 20) batches of batteries studied (figure (4)). Analysis of the material

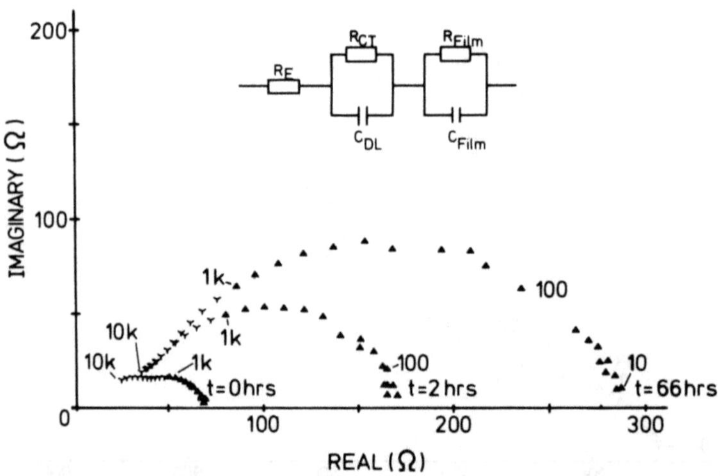


Figure 3. Impedance spectra for a lithium electrode in the battery electrolyte at the rest potential (E = 0) as a function of time (hrs.); electrode area = 0.785 $cm^2$

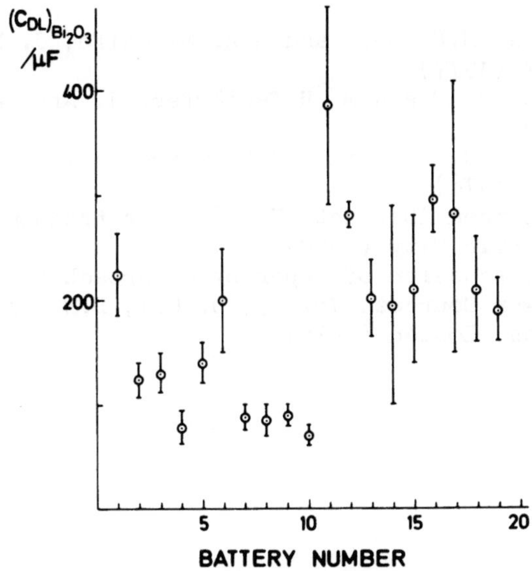


Figure 4. The capacitance of the $Bi_2O_3$ electrode in the twenty cells tested.

used in each batch showed that there was an approximate 2 % incorporation of aluminum oxide into the cathode material used in the second batch. It seems, therefore, that this capacitance is highly sensitive to the exact nature of the electrode and further shows the usefulness of this technique.

## SUMMARY

It has been shown that the a.c. impedance technique is capable of providing useful information in projects connected with the development of new battery systems. In particular, measurements carried out on a lithium/bismuth oxide cell have led to a number of interesting conclusions.

Firstly it was deduced from measurements on a three electrode system that passivation of the lithium anode occurs in this battery system on storage. Secondly, on open circuit, the bismuth oxide cathode behaves as a capacitance, the capacitance value being very dependent on the exact nature of the material used. Finally the resistance of this electrode is not negligible and could cause significant polarisation on discharge.

REFERENCES

1. R.D. Armstrong, M.F. Bell and A.A. Metcalfe, J. Electroanal. Chem. 77, 287 (1977)
2. U. v.Alpen, K. Graf and M. Hafendörfer, J. Applied Electrochem. 8, 557 (1977)
3. P. Fiordiponti, G. Pistoia and C. Temperoni, J. Electrochem. Soc. 125, 14 (1978)
4. A.N. Dey, Extended Abstracts No. 62, Electrochem. Soc. Fall Meeting, Atlantic City (1970)
5. P. Bro, (in discussion of paper by J. Broadhead and F. A. Trumbore) Power Sources, Vol. 5, p. 661, D.W. Collins/ed., Academic Press, London (1975)

ENHANCED IONIC MOTION IN SOLID ELECTRODE MATERIALS

W. Weppner

Max-Planck-Institut für Festkörperforschung
D-7000 Stuttgart-80, FRG

ABSTRACT

High power density batteries require rapid equilibration of compositional inhomogeneities within the mixed ionic-electronic conducting electrodes. Large enhancement of the ionic motion may be observed for semi-conducting materials with a small number of highly mobile electronic species. Metallic conduction is a disadvantage. In order to meet the various conflicting requirements, the coulometric titration curve should show large slopes at completely charged and discharged states.

INTRODUCTION

Good performance of electrode materials requires fast compositional equilibration upon the removal or addition of electroactive species at the interfaces with the electrolyte during the discharge and charge processes. Polarization causes loss of the chemical energy converted into electrical energy in favor of the formation of heat. The problem is particularly significant for the development of high power density batteries. However, in addition, fast responding electrochromic displays and memory elements need rapid equilibration of inhomogeneities within their mixed conducting electrodes.

A solution is presently sought in various laboratories by inverting the traditional concept of batteries by using solid electrolytes and liquid electrodes. Liquid systems as electrodes are generally considered to be of advantage compared to solids with regard to high diffusion rates.

It is the purpose of this contribution to show that solid mixed conducting materials may have even orders of magnitude larger effective diffusion coefficients than liquids. Equilibration may occur as fast as by diffusion in gases. These solid materials will be called "fast solid electrodes (FSE)". They are of interest in conjunction with both liquid or solid electrolytes. The conditions for enhanced ionic motion in solid mixed conductors will be derived. At the same time, the aspect of high energy storage capacity will be taken into account.

## INNER ELECTRIC FIELD OF MIXED CONDUCTING ELECTRODES

The transport of the ionic species in solid electrodes occurs by chemical diffusion. Compositional inhomogeneities which are built up by the removal or addition of electroactive species equilibrate by the simultaneous movement of the mobile ionic and electronic species. The net charge flux is balanced to the current drawn by the external circuit:

$$q \sum_k z_k j_k = i \tag{1}$$

$i$, $q$, $z_k$ and $j_k$ are the current density, the elementary charge, charge number and flux density of species k, respectively. All species are driven by the gradients of their electrochemical potential $\eta_k$ which may be separated into a diffusional part due to the gradient of the chemical potential $\mu_k$ and an electrical part due to the gradient of the electrostatic potential $\phi$. The flux densities are in terms of these quantities:[1]

$$j_k = -\frac{\sigma_k}{z_k^2 q^2} \operatorname{grad} \eta_k = -\frac{\sigma_k}{z_k^2 q^2} \operatorname{grad} \mu_k - \frac{\sigma_k}{z_k q} \operatorname{grad} \phi \tag{2}$$

$\sigma_k$ is the partial ionic conductivity of species k.

In most cases, one type of ionic species ($A^{z+}$) is much more mobile than the other ones. Consequently, only these ions have to be considered beside the electrons or holes in the sum of eq. (1). Insertion of eq. (2) for k = ions $A^{z+}$ and electrons e into eq. (1) and solving for the electrostatic potential gradient yields:

$$\frac{d\phi}{dx} = \frac{t_e}{q}\frac{d\mu_e}{dx} - \frac{1-t_e}{zq}\frac{d\mu_{A^{z+}}}{dx} - \frac{i}{\sigma_e + \sigma_{A^{z+}}} \tag{3}$$

where $t_e = \sigma_e/(\sigma_e + \sigma_{A^{z+}})$ is the electronic transference number. The slower species become enhanced in this inner electric field while the faster ones will be reduced in their velocities in order to maintain local electroneutrality.

## ELECTRONIC PARAMETERS FOR ENHANCED IONIC MOTION

Eq. (3) will be analyzed with regard to the possible action of the inner electric field to enhance the motion of the ions, i.e., the equilibration of the electrode. The first term on the right hand side of eq. (3) results from the tendency of the electrons to move down their activity gradient. Since $d\mu_e$ has the same sign as $d\mu_{A^{z+}}/z$, it represents an electric field which drives the $A^{z+}$ ions forward in the direction to enhance the equilibration of the electrode. In order to make this influence large, it is first necessary to have an electronic transference number $t_e$ close to one. Secondly, it is favorable that the electronic concentration $c_e$ be small since this results in large changes of the logarithm of $c_e$ and therefore large variations of the chemical potential of the electrons according to $d\mu_e = kT\, d \ln c_e$ assuming ideal solution behaviour or by other reasons a constant activity coefficient. In order to make this conditon compatible with the previous one ($t_e \simeq 1$), it is necessary to have a relatively low concentration of highly mobile electronic species. The most favorable materials for electrodes are semi-conducting. The often stated requirement for metallic conduction is in fact a disadvantage for fast solid electrodes.

The next term in eq. (3) describes the electric force to prevent the motion of the ionic species under the influence of their own activity gradient. This retarding electric field contribution has to be kept small. That can be done by the same condition that makes the first term large, $t_e \simeq 1$. In addition, the change of the chemical potential of $A^{z+}$ may be kept small by a large disorder in the sublattice of the mobile ionic species, which is also generally of advantage for the ionic motion.

The last term in eq. (3) is the electrostatic potential drop due to the net charge flux in the electrode (IR-drop). It has the proper sign in order to enhance the motion of the ions. However, this contribution lowers the overall potential drop across the galvanic cell and is not taken into consideration for enhancing the motion of the ions.

## SHAPE OF THE COULOMETRIC TITRATION CURVE

Large enhancement of the ionic motion in mixed conducting materials may also be revealed by the shape of the coulometric titration (CT) curve.[1,2-4] The steady state equilibrium cell potential E is plotted in these diagrams as a function of the content of electroactive species which is determined from the charge flux through the cell.

The (thermodynamic) enhancement factor W is defined by the ratio of the chemical diffusion coefficient and the diffusivity (component diffusion coefficient) of the mobile ionic species. It is related to the change of the activity $a_{A*}$ of the mobile species A in their neutral state with regard to their concentration according to: [1,4]

$$W = t_e \; \partial \ln a_{A^*} / \partial \ln c_{A^*} \tag{4}$$

The quotient may be readily determined experimentally from the slope of the CT curve of the electrode material. The change of ln $a_A$ is given by the change of the cell voltage E multiplied by $z_A q/kT$, and the change in the concentration $c_A$ is $N_A\, d\delta/V_M$ where $N_A$, $\delta$ and $V_M$ are Avogadro's number, the stoichiometric number of A, and the molar volume, respectively. This yields

$$W = -\frac{t_e\, z_A\, q}{kT}\,\frac{dE}{d\delta} \tag{5}$$

It may be seen again that a large enhancement factor requires an electronic transference number close to one. In addition, large values for W exist at compositions that show large slopes in the CT curve. This behaviour is, however, in contradiction to the requirement of high energy densities of batteries since the area underneath the CT curve has to be large for a large energy storage capacity[5]. In addition, the cell voltage will vary considerably during discharge.

The apparent contradiction may be overcome by selecting materials which have shapes of the CT curves as schematically shown in Fig. 1.

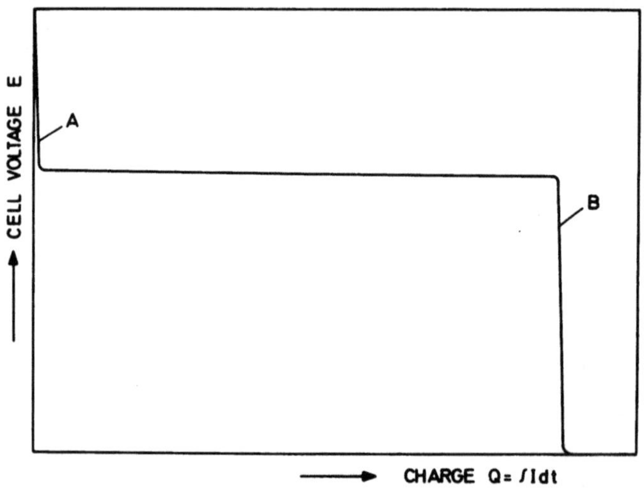


Fig. 1 Schematic representation of an ideal coulometric titration curve. The o.c. cell voltage is plotted versus the charge passed through the galvanic cell which is proportional to the change of composition. Large slopes of the titration curve should be observed at completely charged (A) and discharged (B) states.

Sharp decreases of the voltage should be observed at complete charge (A) and discharge (B). The intermediate range should be comparatively flat and may be, for example, the single-phase region of a compound with a wide range of stoichiometry or it may also be the equilibrium between 2 or more phases. For explanation, let us assume that such an electrode has a homogeneous composition somewhere in the middle of the flat region when a current through the cell is turned on. As a result of the small enhancement factor at the starting composition, the concentration of the electroactive species will increase (or drop in the case of an opposite current) rapidly at the interface with the electrolyte. As soon as a composition is reached, however, at which a fast change of the voltage is observed as indicated by B in Fig. 1 the diffusion will be enhanced and further equilibration will occur readily without increasing the polarization. In spite of the large change of composition at the interface, the cell voltage has only varied a little. The decrease of the voltage compared to the o.c. (thermodynamic) value will be the smallest when the intermediate range of the CT curve is most flat and the subsequent decrease is steep. A similar consideration may be made for the charging process. Again starting at an intermediate composition and turning on the current, the concentration of the mobile species will be reduced until a steep increase of the cell voltage is observed as in point A of Fig. 1. In this case, only a somewhat larger cell voltage than the thermodynamic one has to be applied for rapid charging of the battery.

## EXPERIMENTAL VERIFICATION

A few examples will be given in order to show that the required conditions for enhanced ionic motion in electrodes are not artificial and can indeed be realized experimentally. Many systems may already be indicated by considering reported literature data. For instance, we may think of Sb or Bi as cathodes in lithium batteries. Region A corresponds to the solid solution of Li in Sb or Bi and region B corresponds to the single-phase compounds $Li_3Sb$ or $Li_3Bi$. The flat region comes from the 2-phase equilibria $Li/Li_2Sb$ and $Li_2Sb/Li_3Sb$ or Li/LiBi and $LiBi/Li_3Bi$[4,5]. The enhancement factors of $Li_3Sb$ and $Li_3Bi$ may be as large as 70000 and 370, respectively. By contrast, it is much more difficult to match the above requirements on the shape of the CT curve for electrode systems which do not show phase changes during the entire discharge process (e.g., some intercalation compounds with chemical diffusion coefficients as low as $10^{-9}$ $cm^2$ $sec^{-1}$ at ambient temperature in the case of $Li_xTiS_2$). In the case of systems which include phase changes, the sharp voltage drops may often come from narrow single phase regions.

Another example of a material with a large enhancement factor is $\alpha$-$Ag_2S$. W can be calculated to be about 10000. This results in an effective diffusion coefficient of 0.47 $cm^2$ $sec^{-1}$ at 200 °C,[6] which is much higher than the values found in most liquids.

## CONCLUSIONS

High inner electric fields are the origin for large enhancement of the motion of ions in mixed conductors. They are produced by a sma number of very mobile electronic species. Effective diffusion coefficients may be larger in this way in solids than in liquid systems. Th condition of metallic conduction often demanded earlier for electrode is of disadvantage. Favorable conditions for enhanced ionic motion ma be derived from the coulometric titration curve. The fast solid electrodes are of interest both in the traditional concept for batteries with liquid electrolytes and in all-solid-state batteries using solid electrolytes.

## REFERENCES

1. W. Weppner and R.A. Huggins, Determination of the Kinetic Parameters of Mixed-Conducting Electrodes and Application to the System $Li_3Sb$, J. Electrochem. Soc. 124:1569 (1977)
2. C. Wagner, Investigations on Silver Sulfides, J. Chem. Phys. 21: 1819 (1953)
3. W. Weppner and R.A. Huggins, Electrochemical Methods for Determini Kinetic Properties of Solids, Ann. Rev. Mater. Sci. 8:269 (1978)
4. W. Weppner and R.A. Huggins, Electrochemical Investigation of the Chemical Diffusion, Partial Ionic Conductivities, and Other Kinetic Parameters in $Li_3Sb$ and $Li_3Bi$, J. Solid State Chem. 22:297 (1977)
5. W. Weppnerand R.A. Huggins, Thermodynamic Properties of the Intermetallic Systems Lithium-Antimony and Lithium-Bismuth, J. Electr chem. Soc. 125:7 (1978)
6. W.F. Chu, H. Rickert and W. Weppner, Electrochemical Investigation of the Chemical Diffusion in Wüstite and Silver-Sulphide, W. van Gool, ed., North Holland Publ. Comp., Amsterdam (1973), p. 181

# DYNAMIC CONSIDERATIONS IN THE DESIGN OF BATTERIES WITH INTERCALATION CATHODES

S. Atlung, K. West, and T. Jacobsen

Fysik-Kemisk Institut, The Technical University of Denmark, Building 206, DK 2800 Lyngby, Denmark

The performance of a battery at a given load is to a considerable extent determined by the rate capability of the electrochemical system relative to the applied load. This can conveniently be expressed as a ratio between time constants: Θ/T, where T is the stoichiometric discharge time at the required load and Θ is a characteristic magnitude for the battery, depending on the transport and electrochemical rate constants and the geometry of the electrode configuration.

It is reasonable to assume that most battery systems based on intercalation electrodes will be transportcontrolled, i.e. the rate determining processes will be electrodiffusion of the intercalated ion in the electrolyte and the solid electrode. Under the assumption that the host substance is electronically conductive and the counter-ions in the electrolyte are mobile the concentration changes caused by the ion transport are depicted in Fig. 1.

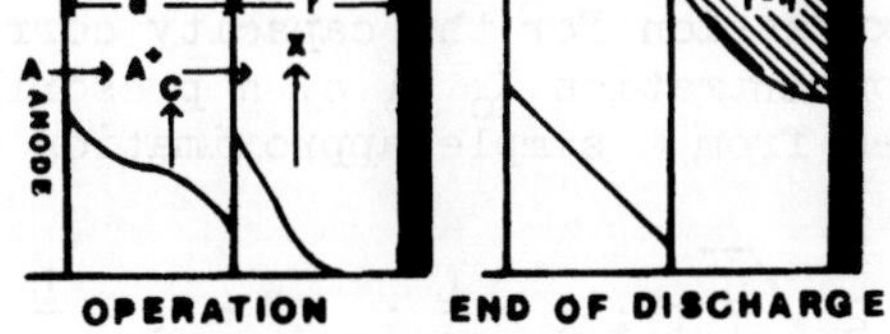


Fig. 1. Concentration Profiles.

## The Discharge End Point

If the transfer processes at the interface are fast, compared to the mass transport, the working potential of the cathode is determined by the interfacial ion concentrations, e.g. by the type of relation proposed by Armand (1)

$$\pi - \phi = E_o - \frac{RT}{F}\left( \ln \frac{X^*}{1-X^*} + f \cdot X^* - \ln c^* \right)$$

where X is the relative concentration of the intercalated ion in the cathode: $X = c/c_{max}$, which corresponds to the local degree of discharge, c is the electrolyte concentration and the asterisk signify interfacial values. f is an interaction parameter which dominates the potential in the middle part of the discharge curve, but as $X^*$ approaches unity, the logarithmic term causes the potential to change very fast to large negative values. At the same time the negative concentration gradient in the solid, which is necessary for the trans port into the solid, is forced to diminish in value and the cathode cannot accept the discharge current. At this point the battery is discharged and it is seen from figure 1 that an amount of cathode capacity corresponding to the hatched area has not been utilized. This is the main reason for the decrease in battery capacity caused by the load.

## Utilization of the Cathode

A mathematical treatment of the situation based on the solution of the partial differential equations for the diffusion in the cathode has been given elsewhere (2). The resulting expression for the surface concentration is:

$$X^* = \tau + \frac{L}{A}\left( \frac{1}{B} - 2 \sum_i \frac{\exp(-\alpha_i^2 \tau / L)}{\alpha_i^2} \right)$$

where L is the ratio between the time constants for diffusion $\Theta = r^2/D$ and discharge T: $L = \Theta/T$ and $\tau$ a dimensionless time variable $\tau = t/T$, which is equal to the total degree of discharge. A, B and $\alpha$ depend on the cathode particle shape. For flat particles (thickness 2r) the values are 1, 3 and $i \cdot \pi$, respectively.

This relation connects the surface concentration and consequently the potential with the capacity withdrawn from the battery.

An explicit expression for the capacity corresponding to a given critical surface concentration $X_o$ (i.e. a prescribed end point potential) can be derived from a simple approximation valid within a few percent:

$$\tau < \frac{L}{\pi} : \ X^* = 2\sqrt{\frac{L}{\pi}\tau} \ ; \ \tau > \frac{L}{\pi} : \ X^* = \tau + \frac{L}{\pi}$$

Solved for $\tau$ the utilization $\eta$ for a critical surface concentration $X_o$ corresponding to the end point potential is:

$$L < \frac{\pi}{2} X_o \; : \; \eta = X_o - \frac{L}{\pi} \; ; \quad L > \frac{\pi}{2} X_o \; : \; \eta = \pi X_o^2/4L$$

The two expressions are valid for what may be termed normal and heavy loads, respectively. $L = \pi/2$ gives $\eta = 0.5\ X_o$.

Packaging Density and Specific Energy

A large $\eta$ corresponds to a low value for L, which requires thin electrodes: $r = \sqrt{LD_sT}$. This, however, is in conflict with the desire to construct a battery with a high "packaging density", w, which is the ratio between the weight of active materials and the total battery weight. The product $\eta \cdot w$ indicates the effectiveness of a given battery design, and multiplied with the stoichiometric energy density the final energy density in Wh/kg can be found. E.g. for lead acid batteries w = about 0.5, $\eta$ = about 0.3 and as the stoichiometric energy content is 178 Wh/kg, a specific energy of 30 Wh/kg is obtained.

For large batteries the electrolyte and the electrode supports (current collectors) constitute the major part of the inactive weight. Using reasonable estimates for the thickness of these components, w can be calculated as a function of r. Assuming a value for the solid state diffusion constant $D_s$, L and then $\eta$ can be calculated for different values of T and the optimum value for the product $w \cdot \eta$ can be estimated. In Fig. 2 the result of these calculations for the $Li/TiS_2$ system ($D_s = 10^{-8}$ $cm^2\ s^{-1}$, $X_o = 1$) is shown for a 2 and a 5 hour load.

Electrolyte Limitations

The discharge causes a decrease in the concentration of the active ion at the electrolyte side of the interface. When this concentration approaches zero, electrolyte limitation occurs. The limiting current depends on electrode spacing, the tortuosity of the separator (3), the cation diffusion constant and the initial electrolyte concentration. However, for times shorter than the time constant for diffusion in the electrolyte considerably larger currents can be drawn. It is worth noting that whereas the behaviour in the steady state depends on the cation mobility only, the mobilities of both ions enter in the time constant and the development of concentration and potential for short times.

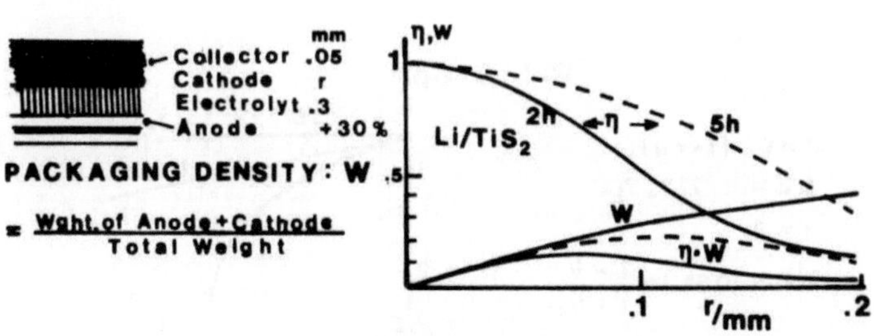


Fig.2. Utilization, Packaging Density and $\eta \cdot w$

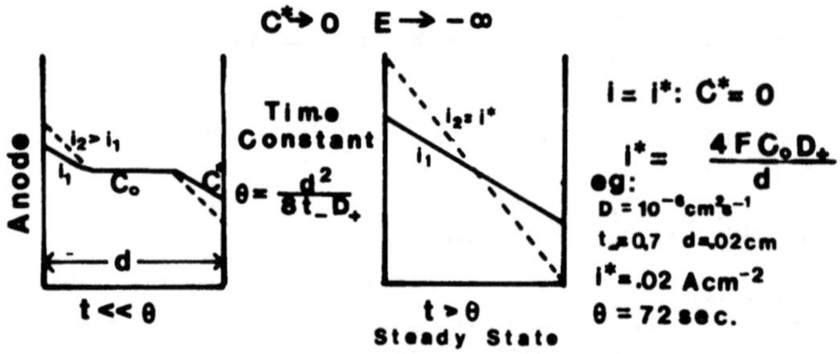


Fig. 3. Limiting Current in Electrolyte

In Fig. 3 the concentration profiles are depicted and some of the relevant formula given. The limiting current phenomena place a restriction on the combined thickness of electrode and electrolyte because the current density is proportional to electrode thickness:

$$r \cdot d < 4TD_+C_oV_m/n$$

The potential loss across the electrolyte can for times considerably shorter than the time constant be calculated as an "I·R" drop from the electrolyte conductivity. For times comparable to the time constant the concentration changes cause a considerably larger potential loss. In Fig. 4 these phenomena are illustrated.

Solid ion conductors behave differently because concentration changes cannot occur. There is no time dependent phenomena and no limiting current ($t_- = 0$, $\Theta \to \infty$). Likewise the potential loss is a real "I·R" drop.

Porous Cathodes

It is obvious that for the plane electrode configuration only low values of $\eta \cdot w$ can be obtained for loads in the 2 hour range. Using a porous electrode it is, however, possible at the same time to choose a cathode thickness large enough to obtain a good packaging density, and use a small particle thickness in order to obtain a high utilization. However, the limiting phenomena due to the ion transport in the pores limit the utilization in the bottom of these. Work is in progress to study these phenomena (4) and some preliminary results will be given here.

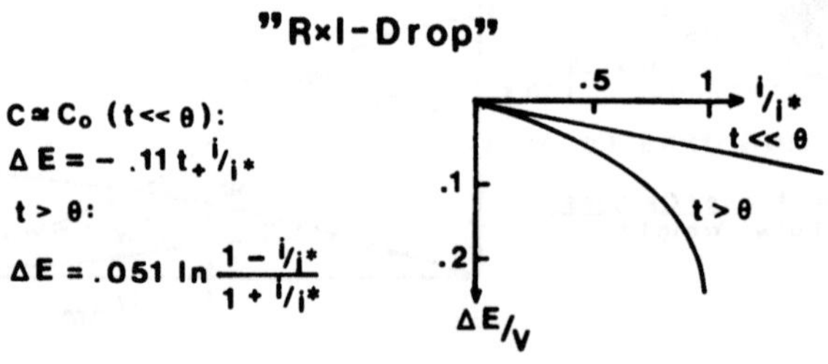


Fig. 4. Potential Loss in Electrolyte

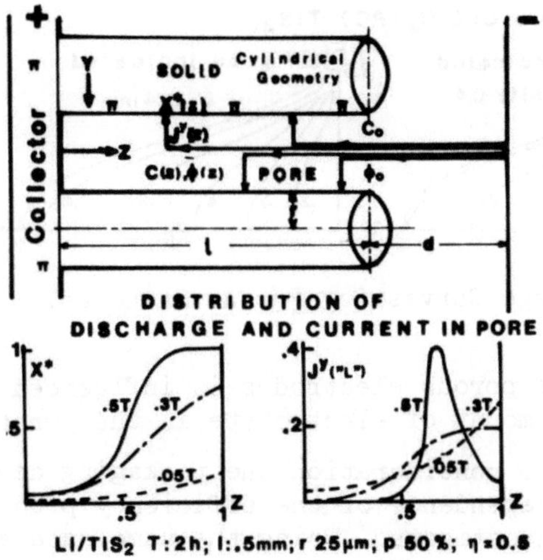


Fig. 5. Porous Cathode Model and Discharge and Current Distribution

The model used for analysis of the behaviour of porous cathodes is shown on Fig. 5. It consists of long cylinders of electronically conducting cathode material in contact with a current collector. The space between the cylinders constitutes the electrolyte filled pores.

The ionic fluxes in the pores are determined by the concentration and electrical potential gradients along the pore. This last adjusts itself to such a value that there is a surplus flux of cations, corresponding to the discharge current, which enters the cathode particle. This in turn determines the value of $X^*$ and thereby the local electrode potential $\pi-\phi(z)$. As $\pi$ is constant in the cathode particle this gives the feedback to adjust the potential in the pore ($\phi$).

The resulting system of partial differential equations are solved by an implicit finite difference procedure developed by Brumleve and Buck (5).

A typical behaviour of a porous cathode is illustrated by the discharge and current profiles shown in the lower part of Fig. 5. Only the outer half part of the cathode has taken part in the discharge and at the end of the discharge the current distribution is very uneven. This behaviour is due to a low electrolyte concentration (not shown) in the inner part of the pore.

Discharge curves calculated for different values of cathode thickness and porosity are shown in Fig. 6. The large influence of these "design parameters" is remarkable and should be kept in mind when porous electrode structures are used for evaluation of unknown electrode materials.

Even if a comparatively large cathode thickness can be used the

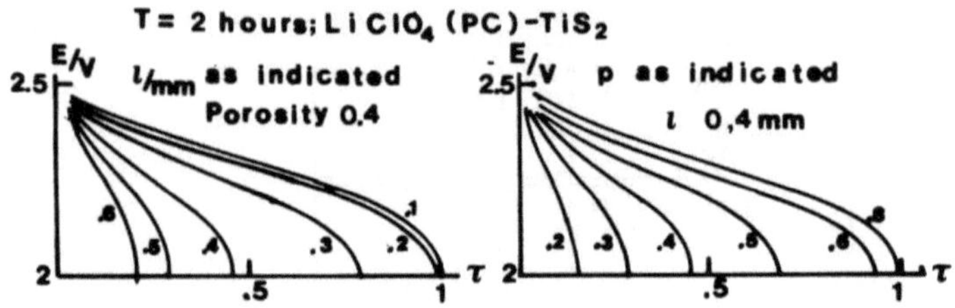


Fig. 6. Discharge Curves for Porous Cathodes. T = 2 hours.

packaging density of porous electrodes is influenced to a considerable degree by the amount of electrolyte in the pores.

Taking this into consideration the packaging density w can be calculated and the dependence of the efficiency product $\eta \cdot w$ on the design parameters investigated. Using the same data as in the calculations for plane electrodes (Fig. 2) the results from the discharge curves above are shown in Fig. 7. An overall efficiency of 25 - 30 % can be estimated for the $Li/TiS_2$ system in agreement with experimental results (6). The improvement over the plane electrode configuration is 1.5 - 2 times. However, for cathode materials with values of $D_s$ inferior to $10^{-8}$ $cm^2$ $s^{-1}$ the improvement due to the porous structure will be larger.

Solid ion conductors used as electrolyte in "porous" electrodes will presumably allow the use of thicker cathodes with a high utilization, but most probable at the expence of a lower discharge voltage. These problems are under investigation.

UTILISATION, PACKAGING DENSITY & η x w

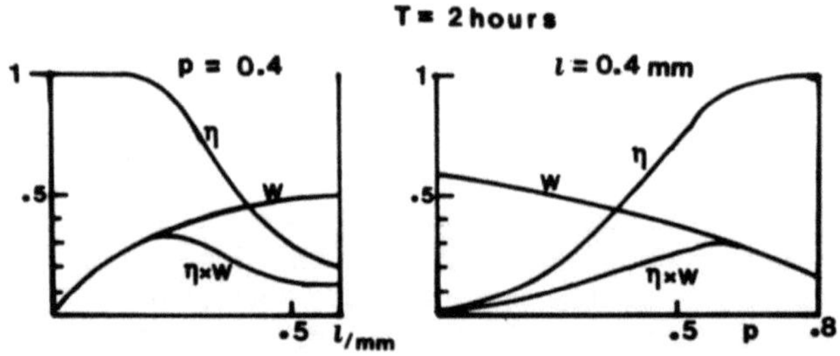


Fig. 7. Results from the Discharge Curves

Acknowledgement. This work was supported by the Danish Government Program for Energy Conservation and (in part) by EC Contracts 315, 316-78 EE-DK, UK.

References:

1) M.Armand Thesis, University of Grenoble (1978)
2) S.Atlung et al. J.Electrochem.Soc. 126 no.8, 1311 (1979)

3) S.Atlung in "Progress in Batteries and Solar Cells" 2, 96 (1979)
4) K.West et al. Extended Abstract no.174, 156' Meeting, the Electrochemical Society. Los Angeles 1979
5) T.R.Brumleve & R.P.Buck, J.Electroanal.Chem. 90. 1 (1979)
6) L.H.Gaines et al. 11' Intersoc.Energy Conversion Conference (1976)

# STRAIN EFFECTS IN INTERCALATION SYSTEMS

W.R. McKinnon and R.R. Haering

Department of Physics
University of British Columbia
Vancouver, B.C., Canada, V6T 1W5

## INTRODUCTION

In this paper we discuss the effects of host lattice strain on the behavior of intercalation systems. In particular, we discuss the role of lattice strain in two kinds of Li intercalation systems: Li/$Li_xMX_2$ cells, where $MX_2$ is a layered structure, such as a transition metal dichalcogenide; and Li/$Li_xMO_2$ cells, where $MO_2$ is a channeled structure, such as a rutile or distorted rutile transition metal dioxide. We shall discuss the main physical principles underlying the strain effects, quoting the theoretical results without proof. Details of our calculations will be published elsewhere.

The voltage of a Li intercalation cell is a direct measure of the energy difference between a Li atom in the Li metal anode and a Li atom located in one of the intercalation sites of the cathode. It is well known that this voltage depends in general on x, the Li content of the $Li_xMX_2$ cathode, which suggests that Li-Li interactions play a role in determining the magnitude of the voltage. It is therefore instructive to analyze the dependence of the voltage, V, on x in terms of a Li lattice gas model and to relate special features of the V(x) curve to the behavior of this lattice gas.[1,2]

Any attempt to describe a Li intercalation cell in terms of a lattice gas model requires, from the outset, a knowledge of the Li-Li interactions. The magnitude and the sign of the various interactions (i.e. nearest neighbor, second neighbor, etc.) play a crucial role in determining phase transitions, ordering, staging etc. of the Li lattice gas. In this paper we derive the strain mediated interactions from first principles.

THEORY

When a Li atom is intercalated into a host lattice, a strain field is established which in turn can interact with other intercalated Li atoms. The resulting strain-mediated Li-Li interaction is conveniently discussed in terms of a short range and a long range part. In general, the short range part of the interaction energy falls off like $r^{-3}$ with distance and is attractive in some directions and repulsive in others. The angular average of the short range interaction is zero. The long range part of the interaction depends on the boundary conditions which apply at the host lattice surface. For a free surface, the long range interaction is attractive, whereas it is repulsive for a clamped surface.

Long range, strain-induced attractive interactions are believed to be responsible for the phase transitions observed in some metal-hydride systems.[3] In addition, the short range forces arising from the host lattice strain field have recently been invoked to account for the staging observed in graphite intercalation compounds.[5]

In discussing the strain induced Li-Li interaction we treat the host lattice as an isotropic elastic continuum with Young's modulus Y and Poisson's ratio $\nu$. Our model can readily be extended to include the effects of elastic anisotropy of the host lattice. The effect of an intercalation atom on the host lattice is described by an appropriate set of body forces, $\underline{f}$. For a particular atom, the net force, $\int \underline{f} dv$, and the torque, $\int (\underline{r \times f}) dv$, both vanish. For a layered lattice, it is found experimentally that the dominant effect of intercalation is a change of the crystallographic c-axis. For a channeled host lattice, the dominant effect is a change perpendicular to the channel direction (Fig. 1). It is convenient to characterize the force field $\underline{f}$ by its first moment tensor $P_{ij}$ defined by

$$P_{ij} = P_{ji} = \int x_i f_j dv \tag{1}$$

$P_{ij}$ is often called the elastic dipole tensor, in analogy with the dipole moment of an electrostatic charge distribution with zero net charge. $P_{ij}$ may be obtained from the observed strains, $\overline{\varepsilon_{ij}}$, associated with the intercalation through the equation

$$\overline{\varepsilon_{ij}} = s_{ijkl} P_{kl} \overline{\rho} \tag{2}$$

where $s_{ijkl}$ is the compliance tensor and $\overline{\rho}$ is the average concentration of intercalated atoms. For layered host lattices ($\overline{\varepsilon}_{33} \neq 0$, $\overline{\varepsilon}_{ij} = 0$ otherwise) we find:

$$P_{33} = Y \frac{(1-\nu)\overline{\varepsilon}_{33}}{(1+\nu)(1-2\nu)\overline{\rho}} \quad ; \quad P_{11} = P_{22} = P_{33} \frac{\nu}{(1-\nu)} \tag{3}$$

For a channeled host lattice ($\overline{\varepsilon}_{11} = \overline{\varepsilon}_{22} \neq 0$, $\overline{\varepsilon}_{ij}$ = otherwise) we find

$$P_{11} = P_{22} = Y \frac{1}{(1+\nu)(1-2\nu)} \frac{\overline{\varepsilon}_{11}}{\overline{\rho}} \; ; \qquad P_{33} = 2\nu P_{11} \tag{4}$$

In terms of $P_{ij}$, the interaction energy between two intercalated atoms located at $\underline{r}_1$ and $\underline{r}_2$ may then be written:

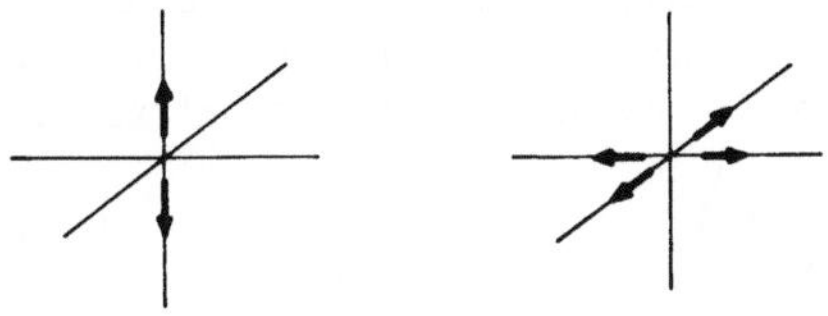


Fig. 1 Force densities used to model the effect of an intercalated atom on the host lattice. (a) Force density appropriate for layered host lattices. (b) Force density appropriate for rutile-related host lattices. In both cases, the diagram assumes $\nu = 0$ (see text).

$$W_{12}(\underline{r}_1, \underline{r}_2) = - P_{ij}\varepsilon^1{}_{ij}(\underline{r}_2) \tag{5}$$

where $\varepsilon^1{}_{ij}(\underline{r}_2)$ is the strain at $\underline{r}_2$ due to the atom at $\underline{r}_1$. It is convenient to divide $W_{12}$ into two parts:

$$W_{12}(\underline{r}_1, \underline{r}_2) = W^{\infty}(\underline{r}) + W^{I}(\underline{r}_1, \underline{r}_2) \tag{6}$$

where $\underline{r} = \underline{r}_1 - \underline{r}_2$. $W^{\infty}(\underline{r})$ is the interaction in an infinite host medium and $W^I$ represents the additional interaction in a finite medium which arises from image forces which are required to satisfy the boundary conditions at the surface. $W^{\infty}(\underline{r})$ may be expressed in terms of the infinite medium Green's function $G_{ik}(\underline{r})$:

$$W^{\infty}(\underline{r}) = P_{ij}P_{mk} \frac{\partial^2 G_{ik}}{\partial x_j \partial x_m} \tag{7}$$

For an isotropic medium, $G_{ik}$ has been calculated by Love.[5] The result is:

$$G_{ik}(\underline{r}) = \frac{(1+\nu)}{8\pi Y(1-\nu)} \left\{ \frac{x_i x_k}{r^3} + (3 - 4\nu) \frac{\delta_{ik}}{r} \right\} \tag{8}$$

Using equations (3), (4), (7) and (8) we then find:

$$W^{\infty}(\underline{r}) = \frac{Y}{8\pi r^3(1-\nu^2)} \left(\frac{\varepsilon_o}{\rho_o}\right)^2 \left[15 \cos^4\theta - 6 \cos^2\theta - 1\right] \tag{9}$$

$$W^{\infty}(\underline{r}) = \frac{Y}{8\pi r^3(1-\nu^2)}\left(\frac{\varepsilon_o}{\rho_o}\right)^2\left[15\cos^4\theta - 6(3+2\nu)\cos^2\theta+3+4\nu\right] \quad (10)$$

The calculation of $W^I$ presents greater difficulties. In general $W^I$ depends on the shape of the host lattice and lacks translational invariance, i.e. depends on where in the host lattice the atoms are located. The only case which has been treated to date corresponds to the interaction between two "dilation centres" ($P_{ij} = P\delta_{ij}$) in a spherical isotropic host lattice (volume V), one of the centres being located at the centre of the sphere.[6] The surface of the sphere is either free or clamped. One then finds:

$$W^I = -\frac{2Y}{v(1-\nu)}\left(\frac{\varepsilon_o}{\rho_o}\right)^2 \qquad \text{(free)} \qquad (11)$$

$$W^I = +\frac{Y(1+\nu)}{v(1-\nu)(1-2\nu)}\left(\frac{\varepsilon_o}{\rho_o}\right)^2 \qquad \text{(clamped)} \qquad (12)$$

We note that the interaction is inversely proportional to the sample volume v and is attractive for a free surface and repulsive for a clamped one. In contrast to $W^{\infty}(\underline{r})$, the angular average of $W^I$ is not zero. Remarkably, the magnitude of $W^I$ is independent of the separation between the two dilation centres.

DISCUSSION

Fig. 4 gives a schematic summary of the Li-Li interaction energies $W^{\infty}(\underline{r})$ in layered and channeled host lattices. It is of interest to estimate the magnitude of these effects. As an example we consider the $Li_xMoO_2$ systems. The $MoO_2$ host lattice is monoclinic (distorted rutile). The intercalation sites lie along chains in the direction of the a-axis (the pseudotetragonal c-axis) spaced by a/4 = 1.4 A°. The chains are arranged in a square lattice separated by $b/\sqrt{2}$ (b = 4.856 A°). During intercalation ($0 \leq x \leq 1$) the a lattice parameter stays nearly constant while b increases by 0.34 A°.[7] If we apply equation (10) with $Y \sim 10^{12}$ dynes/cm$^3$ and $\nu \sim 1/3$ we obtain

$$W^{\infty}(\underline{r}) = \frac{5.6\ kT}{r^3}\ (15\cos^4\theta - 22\cos^2\theta + 13/3)$$

where kT = 25.7 meV and r is measured in A°. Along the chains, the interaction is attractive and equal to -5.4 kT, -0.7 kT, -0.1 kT for first, second and third nearest neighbors respectively. Perpendicular to the chains the interaction is repulsive, the sequence for $\theta = 90°$ being +0.6 kT, +0.2 kT, +0.1 kT... These energies are large enough to produce superlattice ordering or staging effects at ambient temperatures.

The main effect resulting from the finite size of the host lattice is an additional interaction whose range is of the order of the size of the lattice. Because of its long range, this interaction may always be treated by mean field theory in a lattice gas model of intercalation. Its inclusion will produce a term in the free energy of the system which is quadratic in the Li concentration If the average interaction of a Li-Li pair is attractive (see eqn. (11)), the result will be a phase separation into Li-rich and Li-poor regions. Coexisting phases result in a constant voltage for the intercalation cell (Fig. 5). The behavior of the system then depends critically on the coherency stresses at the phase boundaries. These stresses arise because there is in general a lattice parameter mismatch at the phase boundary. Energy losses

2

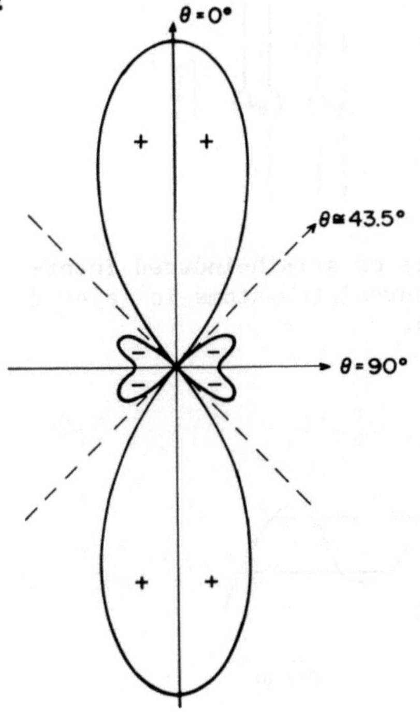


3

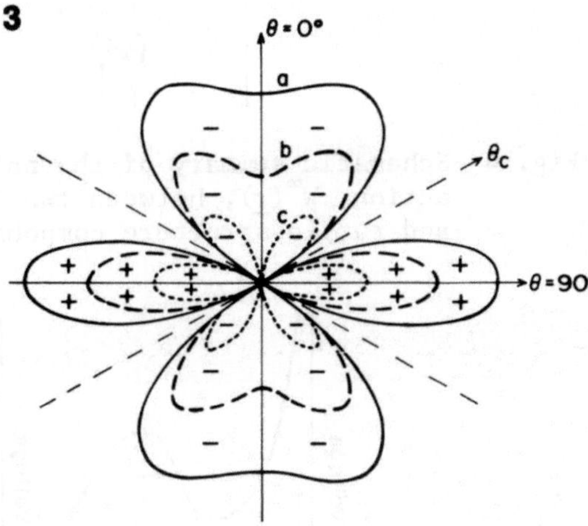


Fig. 2 Polar plot showing the angular variation of the strain-induced interaction $W^{\infty}(\underline{r})$, between two intercalated atoms in layered compounds. The interaction is attractive within a given layer (θ=90°) and repulsive perpendicular to the layer (θ=0°). The critical angle at which the interaction changes sign is about 43.5° for all Poisson ratios.

Fig. 3 Polar plot, similar to Fig. 2, for rutile-related compounds. The interaction is attractive along the intercalation channel (θ=0°) and repulsive perpendicular to the channel (θ=90°). Three curves are shown corresponding to different Poisson ratios, (a) $\nu = \frac{1}{2}$, $\theta_c \sim 60.3°$; (b) $\nu = 1/3$, $\theta_c \sim 61.0°$; (c) $\nu = 0$, $\theta_c \sim 63.4°$.

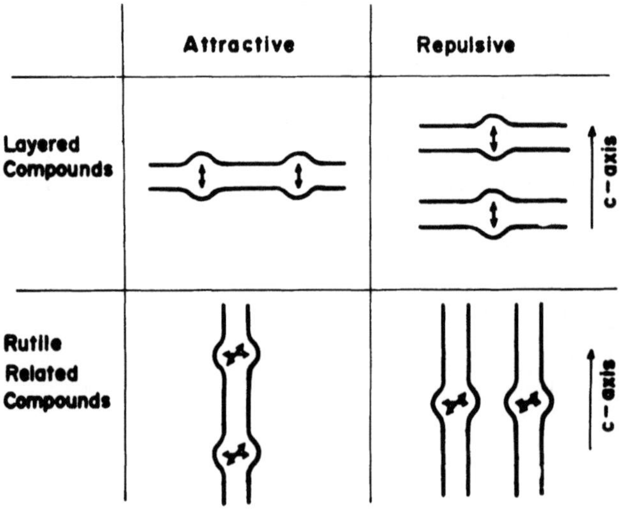


Fig. 4 Schematic summary of the nature of strain-induced interaction, $W^{\infty}(\underline{r})$, between two intercalated atoms in layered and rutile structure compounds.

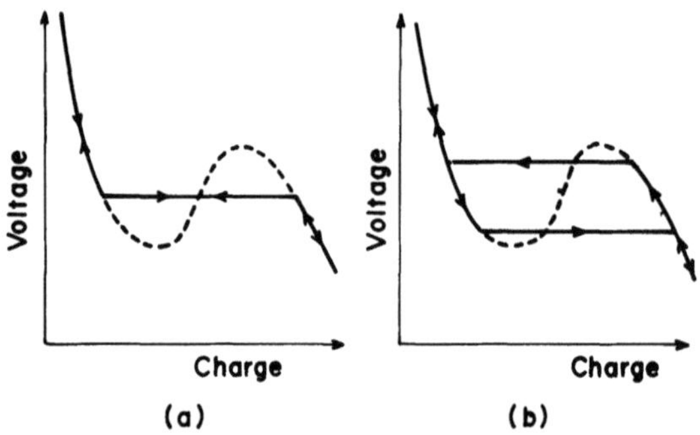


Fig. 5 Schematic discharge curve of an intercalation cell showing condensation due to attractions between the intercalating atoms; (a) Maxwell construction, (b) real system showing hysteresis.

and hence hysteresis in the charge-discharge cycle of the cell may result when the phase boundaries move (Fig. 5). Hysteresis in intercalation systems is to be expected whenever coexisting phases are present.

REFERENCES

1. A.H. Thompson, Phys.Rev.Lett. 40, 1511 (1878).
2. A.J. Berlinsky, W.G. Unruh, W.R. McKinnon and R.R. Haering, Solid State Comm. 31, 135 (1979).
3. H. Wagner, Hydrogen in Metals I, G. Alefeld and J. Volkl Ed. Springer Verlag, Berlin (1978).
4. S.A. Safran and D.R. Hamann, Phys.Rev.Lett. 42, 1410 (1979).
5. A.E.H. Love, A Treatise on the Mathematical Theory of Elasticity, 4th Ed., Dover, New York (1944).
6. G. Liebfried and N. Breuer, Point Defects in Metals I, Springer Verlag, Berlin (1978).
7. U. Sacken and R.R. Haering (to be published).

# AMORPHOUS CATHODES FOR LITHIUM BATTERIES

M. Stanley Whittingham, Russell R. Chianelli and Allan J. Jacobson

Corporate Research
Exxon Research and Engineering Company
P. O. Box 45, Linden, New Jersey 07036 U.S.A.

## INTRODUCTION

Interest in ambient temperature secondary lithium batteries has focused recently on the use of crystalline layered compounds as cathode materials (1). In particular the dichalcogenide, $TiS_2$, has been found to possess many of the characteristics desired of a secondary electrode, such as a continuous non-stoichiometric phase, $Li_xTiS_2$ for $0<x<1$, high electron conductivity, and high free energy of reaction. Many of the other dichalcogenides show similar but generally inferior electrochemical characteristics. However, amongst the group VIB dichalcogenides, crystalline molybdenum disulfide only intercalates about 0.1 Li/Mo and the tungsten compounds react directly to the lithium chalcogenide, $Li_2S$ or $Li_2Se$. This behavior is just that expected from thermodynamic calculations. Ideally to achieve optimum energy storage capacity, either volumetric or gravimetric, reaction with more than one alkali metal per transition metal is desired without breaks in the discharge curve as observed with $VSe_2$. Although some crystalline trichalcogenides such as $TiS_3$, $MoO_3$ and $NbSe_3$ have been studied only one, $NbSe_3$, shows significant reversibility under deep discharge conditions and it contains heavy, expensive and toxic components; moreover, its gravimetric energy density under load is less than that of $TiS_2$. This paper discusses some initial work on a new class of cathode materials, amorphous transition metal chalcogenides. As examples, the properties of $MoS_2$, $MoS_3$ and $V_2S_5$ will be described (2).

## SAMPLE PREPARATION

Amorphous $MoS_2$ was prepared (2, 3) by adding with stirring $MoCl_4$ and $Li_2S$ in stoichiometric proportions to dry tetrahydrofuran. A black precipitate separates out, and the LiCl formed is removed by repeated washing with tetrahydrofuran and ethyl acetate. The $MoS_2$ formed was then dried of excess solvent by heating in vacuo at 150°C for 5 hours; this product loses less than 5 wt.% on heating to 500°C, and this weight loss is possibly associated with a little excess sulfur

Molybdenum trisulfide was prepared by the decomposition, thermal or chemical, of ammonium thiomolybdate. This in turn was prepared by passing hydrogen sulfide through a solution of ammonium paramolybdate in aqueous ammonia at ambient temperature for four hours. Red crystals of $(NH_4)_2MoS_4$ separate out during the reaction. The crystals were removed by vacuum filtration, washed and dried. Their composition was checked by both thermogravimetric and x-ray analysis. The thermogravimetric data is shown in Fig. 1. The weight loss at the center of the intermediate plateau, 325°C, was 26.24% which is within experimental error of that expected, 26.16%, for formation of $MoS_3$. The final product of the thermal decomposition at 1000°C was $MoS_2$ as shown both by x-rays and the overall weight loss; observed 38.88%, calculated 38.48%. Large samples of $MoS_3$ were prepared from $(NH_4)_2$ $MoS_4$ either on heating in helium at 300°C, or by acid treatment at ambient temperatures (2).

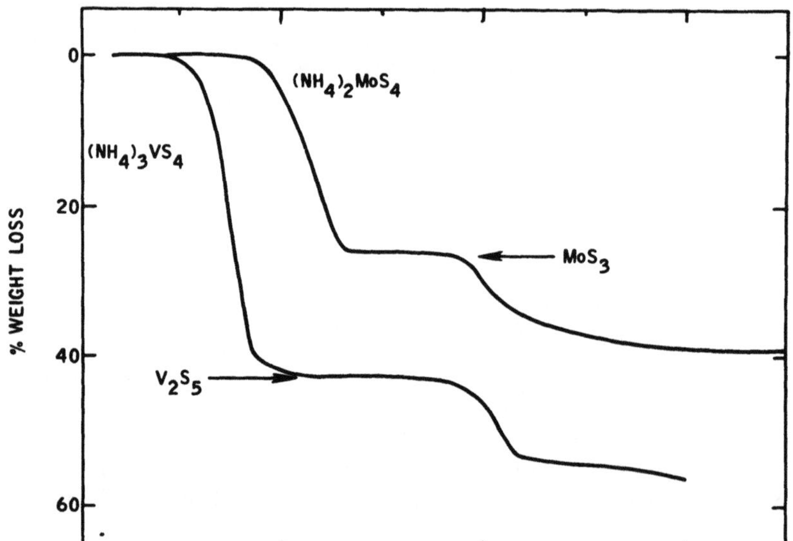


Fig. 1. Thermogravimetric Analysis of $(NH_4)_3VS_4$ and $(NH_4)_2Mo_{S4}$ at 10°C/min.

Divanadium pentasulfide was prepared in an analogous way through the decomposition of $(NH_4)_3VS_4$ (2). The thermogravimetric data is compared with that of the molybdenum compound in Fig. 1. The first weight loss corresponds to the formation of $V_2S_5$, 44.5% observed and 43.8% calculated, and the second to the formation of $VS_{1.53}$. Large samples of $V_2S_5$ were prepared by heating $(NH_4)_3VS_4$ in helium at 275°C.

REACTIONS WITH ALKALI METALS

All the samples of the amorphous sulfides were reacted with a 1.6 molar solution of n-butyl lithium in hexane; to slow down the reaction the solid phase was initially mixed with an excess of hexane. Molybdenum disulfide whether crystalline or amorphous was found to consume the equivalent of 1.5 Li/Mo; molybdenum trisulfide reacted with four lithium and divanadium pentasulfide with six lithium. All the x-ray diffraction patterns indicated amorphous products, except for the crystalline $MoS_2$ where there was some evidence for the formation of lithium sulfide. In the case of molybdenum trisulfide, reaction with sodium and potassium naphthalide was also carried out; both sodium and potassium were consumed to the extent of 4 alkali per molybdenum. Thus in all cases these reactions are suggestive of reduction of the transition metal to the +2 oxidation state. That $MoS_2$ did not react with two lithiums is probably a kinetic effect, possibly related to blockage of the surface with the product of reaction in the crystalline case; also thermodynamic factors, such as too low a free energy of reaction for n-butyl lithium, cannot be ruled out.

The electrochemical measurements were made using Princeton Applied Research Potentiostats in the constant current mode and with voltage limiting devices. Cathodes were prepared by hot pressing the sulfide material admixed with Teflon into an expanded metal grid. These cathodes were then surrounded by polypropylene separators and a sheet of lithium and immersed in a 2 molar solution of lithium perchlorate in dioxolane.

The data for molybdenum disulfide is shown in Figs. 2 & 3 (4). In Fig. 2 the capacities of $MoS_2$ cells are shown as a function of the temperature at which the $MoS_2$ was heated. Very clearly the low temperature $MoS_2$ has a substantially larger capacity for electrochemical reaction with lithium than the high temperature crystalline $MoS_2$. Whereas the two lowest temperature samples show a smoothly varying voltage/composition profile, the break in the curve of the 400°C sample is suggestive of the presence of two phases. In addition x-ray diffraction studies indicated that the amorphous $MoS_2$ found at low temperatures begins to crystallize around 300°C but is not complete until 800°C. Also a very large increase in the surface area, found on heating the material to around 300°C, is suggestive of a radical reorganization of the crystal structure. It therefore appears that $MoS_2$ exists in two bonding modifications, the low temperature form

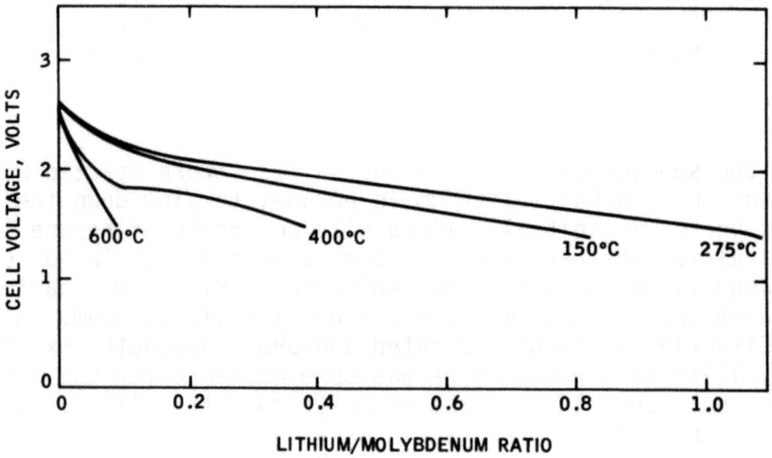


Fig. 2. Effect of Heat Treatment on $MoS_2$ in Lithium Cells (from 4)

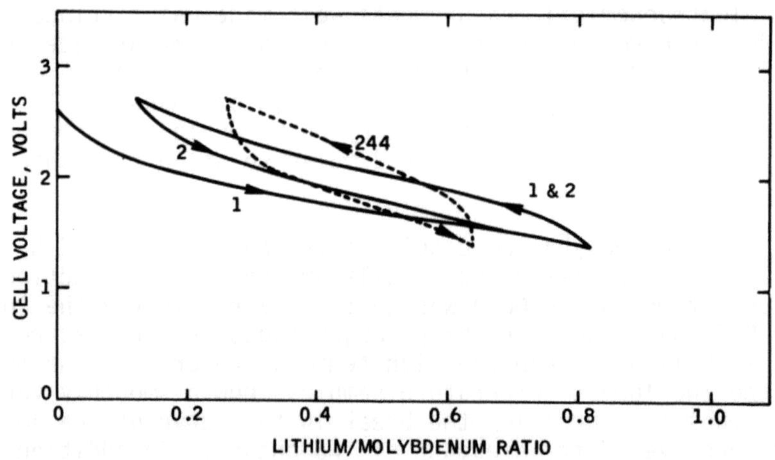


Fig. 3. Cycle Behavior of 150°C Dried $MoS_2$ in Lithium Cells (from 4)

being much the more active in electrochemical cells. This form of $MoS_2$ is also highly reversible as shown for the 150°C treated sample in Fig. 3 (4). Even after 244 discharge/charge cycles the capacity exceeded 50% of that of the 2nd discharge. The 400°C sample showed similar high reversibility but at a much lower capacity, 0.2 - 0.1 Li/Mo. The sloping discharge/charge data of Fig. 3 is very similar to that exhibited by the $Li/TiS_2$ reaction (1), suggesting that a similar non-stoichiometric reaction is taking place. Confirmation of the reaction type will have to wait elucidation of the bonding within the $MoS_2$ and $Li_xMoS_2$ phases.

The discharge behavior of $MoS_3$ in lithium cells is also effected by heat treatment as shown in Fig. 4. A large increase in surface area is also observed in this instance, and presumably the highest temperature samples are the same in both cases. The degree of reaction with lithium of the $MoS_3$ in these cells is only slightly less than found with n-butyl lithium; the difference can be attributed to polarization and possibly electrical resistance effects. The precise degree of reaction is dependent on the method of preparation of the molybdenum trisulfide, as is also the ease of reversibility of the reaction; samples formed by thermal decomposition of the thiomolybdate tend to have slightly lower initial capacities but higher capacities after repetitive cycling than samples formed by chemical decomposition. The rate capability of $MoS_3$ is indicated in Fig. 5. This data was collected by discharging at the highest rate, allowing the cell to rest for 10 minutes, then discharging at a lower rate; this was repeated until the lowest rate was reached (5). Fig. 5 indicates that $MoS_3$ can sustain several $ma/cm^2$ and could make an intermediate rate battery; it does not, however, at the moment have the high rate capability of $TiS_2$. The addition of the conductive diluent carbon did not significantly effect the capacity for lithium on the first discharge.

Electrochemical tests in a sodium cell confirmed the activity predicted by the chemical studies. A cell was constructed as for the lithium studies but as electrolyte a 1 molar solution of $Na(NC_4H_4)Et_3B$ in dioxolane was used; this is one member of a new class of electrolytes (6). The discharge behavior of this cell is compared with that of the lithium cell in Fig. 6. The capacity found electrochemically is significantly less ∿1 Na/Mo, than that found chemically; this is probably associated with the 1.30 volt cut-off used in the cell tests. In initial tests these sodium cells were found to be reversible but not to the same degree as for lithium; further studies are underway to ascertain the inherent reversibility of the molybdenum trisulfide cells.

The cycling data obtained for $V_2S_5$ cells in lithium cells is shown in Fig. 7. The discharge data are seen to be smoothly varying over the whole composition range and there is a definite though broad

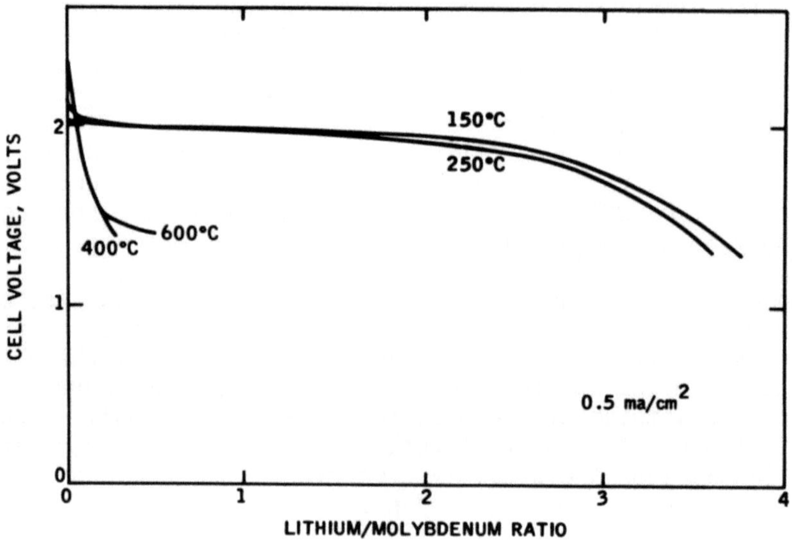


Fig. 4. Effect of Heat Treatment of $MoS_3$ in Lithium Cells ($0.5\ ma/cm^2$)

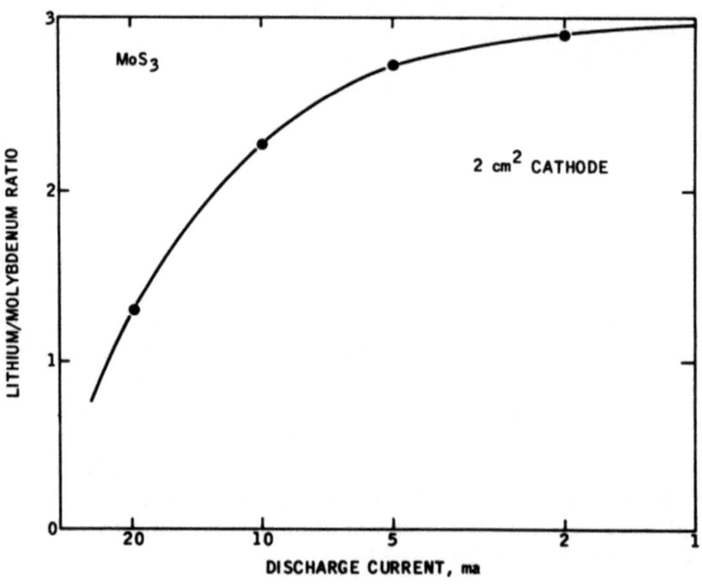


Fig. 5. Effect of Current Drain on Capacity of Li/$MoS_3$ Cells

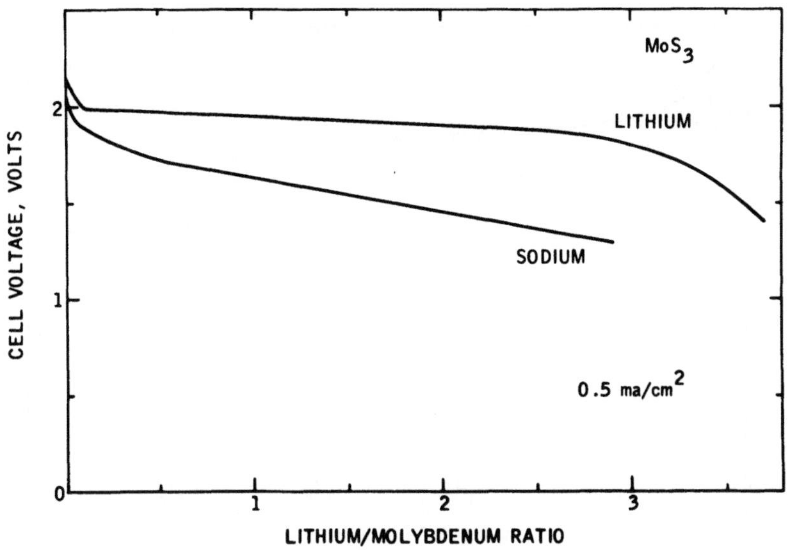


Fig. 6. Reaction of $MoS_3$ in Lithium and Sodium Cells

end point at the end of the discharge; addition of carbon made little difference to the discharge capacity. Increasing the discharge rate from 0.5 ma/cm2 to 2.5 ma/cm2 cut the capacity by less than 20%; even at 8 ma/cm2 the capacity was above 50% of the 0.5 ma/cm2 capacity indicating that just as for $MoS_3$ moderate discharge rates could be obtained in batteries. As Fig. 7 shows there is a marked reduction in capacity on the second discharge cycle. The capacity continues to decrease with repeated cycling but eventually levels out to about one third of the initial capacity, ~0.8 Li/V, as shown in Fig. 8. The fall in capacity is accompanied by a noticeable increase in the voltage profile suggesting that the cathode material is changing in composition; the mean discharge voltage is 1.96, 2.05, 2.12, and 2.28 volts on cycles 1, 2, 12 and 30. The discharge/charge curve separation also changes, markedly decreasing - 340, 250, 130, and 100 mV on the above cycles; this again is suggestive of cathode material changes. This phase could conceivably be an amorphous form of $VS_2$, but there is no definitive structural evidence available yet.

## DISCUSSIONS AND CONCLUSIONS

The higher capacity of the amorphous over crystalline molybdenum disulfide is presumably associated with the disordered structure which prevents the decomposition to lithium sulfide found for the latter (7). Ongoing structural studies on the di- and tri-sulfides using EXAFS and radial distribution studies (8) clearly show that there are marked differences between the amorphous and crystalline

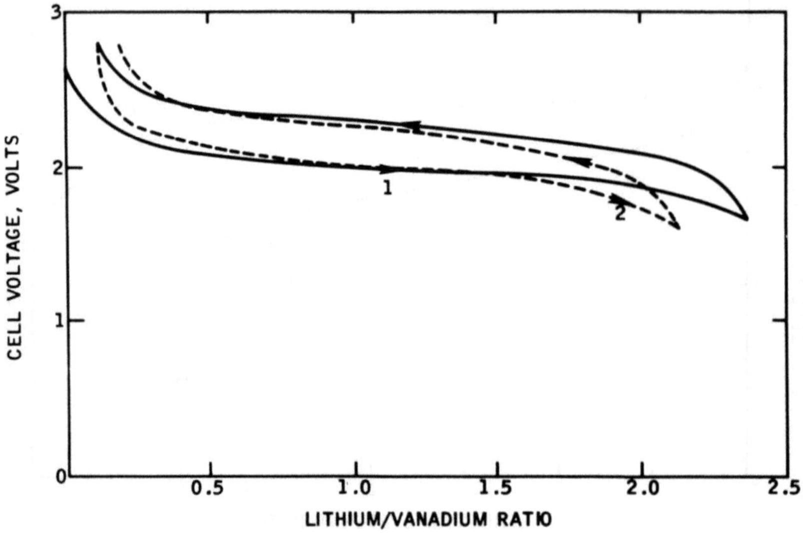


Fig. 7. Early Cycle Behavior of $V_2S_5$ in Lithium Cells

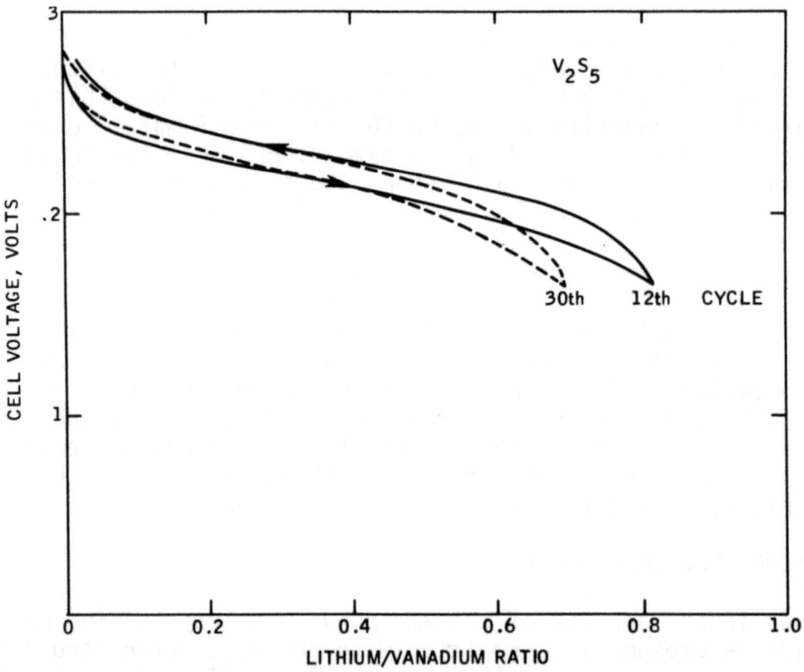


Fig. 8. Later Cycle Behavior of $V_2S_5$ in Lithium Cells

compounds. In both amorphous molybdenum compounds, and in the analogous tungsten materials, there is clear evidence for a Mo-Mo bond of 2.8Å length just as in many molecular complexes. Ongoing investigations are aimed at determining what happens to this bond during reaction with lithium, and the nature of the sulfur bonding. In the case of $V_2S_5$, a possible explanation for the apparent behavior is based on the microcrystalline model of Diemann (9); this model would allow the removal of edge sulfur atoms, possibly as lithium sulfide, followed by polymerization giving an amorphous material of composition approaching $VS_2$. Additional data is now being collected on amorphous $VS_2$ (3), formed as for $MoS_2$, to confirm this hypothesis. This material behaves very differently to crystalline $VS_2$, where a number of intermediate phases are found and the capacity is less than 0.5 Li/V (10).

The energy storage capacities of these amorphous cathode materials is very high. Thus on primary discharge to 1.4 volts $V_2S_5$ reacts with five lithium at a mean voltage of 1.93 volts giving an energy density of 1 kW hr/kg; this drops to around 300 W hr/kg after extended cycling. For $MoS_3$ the initial energy density on discharge is also 1 kW hr/kg. The corresponding theoretical values, based on the n-butyl lithium reactions, are 1.2 and 1.1 kW hr/kg; these may be compared with 0.48 and 0.8 for $TiS_2$ (1) and $V_6O_{13}$ (11) respectively. Of these $TiS_2$ is the only cathode that can sustain extended cycling at high rates, ∿10 ma/cm$^2$, and has the highest reported actual energy density after extensive cycling. Whether these amorphous sulfides, or corresponding oxides, can show the same characteristics, whilst maintaining their initial high energy densities, has still to be proved.

REFERENCES

1.) M. S. Whittingham, Progress in Solid State Chemistry 12, 41 (1978).
2.) A. J. Jacobson, R. R. Chianelli and M. S. Whittingham, U.S. Patents 4,144,384 and 4,166,160 and J. Electrochem. Soc., Mat. Res. Bull., and J. Less Common Metals, in press.
3.) R. R. Chianelli and M. B. Dines, Inorg. Chem., 17, 2758 (1978).
4.) A. J. Jacobson, R. R. Chianelli and M. S. Whittingham, J. Electrochem. Soc., 127, (1980).
5.) B. M. L. Rao devised this technique.
6.) L. P. Klemann and G. H. Newman, U.S. Patent, 4,117,213.
7.) M. S. Whittingham and F. R. Gamble, Mat. Res. Bull., 10, 363 (1975).
8.) S. P. Cramer, K. Liang et al., in press.
9.) E. Diemann and A. Müller, Z. anorg. allg. Chem., 444, 181 (1978).
10.) D. W. Murphy, J. N. Carides, F. J. DiSalvo, C. Cros and J. V. Waszczak, Mat. Res. Bull., 12, 825 (1977).
11.) D. W. Murphy, P. A. Christian, F. J. DiSalvo, and J. N. Carides, J. Electrochem. Soc., 126, 497 (1979).

# A STUDY OF ELECTROINTERCALATION AND DIFFUSION OF SILVER IN TANTALUM DISULPHIDE USING SOLID STATE CELLS

F. Bonino*, M. Lazzari* and C.A. Vincent

Department of Chemistry, University of St. Andrews, St. Andrews, Fife, Scotland

## INTRODUCTION

The theoretical and practical importance of layered dichalcogenides as hosts for the intercalation of atomic and molecular species has recently been pointed out by various authors. In view of possible applications, a screening of the systems based on the determination of the diffusion coefficient of the relevant species seems useful. In this respect electrochemical techniques offer considerable advantages, primarily because of the short time scale of the experiments.

A comparison of two experimental methods for the determination of the diffusion coefficients in solid state cells was the main aim of the present study. The methods were applied to a model system, namely to the diffusion of silver in tantalum disulphide. The study covered a range of composition in which no phase changes occur.

## EXPERIMENTAL TECHNIQUE

Tantalum disulphide was prepared by reacting the elements in the stoichiometric ratio in evacuated quartz ampoules at 900°C for 6 days and then slowly cooling to room temperature. X-ray powder diffraction patterns indicated that the material was mostly 2H-$TaS_2$ with traces of 1T-$TaS_2$ and 4H-$TaS_2$.

The iodotungstate silver solid electrolyte of composition $0.2Ag_2WO_4 \cdot 0.8AgI$ was used. It was prepared in a vitreous state by

*Permanent address: Centro Studio Processi Elettrodici del C.N.R., Politecnico di Milano, Milan, Italy.

melting in a pyrex tube the mixture of the two silver salts and by quenching the melt between two metal plates of high thermal capacity. The glass disks obtained were then finely ground: the powdered electrolyte was used to assemble the cells. The latter were formed by pressing at 75-100 MPa the various components in a suitable composite die (1). In the final cylindrical pellet the working electrode ($TaS_2$) was in the centre, surrounded first by the electrolyte and then by the annular reference electrode (Ag-electrolyte mixture). On top of this pellet was pressed first a layer of electrolyte and finally a layer of silver to act as counter electrode. This configuration, schematically shown in Fig. 1, is optimized as far as the electric field distribution and the ohmic drop are concerned.

METHOD 1

In this technique a short galvanostatic pulse was applied to the cell and the voltage relaxation was recorded after switching off the current. A similar method has been applied to several systems and in particular for determining the diffusion coefficient of $Li^+$ and $Na^+$ in $TiS_2$ (2) and $TaS_2$ (3).

The relationship between overvoltage $\eta$ and time during the relaxation is given (for our cell configuration and a one electron process) by

$$\eta = \frac{RT}{F} \ln \left( 1 + \frac{i\tau}{F (\pi Dt)^{1/2} c_o} \right) \qquad (1)$$

where: i is the current density (A $cm^{-2}$); $\tau$ the pulse duration (s); $c_o$ the initial equilibrium concentration of silver in $TaS_2$ (mol $cm^{-3}$); D the diffusion coefficient of silver ($cm^2\ s^{-1}$); and the other symbols have their usual meaning. Eq. 1 is deduced assuming that the recovery of the voltage is governed by the mass transport of the metal within the cathode. Furthermore, the current pulse must be short, i.e. it must approximate to an instantaneous planar source at the electrode/electrolyte interphase. A plot of $[\exp(\eta F/RT) - 1]$ vs $t^{-1/2}$ should be a straight line with slope

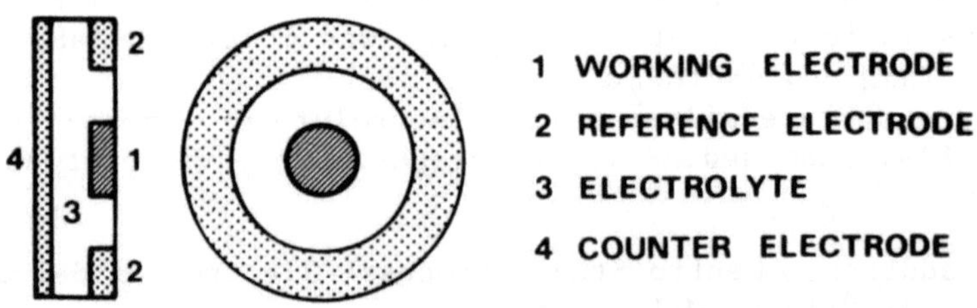


Fig. 1. Schematic diagram of the cell

$$\frac{i\tau}{F\ (\pi D\ )^{1/2} c_o}$$

from which the diffusion coefficient can be obtained.

METHOD 2

In the second technique a galvanostatic pulse of short duration (<1 s) in the range 100-500 μA $cm^{-2}$ was applied to the cell and the resulting overvoltage/time curve was analyzed. This technique has been applied by Bottelberghs (4) to the study of the diffusion of silver in solid $Na_2WO_4$ and by Scholtens (5) to the study of the diffusion of silver in $Ag_xV_2O_5$. By means of operational amplifiers the open circuit voltage of the cell could be partially offset so that the highest sensitivity range (0-10 mV) of a digital signal analyzer could be used to record the transient. The content of the memory was transferred to a punched tape for the subsequent calculation steps. The diffusion coefficient of silver in $TaS_2$ can be deduced from the following equation (4):

$$\eta = \frac{RT}{F} \ln \left(1 + \frac{2\ i\ t^{1/2}}{c_o\ F\ \pi^{1/2} D^{1/2}}\right) \qquad (2)$$

where the symbols have the same meaning as in Eq. 1. Eq.2 is applic-

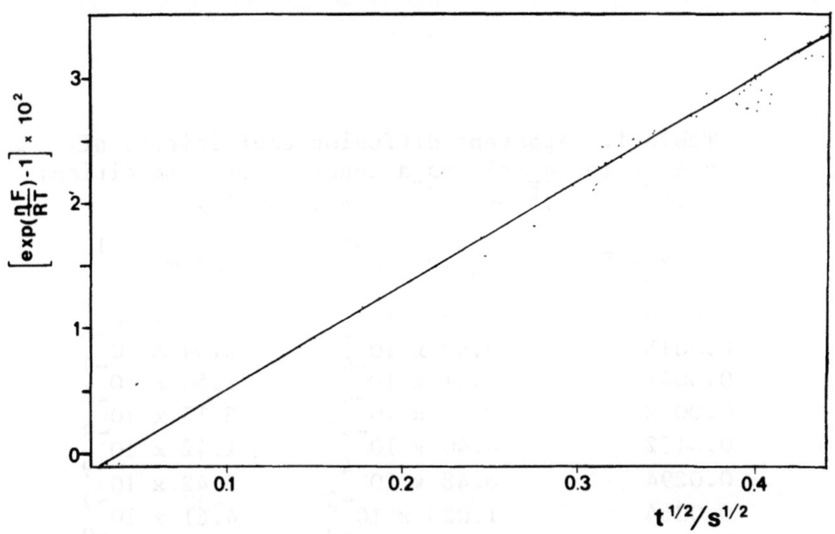


Fig. 2. Plot $[\exp(\frac{\eta F}{RT}) -1]$ vs $t^{1/2}$ for a typical experiment according to Method 2; 48.3°C; 0.4 mA $cm^{-2}$; pulse duration 0.2 s; Co $0.67 \times 10^{-3}$ mol $cm^{-3}$

able if $RT/F >> \eta$ and the electromigration of silver can be ignored. Plotting $[\exp(\eta F/RT) - 1]$ vs $t^{1/2}$ a straight line with a slope $2i/c_oF\pi^{1/2}D^{1/2}$ is obtained, from which the diffusion coefficient can be calculat In Fig. 2 the plot of $[\exp(\eta F/RT) - 1]$ vs. $t^{1/2}$ foı a typical experiment is reported.

## CONCLUSIONS

Only preliminary conclusions can be given in the present paper; a more detailed analysis of the results obtained will be presented elsewhere.

As far as a comparison between the methods is concerned, we found that the results obtained by Method 2 were less influenced than those obtained by Method 1 by experimental factors such as the pulse duration and the current density. These parameters played a more important role at low concentration of silver; as the concentration increased, the results obtained by the two methods tended to converge.

The major problem in this kind of determination is clearly the evaluation of the true area of the interphase electrode/electrolyte. We compared the results obtained with cathodes formed by pure $TaS_2$ and by $TaS_2$ mixed with the electrolyte (1:1 by weight). With the latter configuration the apparent values of D were reproducibly higher by a factor of 2.3-2.4 than with pure $TaS_2$. The difference is limited and this result suggests that diffusion experiments per-

Table 1. Apparent diffusion coefficients of silver in $Ag_xTaS_2$ as a function of composition; 48.3$^oC$; 0.4 mA cm$^{-2}$; pulse 0.2 s

| x in $Ag_xTaS_2$ | $c_o$/mol cm$^{-3}$ | D/cm$^2$ s$^{-1}$ |
|---|---|---|
| 0.0010 | $0.29 \times 10^{-4}$ | $2.33 \times 10^{-6}$ |
| 0.0015 | $0.43 \times 10^{-4}$ | $1.04 \times 10^{-6}$ |
| 0.0042 | $1.21 \times 10^{-4}$ | $1.50 \times 10^{-7}$ |
| 0.0072 | $2.07 \times 10^{-4}$ | $5.17 \times 10^{-8}$ |
| 0.0152 | $4.40 \times 10^{-4}$ | $1.42 \times 10^{-8}$ |
| 0.0294 | $8.48 \times 10^{-4}$ | $5.42 \times 10^{-9}$ |
| 0.0354 | $1.023 \times 10^{-3}$ | $4.61 \times 10^{-9}$ |
| 0.0475 | $1.373 \times 10^{-3}$ | $3.05 \times 10^{-9}$ |
| 0.0596 | $1.723 \times 10^{-3}$ | $2.17 \times 10^{-9}$ |

Table 2. Activation energy for the diffusion process of silver in $Ag_xTaS_2$ as a function of composition. Temperature range 48-92 °C.

| x in $Ag_xTaS_2$ | $c_o$/mol $cm^{-3}$ | $E_a$/kJ $mol^{-1}$ |
|---|---|---|
| 0.0039 | $0.111 \times 10^{-3}$ | 48.5 |
| 0.0160 | $0.461 \times 10^{-3}$ | 45.7 |
| 0.0384 | $1.111 \times 10^{-3}$ | 48.2 |
| 0.0505 | $1.460 \times 10^{-3}$ | 48.3 |
| 0.0566 | $1.635 \times 10^{-3}$ | 50.7 |

formed on solid state cells in reasonably close experimental conditions are directly comparable.

In Table 1 the values of apparent diffusion coefficient of silver in $Ag_xTaS_2$ as a function of composition, determined by Method 2, are reported. They show a regular decrease when the concontration, i.e. the occupancy of available sites within the Van der Waals gap, is increased. However, the activation energy for the diffusion shows no significant changes within the range of composition investigated (Table 2). As expected, the apparent diffusion coefficients in the highly diluted range here investigated are of several orders of magnitude higher than those found for silver in a onesional tunnel material such as $Ag_xV_2O_5$ (5).

ACKNOWLEDGEMENTS

We thank the Science Research Council, British Council and Consiglio Nazionale delle Ricerche for support.

REFERENCES

1. G. Razzini, unpublished results.
2. D.A. Winn, J.M. Shemilt and B.C.H. Steele, Titanium disulphide: a solid solution electrode for sodium and lithium, Mat. Res. Bull. 11:559 (1976).
3. S. Basu and W.L. Worrell, Electrochemical determination of the chemical potential of lithium in $Li_xTaS_2$ at 300°K, in Proc, Symp. on Electrode Materials and Processes for Energy Conversion and Storage, ECS (1977).
4. P.H. Bottelberghs, Thesis, University Utrecht (1976).
5. B.B. Scholtena, Diffusion of silver in silver vanadium bronzes, Mat. Res. Bull. 11:1533 (1976)

# SECTION III

# STUDY GROUP REPORTS

## STATE-OF-THE-ART AND NEW SYSTEMS

This study group had the responsibility of discussing and evaluating the state-of-the-art in battery development and projecting what new systems should be studied as candidates for future development. The systems studied are typically oriented towards four general areas of application:

- Load Levelling
- Electric Vehicles
- Portable Applications
- Solar and Wind Energy Storage

The first of these applications is predominantly a U.S. oriented application while the other three are more universal in their applicability. Performance requirements for these applications dictate a need for higher energy density, higher power density, longer cycle life, reliability, lower cost and use of available (nonstrategic) materials.

The state of the art, baseline, performance of battery systems is given below and includes present and projected (1995) performance of systems which are in production, in specialized use, (military, aerospace, etc.) or are near commercialization. Energy densities that are strongly rate dependent are quoted at the 5 hour rate. Where battery system design influences cycle life (eg. plate thickness in $Pb/PbO_2$ cells), a mean value is quoted for the system. The values given in Table 1 are "hard-nosed" and reflect a realistic mean value of energy density, cycle life and turn around efficiency.

The performance values of systems in this table are arranged approximately in order of commercial realization and/or development.

Table 1

Present and Projected Performance of Baseline Systems

| System | | Energy Density WH/kg at 5 hr rate | | Cycle Life at 80% depth of discharge | | Turn Around Energy Efficiency % |
|---|---|---|---|---|---|---|
| | | Now | 1995 | Now | 1995 | |
| $PbO_2 \mid H_2SO_4 \mid Pb$ | Cell | 40 | 55 | 1000 | 1000 | 70 |
| | Battery | 35 | 48 | 1000 | 1000 | |
| NiOOH\|KOH\|Cd | Cell | 35 | 50 | 1500 | 1500+ | 70 |
| | Battery | 35 | 50 | 1500 | 1500+ | |
| NiOOH\|KOH\|Fe | Cell | 25 | 65* | 3000 | 1000+* | 50 |
| | Battery | 25 | 65 | 3000 | 1000+ | |
| NiOOH\|KOH\|Zn | Cell | 60 | 70 | 200 | 500 | 65 |
| | Battery | ? | (50) | ? | (500) | |
| $Cl_2(H_2O) \mid ZnCl_2 \mid Zn$ | Cell | 80-100 | ? | (1000) | 1000+ | 50-70 |
| | Battery | ? | ? | ? | ? | |

* Extra energy density acquired at the expense of cycle life

| System | Energy Density WH/kg at 5 hr rate: Now | Energy Density WH/kg at 5 hr rate: 1995 | Cycle Life at 80% depth of discharge: Now | Cycle Life at 80% depth of discharge: 1995 | Turn Around Energy Efficiency % |
|---|---|---|---|---|---|
| $O_2$ \| KOH \| $H_2$ (as a closed system) | 80-100 | 140-200 | - | 20,000 hrs. | - |
| NiOOH \| KOH \| $H_2$ | 60 | 80 | 1000+ | - | 65 |
| $Cl_2$ \| NAFION \| $H_2$ | - | 80-100 | - | - | 70 |
| $O_2$ \| KOH \| Fe | 40-60 | 70 | 500 | 1000 | 50 |
| $O_2$ \| KOH \| Al | | ~800 | Mechanically Rechargeable | | <20 |
| $O_2$ \| KOH \| Zn | 100 | - | 10,000 hrs. | | 50 |
| $O_2$ \| $H_2O$ \| Li | 800-1000 | - | Mechanically Rechargeable | | ~30 |

Values given in parentheses are either preliminary results or are tentative projections.

Recommendations for Research

The lead acid battery system which has been known for more than a century and is by far the most widely used secondary system still has a lack of fundamental understanding of the processes involved in energy storage. Over the last 10-15 years, and in particular the last several years, fundamental studies of the kinetics of discharge of $PbO_2$ have been conducted. Techniques used to investigate the properties and in particular, to elucidate the mechanism of loss of activity during charge and discharge include SEM, energy dispersive X-ray analysis, NMR, X-ray, DTA, DSC, etc. These studies of oxide characterization have typically been covered by many different organizations, each using its own special diagnostic technique. To better assist in understanding the complete picture of the mechanisms involved in $PbO_2$ cycling and degradation, it is proposed that a coordinating entity should be set up to organize and set up the "matrix" of investigative techniques to be focussed on the problem. This entity would be responsible for assisting in the selection of the starting active material, the selection of cycle regimes and most important, the assignment of analytical tasks to the organizations experienced in the specific techniques. A concerted effort like this should have a far better chance of elucidating the mechanism of loss of activity of $PbO_2$ in the positive plate.

$PbO_2$ was taken as the example of a material needing a coordinated approach to fundamental R & D. This idea, however, can and should be extended to cover the negative expander in the Pb electrode and even further to the problems of dendrite formation and morphological change that takes place at zinc electrodes in Ni/Zn systems.

In other areas of aqueous energy storage system research, work on ion exchange membranes such as NAFION should be encouraged, in particular with respect to life and identification of degradation mechanisms in the various redox systems presently in early development.

In new exploratory systems, the opportunities are many and varied and would be extremely lengthy to outline in detail. By way of example, the list below shows some opportunities for research and development in a variety of proposed systems. The listing is arranged in a generalized mechanistic format followed by an example of a known system which approximates the format:

## Exploratory Systems

Nomenclature: D = Displacement or Stripping Electrode; I = Insertion Electrode; S = Solid Electrolyte; A = Amorphous: glass, polymer, composite; L = Liquid; G = Gas; E = Metal Current Collector; R = Redox in Aqueous Electrolyte.

### Batteries

A. Proven Approaches

1. D/L/D : $Ag_2O/KOH/Zn$
2. L/S/L : $(Na_xS + C)/Na\text{-}\beta/Na$
   Dissolved $S(?)/Li\beta''/Li(NH_4)_3$
3. I/L/D : $V_6O_{13}/Li^+$- nonaq/Li
4. I/L/I : $NiOOH/H^+$-aq/Hydride
   $Li_xCoO_2/Li^+$-nonaq/$Li_xTiS_2$

B. Cells with Solid/Solid Interfaces

1. I/A/D : $TiS_2/Li^+$-polymer/Li
   $MoS_2/Li^+$-polymer/Li
2. L,G/A/I : $H_2O$,Air/HUP/Hydride
3. I/A/I : NiOOH/NAFION/Hydride
4. I/A/L : $MoS_2/Na^+$-glass/Na

C. Hybrid Electrolytes

1. I/L,S/L : $MoS_2/Na^+$-aq $Na_3Zr_2Si_2PO_{12}/Na$
2. D´/A/D : $Sm/SmF_2$ $F^-$-glass $CaF_2/Ca$

### Electrolysis/Fuel Cells

(Closed & Open Systems)

A. Liquid Electrolyte (Corrosion)

L,G E/L/E G: HCl,$Cl_2$ $Ti_{1-x}Ru_xO_2$/KOH/Pt $H_2$

B. Solid Electrolyte (Cell Design & Materials Mismatch)

L,G E/S/E,G

Low-T: $H_2O,O_2$ $La_{1-x}Sr_{1+x}CaZrO_6$(?)/NAFION/Pt,$H_2$

High T: $H_2O,O_2$ $La_{1-x}Sr_xFeO_{3-4}$(?)/$ZrO_2$(?)/Ni,$H_2$

Redox Systems

E/R/S/R/E: E/$Fe^{2+}$,$Fe^{3+}$/Framework Hydrate/$Cr^{3+}$,$Cr^{6+}$/E

The above outline gives a shopping list of the many possibilities open for evaluation. In a more generalized sense, the list suggests several areas of study that should be identified for specific attention:

a.) Solid-Solid Interfacial Research

How does one maintain good interfacial contact during cycling? What are the mechanisms of electron and/or ion transfer across such material interfaces?

b.) "Insertion Compounds" Research

What factors control the energy and power densities? What factors influence cycle life?

c.) Electrolyte Research

Research should continue to identify new electrolyte materials that have appropriate voltage windows compatible with low losses at typical environmental operating temperatures.

These areas are discussed in more detail in the accompanying reports from the other study groups.

Throughout the discussions, emphasis was often directed towards enhancing the working relationships between the research and development communities. A coordination such as this was outlined early in this report and it is important to remind ourselves that this theme should continue throughout future studies so as to maximize our strategic options for 1995 and beyond.

Acknowledgement

This report was prepared by M. Armand, J. Broadhead, E. J. Casey, S. M. Caulder, J. B. Goodenough, A. LéMehauté, and E. Voss.

# HIGH TEMPERATURE SYSTEMS

## INTRODUCTION

High temperature systems have been defined for the purposes of this report as those types of rechargeable electrochemical cells or batteries operating at temperatures greater than 100°C, but not excluding cells based on molten salt systems with melting points below 100°C (1). Aqueous based systems (largely excluded by this definition) are discussed elsewhere in this Proceedings (2, 3). There are two main types of high temperature systems that have reached an engineering state of development: first, the Na/S based systems that have been described in this Proceedings by Jones (4) and second, the $Li/MS_x$ based systems which have been described by Vissers (5) and Bélanger et al (6) in this Proceedings. The study group identified four specific cell systems, two from each catagory, and then summarized the performances achieved or projected with these cells as Part I of this report. In such a summary, it is essential to recognize the qualifications and limits associated with the comparative numerical estimates. While the group has made a best effort to ensure the accuracy of this general performance summary, it should not be taken as the final word. Another group might well place a different emphasis on the performance characteristics of these systems based on specific application goals. In Part II of the report, problem areas in the engineering development of these four systems were identified and the relative severity of the problems were evaluated. Finally, promising new research areas were discussed and are listed in Part III.

While research of importance to the development of various cell components and high temperature material problems is underway in many parts of the world, systems development is less widespread. Table 1 summarizes the locations that were identified and the high temperature systems being developed. A comparison of the US and European high temperature battery programs in 1976 has been made by Nelson (7).

## I. CURRENT SYSTEMS OVERVIEW

The results of a comparison of four high temperature systems are summarized in Table 2: two sodium/sulfur solid electrolyte systems, those based on the -alumina ceramic and those based on the hollow sodium-ion conductive glass fibres are covered; two solid electrode molten salt electrolyte systems are discussed, the Li-Al/FeS and the Li-Al/$FeS_2$ and in addition some Li-Si/$FeS_2$ results are included. The systems chosen represent those cells for which considerable research and development effort has been expended. The incentive for this investment has been the prospect of high specific energy (>100 Wh/kg v.s. ~40 Wh/kg for a potential improved lead/acid battery), high specific power (>100 W/kg), an extended cycle life at high depth of discharge (D.O.D.) (>1000), and a lower cost per kWh. Development is mainly directed toward applications in load-levelling (L.L.) for electric utilities and for electric vehicles (E.V.) where the size of the battery modules will have a minimum of 10-20 kWh capacity (8,9). The weight/volume penalty for high temperature batteries of this size or larger due to the need for insulation becomes less significant.

In compiling the performance data in Table 2, the authors of this report have tried to choose conservative but realistic numbers for purposes of giving a comparison of the state-of-the-art of these four systems as it is known to them. In addition, the table has been discussed with several others active in the development of these systems.

### Safety

The comments in Table 2 related to safety, the public image of the high temperature systems, and the environmental impact of the systems require some discussion. It is important to note that high energy density (HED) batteries, when compared to other forms of stored chemical energy, eg. gasoline, T.N.T. and liquid propane, used extensively by the public, are not exceptionally concentrated forms of stored energy. Recent experience has indicated in addition, that the Na/S ceramic batteries, in tests by British Rail (10), and the LiAl/FeS battery, in tests by Argonne National Laboratory (ANL) (8), can undergo massive short

Table 1. High Temperature Systems R & D Centres

| Type | Location | Applications |
|---|---|---|
| Li/MS | U.S.A. | |
| | Argonne National Laboratory | EV, Load-levelling |
| | Eagle-Picher Industries Inc. | EV, Military, load-levelling |
| | Gould Inc. | EV, Load-levelling |
| | Rockwell International | Load-levelling |
| | General Motors Corp. | EV |
| | Sandia Laboratories | Military |
| | KDI Score | Military |
| | CANADA | |
| | Hydro Québec | Load-levelling |
| | Defence Research Establishment Ottawa | Military |
| | GERMANY | |
| | Varta-Batterie AG and Battelle Institute | EV |
| | U.K. | |
| | Admiralty Marine Technology Establishment, Harwell | Military |
| | Shell Laboratories | Military |
| Na/S | U.S.A. | |
| | Ford Motor Company | EV, Load-levelling |
| | General Electric | Load-levelling |
| | Dow Chemical Company | Load-levelling |
| | U.K. | |
| | British Rail | Locomotives |
| | Chloride Silent Power Ltd. | Busses and vans |
| | U.K. A.E.A., Harwell | Busses and locomotives |
| | FRANCE | |
| | Laboratoires de Marcoussis C.G.E. | Basic R & D |
| | SAFT | EV |

Table 1 (Cont'd)

| Type | Location | Applications |
|---|---|---|
| Na/S (Cont'd) | GERMANY | |
| | B.B.C. | EV, Load-levelling |
| | JAPAN | |
| | Yuasa Battery Co. | EV, Load-levelling |
| | Toshiba Electric Co. | EV, Load-levelling |
| | BULGARIA | |
| | Central Laboratory of Electrochemical Power Sources, Bulgarian Academy of Sciences, Sofia | EV, Load-levelling |
| | U.S.S.R. | |
| | Institute of Electrochemistry, Academy of Sciences, Moscow | Basic R & D |
| | PEOPLES REPUBLIC OF CHINA | |
| | Program based on ceramic expertise | EV |

Table 2. The Present High Temperature System Performance

| Factor | | Li-Al(Li-Si)/LiCl-KCl/FeS (a) | Na/S($\beta$-$Al_2O_3$ ceramic) | Na/S (glass) | Li-Al(Li-Si)/ LiCl-KCl/$FeS_2$ |
|---|---|---|---|---|---|
| Open Circuit Voltage, OCV | | 1.35 (1.6) | 1.7 | 1.7 | 1.7 (1.9) |
| Specific Energy $E_{sp}$,Wh/kg @ C/4 (b) | theoretical(5) | 460 (690) | 760 | 760 | 650 (900) |
| | goal | 120/110 (130/115) | 250/150 | 220/ - | 170/150 (190/170) |
| | achieved | 110/73 ( 90/ - ) | 120-180/83 | 80/ - | 90/ - (125/ - ) |
| | | | long cycle life can cut $E_{sp}$ by 50% | | |
| Energy Density Eρ,Wh/l @ C/4 (b) | goal | 400/200 (440/220) | 250/150 | | 525/300 (575/220) |
| | achieved | 240/100 (150) | 200/100 | | 200/ - (220/ - ) |
| Peak Specific Power, Eρ, W/kg @ 50% DOD (b) | goal | 125/100 (125/100) | 125/75 | 200/ - | 200/160 (200/160) |
| | achieved | 95/75 ( - / - ) | 100/40 | 160/ - | 100/ - ( 95/ - ) |
| Cycle Life (c) | goal | 1000(3 yrs)/3000(10 yrs) (1000/3000) | 1000/3000 | 1000/3000 | 1000 (3 yrs)/ 3000 (10 yrs) |
| | achieved | 1000/ - (230/ - ) | 1000-1500 max 300-500 av | not avail. | 650/ - (650/ - ) |
| Operating temperature | | 400-450 | 350-400 | 350-400 | 400-450 |
| Cost, 1979 US ($) dollars/kWh (d) | goal EV | 50-60 | 70 | | 50-60 |
| | goal L.L. | 45-55 | 40 | 32 | |
| | achieved | no meaningful estimate possible. | 100 | - | - |
| Self-discharge, % capacity loss/day (6) | | 1 | 0 | 0 | - |
| Reliability | goal | > 10,000 mean cycles between failure (MCBF) | >10,000 MCBF | >10,000 MCBF | - |
| | achieved | > 1000 cycles on engineering cells, but can't project data to battery | ~1000 MCBF | ~1000 MCBF | - |

Table 2 (Cont'd)

| Factor | | Li-Al(Li-Si)/LiCl-KCl/FeS (a) | Na/S($\beta$-$Al_2O_3$ ceramic) | Na/S (glass) | Li-Al(Li-Si)/LiCl-KCl/$FeS_2$ |
|---|---|---|---|---|---|
| Safety, public image, and environmental impact (e) | | Solid electrodes and self-limiting discharge rates on shorting are positive features; no toxic components, gases or products of an accident. | Special design features necessary to reduce risk due to liquid active materials. | Nature of glass capillary system good safety feature but same problem with liquid active material. | Same feature as FeS system |
| Status of System Development (Sept., 1979) | | Batteries in engineering development state. | Batteries in engineering pilot plant stage of commercialization. | Scale-up to engineering size cells. | Cells still undergoing development, no batteries. |
| Demonstration Batteries | EV | 40 kWh battery tests in 1979, 50 kWh and small modules in 1980. | 10-50 kWh batteries. | | |
| | L.L. | 5-6 MWh battery for the B.E.S.T. facility (mid 1980s). | Have been built and tested. | None | None |

**Notes**

(a) Results in parentheses refer to the Li-Si/LiCl-KCl/$FeS_x$ system.

(b) Results are given for cell/battery.

(c) Cycle-life for EV/load-levelling (L.L.); cycle-life depends on several factors including depth of discharge (D.O.D.), rate of charge and discharge, temperature, amount of overcharge, and duty cycle. Range indicated for typical cell performance, 1979, for severe to modest duty.

(d) See text for comments on cost and availability of critical components.

(e) See text for further discussion of these topics.

circuit, or case rupture, or ceramic rupture without explosion or release of gases. With present insulated container designs, it has been found that the external surface temperature for both systems remains within safe limits despite an internal thermal excursion of >1000°C. Since the operating temperature of these systems, 300 - 500°C, is often seen as a possible drawback to the acceptance of this type of system by the public, some discussion of this point is important. In appreciating the operational significance of high temperature battery systems, it should be remembered that the piston chamber temperature of internal combustion engines is about 800°C and that the exhaust manifold of an operating engine glows cherry red (~650°C). By comparison, even the internal operating temperature of the high temperature batteries is low. The surface temperature of the battery case of the first generation batteries developed by ANL was about 25°C; in fact, the total heat loss will be limited to 200W per 20 kWh module by an advanced design insulated battery container (11). A design criterion for high temperature batteries is that the loss in W be not greater than 1% of the capacity of the system in Wh. This seems to be readily achievable with today's technology and materials. This thermal emissivity is almost equivalent to that of the surface of a human body. It is important in developing a relatively novel technology for a mass market to consider the public safety and environmental impact. A balanced view of the new battery technology relative to the old will then emerge and perhaps avoid the development of possible public resistance based on a mistaken understanding at a later stage. The above surface-temperature concepts are felt to be significant in terms of both real safety and in helping to achieve an acceptable public image of high temperature battery systems. Finally, in the selection of materials used in the systems discussed, there need be no toxic or corrosive materials used. The Li/MS systems make use of solid electrodes in which the maximum rate of discharge in the event of a cell short circuit is self-limited by composition changes at the active material/electrolyte interface i.e. off-eutectic freezing of the electrolyte occurs which blocks the discharge reaction. If the case should rupture in an accident, the molten salt electrolyte merely freezes when it reaches ambient temperature. Although the Na/S system is less ideal from a safety viewpoint, having liquid active material and the potential of $SO_2$ and $H_2S$ gas production, the reaction of liquid sodium with liquid sulfur is not explosive, provided that the temperature remains below 1100°C, due to the formation of blocking solid product films at the Na/S interface. It should not be forgotten that the production of high internal pressures from gas evolution or thermal-excursion leading to explosion, or the production of explosive gases such as hydrogen, are characteristic of the aqueous and organic electrolyte battery systems, (but not of the high temperature systems) that should also be of concern to public safety. These hazards are especially serious in the large battery plants being developed for load levelling. From the viewpoint of safety, especially for the Na/S systems, initial application should perhaps emphasize L.L.

Cost

The cost (to the consumer) item in Table 2 requires some further discussion. Since the Na/S systems have received more development funding than Li/MS systems and therefore have reached a more advanced stage of engineering, the estimate of $100/kWh for Na/S is realistic based on a $75-120/kWh range for a plant producing twenty-five 100 MWh batteries/yr for stationary energy storage applications. The major uncertainty is the cost of production of  -alumina ceramic which ranges from 20% to 50% of the cost of raw materials. Another major uncertainty is the charge controller/monitor system for charge equalization between cells in a series stack. However, the raw materials of the Na/S system are abundant and cheap and even if full production were undertaken, there are unlikely to be any supply problems.

The cost of the Li/MS system is less well defined since there are still uncertain engineering costs in several component areas. The price of lithium as $Li_2S$ is projected as $5/kWh, but is dependent on increases in proven resources. This price assumes that $Li_2S$ will be available from a $Li_2CO_3$ production process for use in discharged state cell assembly. There does not appear to be a serious lithium supply problem since full commercialization at the most optimistic level would require only 15% of US proven (or 5% of proven world) resources by the year 2000 and the increased demand for lithium would bring many additional lithium producers on to the market. The concentration of research on the lithium system in North America and the relative lack of European interest is probably a result of the presence of lithium deposits in Canada and the U.S.A. and the absence of European deposits. The most successful separator for the Li/MS cell is a boron nitride felt. This material is being produced by the Carborundum Co. which is developing a process for the factory fabrication of the felt which will reduce the price from about $\$1000/m^2$ felt in 1979 to $\$10/m^2$ projected for 1986. However, the separator cost is estimated to be $4.9/kWh. Powder substitutes for the BN felt are being studied and could reduce this cost item. The cost of the screens and retainers used is significant, estimated as high as $15/kWh. There are programs to develop screenless cells. The cost of feedthroughs including fabricated ceramic parts (but not hermetic seal feedthroughs) is estimated to be $2-3/kWh. It is too early to make a firm estimate of the total cost of Li/MS batteries, however present projections indicate a cost of $60-80/kWh.

## II. PROBLEM AREAS IN ENGINEERING DEVELOPMENT OF PRESENT HIGH TEMPERATURE SYSTEMS

To obtain perspective on the state of development of the high temperature battery systems, the four current systems of Part I above were discussed and a rating assigned to represent the degree of severity of each problem area identified. The results of this exercise of the high temperature study group are presented in Table 3. The table is not meant to indicate an overall comparative rating between the systems, but to identify and rate the severity of specific problem areas.

### Electrical Feedthroughs

In high temperature systems, the electronic conductivity of the feedthroughs and connections is seriously affected by corrosion. Leakage of electrolyte out (in the Li/MS systems) or of $H_2O$ or $O_2$ in, can cause cell failure. Hermetic seals of light weight, corrosion resistance, and mechanical strength have been designed for both Na/S and Li/MS systems. In the case of the Li/MS system the hermetic seal requirement can be relaxed (with some cost savings) if an inert gas blanket is used inside the battery module. The Na/S hermetic seals, based on an -alumina-aluminum compression joint, have been developed with safety and mass production being prime considerations (12).

### Corrosion

The severity of this problem varies with the function and location of the material in the cell. Corrosion problems at the negative electrodes in Li/FeS cells are slight with the Li-Al alloys, but more serious in the Li-Si alloys. Most common materials such as Ni, Inconel (or other stainless steels), Fe, Cu or Co, are useful for retaining screens, frames, and current collectors (c.c.) at the positive electrode of FeS cells having a charge cut-off potential of 1.7V. However, these materials are rapidly corroded in the $FeS_2$ cathode due to the high sulfur activity present between 1.7V and the $FeS_2$ cut-off potential of 2.1V (13). Molybdenum is one of the few acceptable materials from a corrosion consideration for $FeS_2$ cathode current collectors, but it is a high cost material. Choice of current collector and container material for the sulfur electrode of Na/S cells is a problem due to chemical attack by $Na_2S_4$. A wide range of alloys is possible for the sodium compartment of the cell (11).

Table 3. Problem Areas in Present High-Temperature System Engineering (Sept., 1979)

Degree of Problem
1 severe
2 moderate (a)
3 mild
- not a problem

| Problem Area | Li-Al/LiCl-KCl/FeS | Na/S (Ceramic) | Na/S (Glass) | LiAl/LiCl-KCl/$FeS_2$ |
|---|---|---|---|---|
| Feed-throughs | 2 | 2 | 1 | 2 |
| Corrosion -ve electrode | -(LiAl), 2 (Li-Si) | - | - | -(LiAl), 2 (Li-Si) |
| +ve electrode | 3 | 2 | 2 | 1 |
| container | -- | 3 | 2 | - |
| Separator-Solid Electrolyte | | | | |
| Cost | 2 (E.V.), 1 (L.L.) | 1 | 2 | 2 (E.V.), 1 (L.L.) |
| Reliability | 2 | 1 | 1 | 2 |
| Active Material Retention | 2 | - | - | 2 |
| Critical Active Material Cost and Availability | 2 | - | - | 2 |
| Cell-battery Assembly | 2 | 3 | 2 | 2 |
| Self-discharge (b) | 3 | - | - | 3 |
| Freeze-thaw Cycle | 3 (L.L.), 2 (E.V.) | 2 | 1 | 3 (L.L.), 2 (E.V.) |
| Thermal Management (b) | 3 (L.L.), 2 (E.V.) | 3 (L.L.), 2 (E.V.) | 1 | 3 (L.L.), 2 (E.V.) |
| Reliability | 1 | 1 | 1 | 1 |
| Battery Engineering (b) | 2 | 2 | 2 | 2 |
| Charge Control/ Discharge Control | 3 | 3 | 3 | 3 |
| Safety/Environment (b) | 3 | 2 | 2 | 3 |

(a) Areas of degree 1 and 2 require R & D effort.

(b) Area requires further development.

## Separator-Solid Electrolyte

The reduction of the cost of separator material either by development of the BN felt process or by substitution of a cheaper material is necessary for the commercialization of the Li/MS system. In addition, problems remain in the reproducibility of BN felt separator performance. The solid electrolyte -alumina used in Na/S cells has still not achieved a sufficiently reliable performance. However, pilot plant scale operation, with rigorous quality control in mass production, may provide for more reproducible cell performance than the cells constructed from bench-scale produced ceramic separators. The solid electrolyte accounts for about 50% of the cost of a Na/S cell.

## Critical Active Material Cost and Availability

The materials used in the Na/S system are cheap and available in unlimited supply. However, the lithium cost and supply is less certain. As discussed in Part I above, a rise in the demand for lithium should bring more lithium into production. It is estimated that the lithium cost will account for about 50% of the total Li/MS material cost. If the projected commercialization of the Li/MS system is undertaken in the U.S.A., then by the year 2000, 15% of the known U.S. resources (5% of world resources) would be in use. Lithium recycling could be used to reduce demand in the same manner that lead is presently being recycled. Use of the above proportion of known resources would provide storage batteries for 20 million electric vehicles (EV) and 3,500 utility load-levelling plants of 100 MWh capacity (14).

## Cell-Battery Assembly

The present assembly procedure for Li/MS cells now largely conducted in an inert atmosphere in laboratory glove boxes will not be acceptable in a commercially viable manufacturing process. Some manufacturers are now experimenting with dry room cell assembly. The adaptation of the assembly procedure to a dry room process and the automated injection of active material are both necessary developments to reduce cost and ensure a uniform high cell performance. The assembly procedures are more developed in the Na/S system.

## Self-Discharge

There is no self-discharge in the Na/S system. Results indicate that in the Li-Al/LiCl-KCl/FeS cell, self-discharge at operating temperature occurs at 1% per day (6). Energy required to balance heat loss from the system could also be considered a form of self-discharge for high temperature systems.

## Thermal Management-Freeze-Thaw Cycle

Freeze-thaw cycles have occurred during testing of Li/MS cells and in addition, the cells are often shipped in the frozen state after formation cycling. When these cells have had woven fabric BN separators no damage has been evident. However, BN felt separators may be more susceptible to damage and work is required in this area especially for EV applications. Na/S (ceramic) cells can be cooled with care when in the charged state, but serious damage results to the glass fibre type Na/S cells if rapid temperature change occurs.

A more general battery engineering effort is required in the area of thermal management for all the high temperature systems (11). Insulation jackets, heat exchangers, and efficient utilization of the resistive heating is an important area for development. The present state-of-the-art of jacket design can be appreciated from the following. It is estimated that if no heat energy were added to the 40 kWh Mark IA EV battery built by Eagle-Picher for A.N.L. in 1979, it would take about 24 hours before the cells would reach a temperature below 352$^{o}$C (the melting point of the eutectic LiCl-KCl electrolyte). Thermal management should be a more easily solved problem in high temperature systems than in ambient systems for two reasons: first, the components have higher thermal conductivities, and second, there is a larger temperature difference from ambient (~400$^{o}$C) which enables more rapid heat dissipation when necessary.

## Battery Reliability

In the development of a commercially viable (in terms of an acceptable product warranty) high temperature battery, the manufacturers must obtain performance better than 100 years mean-time-between-failure (MTBF) i.e. 1% of the cells fail per year. This has not been achieved for any of the high temperature battery systems to date. A major factor in reaching this goal is the development of highly uniform manufacturing processes. Cells produced in a pilot scale manufacturing facility will be required to define failure modes and to test the feasibility of achieving the reliability goals.

## Battery Engineering

The packaging of cells, connector design, battery insulation feedthroughs, mechanical-thermal reliability and cell uniformity are areas which require effort, but do not present extreme technical problems. In this category is the development of appropriate techniques for charge/discharge control for cell equalization in a series battery.

## Safety/Environment

This subject is of serious concern and has been discussed above. It is recommended that an analysis of all credible accidents for Li/MS and Na/S systems be carried out. Some safety testing has been conducted on both systems, e.g., fully charged multiplate Li-Al/FeS cells at 460°C have been subjected to the impact equivalent to a 50km/hr barrier crash (8).

# III. NEW RESEARCH AREAS

## (i) Li/MS and Na/S

Work is recommended on problems identified as severe in Table 3 to provide more rapid improvements of the systems receiving engineering development. However, the research approach should aim at achieving an understanding of the basic electrochemical engineering parameters as well as trying to solve the practical problems of cell design by an empirical approach.

## (ii) Novel Electrolytes

At this meeting, Mamantov (1) discussed several promising molten salt electrolytes including chloroaluminates and organic halides. The specific conductivities of one of the organic halides being considered for battery applications, eg. alkyl pyridinium chloride - $AlCl_3$ mixtures, are lower than the $AlCl_3$ - alkali chloride melts. However, these organic halides melt at room temperature. The melt $AlCl_3$-NaCl (63-37 mole percent) operated between 180-260°C with a sodium negative electrode, a β"-alumina separator, and tetravalent sulfur as the active cathode material dissolved in the melt, provides a rechargeable cell with an OCV of 4.2V (15). Possible problems with this cell include the stability of ceramics in the melt, cost of the positive current collector, and the high vapor pressure of $AlCl_3$. Advantages are the low cost, low temperature of operation, and the high specific energy and cell voltage. In view of the significant investment being made in developing viable LiCl-KCl based molten salt batteries, it would be prudent to support basic studies on a variety of cells based on alternate molten salt electrolytes.

### (iii) Calcium/Metal Sulfide

Cells of this type will follow the same development path as the Li/MS cells (8), and are of interest because they use inexpensive materials including a LiCl-NaCl-$CaCl_2$-$BaCl_2$ eutectic ($390^oC$) (16) and $FeS_2$ and $NiS_2$ positive electrodes. One of the advantages of these cells is a very low lithium content, but the cells have a lower specific energy density and specific power. There are also problems of separator corrosion due to calcium attack.

### (iv) Mangesium Metal, Aluminum Metal and Negative Electrode Development

Solution of the problem of dendritic growth at the anode on recharging cells made with magnesium or aluminum anodes in various molten salt electrolytes, would help to provide an alternative to the expensive lithium based cells. The higher specific energies of the silicon and boron lithium alloys (see Table 2) suggest that their further development would be a good area for research effort.

### (v) Positive Electrode Development

The intercalation compounds being studied intensively for applications in ambient temperature batteries (17) have recently received attention in molten salts (6). Topochemical compounds showing interesting electrochemical properties as positive electrode materials were recently identified and are now being evaluated in development cells (18, 19). Another promising area of research is that of immobilized active material e.g., the adsorption of active sulfur catholyte on molecular sieves. For low power, high capacity applications, the dissolved catholyte system, Na (melt)/ -alumina/ (sulfur and sodium-sulfur species) (dissolved) has the advantage of a lower operating temperature than the Na/S ceramic cells discussed in parts I and II. This results in lower corrosion rates and the possibility of lighter and cheaper plastic battery components (20).

(vi) Instrumental Research & Development
<u>for High Temperature Applications</u>

Important new data on reactant characterization, reaction rates, and mechanisms at high temperatures would be obtained if more effort were put into development of instrumentation able to operate on systems while at high temperatures e.g., x-ray diffraction equipment and more development of atomic and molecular spectroscopic techniques (IR and laser Raman). The additional use of the ring-disk in molten salt electrochemical studies would be advantageous.

ACKNOWLEDGEMENTS

This report was prepared at the NATO Advanced Study Institute on "Materials for Advanced Batteries" held at Aussois, France, September 9-14, 1979. The high temperature systems study group consisted of W.A. Adams (Chairman), A. Bélanger, H. Bohm, I.W. Jones, G. Mamantov, R. Marassi, S. Sudar, G. Weddigen, and D.R. Vissers.

REFERENCES

1. G. Mamantov. "Molten Salt Electrolytes in Secondary Batteries." These Proceedings.
2. E.J. Casey. "Reflections on Recent Studies of Materials of Importance in Aqueous Electrochemical Energy Storage Systems." These Proceedings.
3. E.J. Cairns. "Requirements of Battery Systems." These Proceedings.
4. I.W. Jones. "Sodium Sulfur Batteries." These Proceedings.
5. D.R. Vissers. "Lithium-Aluminum/Iron Sulfide Batteries." These Proceedings.
6. A. Bélanger, F. Morin, M. Gauthier, W.A. Adams and A.R. Dubois, "Molten Salt Electrochemical Studies and High Energy Density Cell Development." These Proceedings.
7. P.A. Nelson in Proc. of the Symposium and Workshop on Advanced Battery Research and Design, March 22-24, 1976 (Argonne National Laboratory, Illinois) ANL-76-8, p. A-99 (1976).
8. P.A. Nelson et al. "High Performance Batteries for Electric Vehicle Propulsion and Stationary Energy Storage: Progress Report for October 1978 - March 1979." Argonne National Laboratory Report ANL-79-39 (1979).
9. E. Behrin. "Energy Storage Systems for Automobile Propulsion." Lawrence Livermore Laboratory. Rept. UCRL-52553, Volumes 1 and 2 (1978).

10. M.D. Hames, D.G. Hartley and N.M. Hudson. "Some Aspects of Sodium-Sulphur Batteries" in Power Sources 7, Ed., J. Thompson (Academic Press, London) p. 743 (1979).
11. M.M. Farahat, A.A. Chilenskas and D.L. Barney. "Thermal Management of the First 40 kWH Li/MS Electric Vehicle Battery" Extended Abstracts 156th E.C.S. Meeting, Los Angeles. (The Electrochemical Society, Princeton) p. 391 (1979).
12. W. Fischer, W. Haar, B. Hartmann, H. Meinhold and G. Weddigen. "Sodium/Sulfur Battery" in Progress in Batteries and Solar Cells, (IEC Press, Inc. Cleveland, Ohio) (1978).
13. J.A. Smaga, F.C. Marazek, K.M. Myles and J.E. Battles. "Materials Requirements in LiAl/LiCl-KCl/$FeS_x$ Secondary Batteries." Paper #111 in Corrosion/78 (National Association of Corrosion Engineers, Katy. Texas) (1978).
14. A.A. Chilenskas, Argonne National Laboratory, Argonne, Il., private communication, (1979).
15. R. Marassi, G. Mamantov, M. Matsunaga, S.E. Springer and J.P. Wiaux. J. Electrochem. Soc., 126, 231 (1979).
16. T.V. Tsyvenkova, V.I. Vereshchagina and K.V. Gontar. Rus. J. Inorg. Chem., 18, 426 (1973).
17. M. Armand. "Intercalation Electrodes." These Proceedings.
18. M. Gauthier, F. Morin and R. Bellemare (IREQ). "An Electrochemical Study of Li-Al/$FeS_x$ Cells and Construction of a Prototype." D.R.E.O.-N.R.C.C.-E.M.R. Contract No. 2SR-00162 (Ottawa, Canada) (1978). Final Report.
19. M. Gauthier, F. Morin, R. Bellemare, G. Vassort and A. Bélanger (I.R.E.Q.). "Research and Development on LiAl/FeS Batteries." D.R.E.O.-N.R.C.C.-E.M.R. Contract No. 2SD78-00034, (Ottawa, Canada) (1979). Final Report.
29. G. Weddigen. "Electrical Data of Sodium/Sulfur Cells Operating with Dissolved Catholyte" in Proceedings of the Symposium on Battery Design and Optimization, Ed., S. Gross, ELectrochem Soc. 79-1, 436 (1979).

## SOLID ELECTROLYTES

The first section of this report consists of a general discussion about the state of the art and future advances in the field of battery electrolyte development. In the second part, special emphasis will be given to problems which are related to the physics and chemistry of electrolytes. The group has focussed its interest on battery systems which either contain a solid electrolyte or in which an ionically conducting film forms on the electrode by reaction with an aprotic solvent. Aqueous and molten salt electrolytes are excluded, the latter being reviewed in the report of the study group on high temperature systems.

### SECTION I
General Discussion

#### I.1 State of the Art
Solid electrolytes have become increasingly of interest, both technically and scientifically, because of their potential use in high energy density battery systems and gas sensors. Thus, the Na/S battery system employs the sodium ion conductor, ß-alumina resp. ß"-alumina, as a separator. Although there are problems connected with the mechanical stability of the ß-alumina ceramic tubes, this system is in an advanced state of development.

For ambient temperatures, most interest has been devoted to the development of lithium based battery systems, two types of which are commercially available today. The first relies on the reaction between lithium and iodine to form the solid, ionically conducting separator, LiI. This electrolyte has a high resistance. Consequently, the use of this type of battery is restricted to low current applications such as cardiac pacemakers and watches. About 100 000 cells are now produced per year. Similarly, much research has been focussed on cells in which the electrolyte consists of a lithium salt dissolved in a non-aqueous solvent. Many different cells are commerçially available and the total production reaches around 1 million per year. These systems are of interest to this group because lithium metal reacts with the solvent to produce an ionically conducting solid on the electrode surface.

Anion-conducting electrolytes have found application as gas sensors. For example doped zirconia, an oxygen ion conductor, can be used in combustion engines to control fuel injection.

### I.2 Advances Which Lead to Economic, Social and Technical Advantages

As mentioned previously the Na/S battery system is in an advanced stage of development and demonstrates the potential usefulness of solid electrolytes. Because this system may offer the possibility of electrical energy storage at low cost, it may find application for load leveling and for traction leading to reduction in the world's dependence on oil. This is an example of the significant advances to be expected from research into solid electrolyte systems.

As all solid state systems would have minor problems connected with safety and storage, the search for appropriate solid electrolytes and solving the interfacial problems connected with solid-solid contacts remain an important goal of the future. Novel techniques in thin film technology offer the possibility of the in-situ preparation of a solid state battery in a microprocessor.

For many applications, a high energy density at ambient temperature is required and primary and secondary cells containing aprotic solvents are under development to meet this demand.

### I.3 Prediction of Capabilities of Battery Types by 1995

The commercial production of the Na/S system using either ß-alumina or glass separators is to be expected around 1985. High-temperature oxygen sensors are already used in cars. The prediction for the applicability of secondary Li-aprotic solvent systems is that to date's cycling life of 100 cycles with sufficient cycling efficiency should be extended to at least 500 - 1000 cycles. Although safety problems have yet to be solved, the technology should allow for the application of these systems in electric vehicles in the distant future. The near future prediction covers the use of secondary Li-systems in TV cameras, watches, calculators, radios, life-support systems, transceivers and for military use. In the field of primary cells the thionyl chloride system plays an important role, especially for military applications because of its high energy density. The Li-water and Al-water systems may play an important role for any maritime applications. Research in both aqueous and non-aqueous systems has to be focussed on an understanding of the formation of a passivating solid electrolyte film on the anode. Solid proton conductors may offer the possibility to construct new types of fuel cells and electrolyzers which operate at moderate temperatures.

## SECTION II
### Materials Problems

The second part of the report is related to the material problems of some solid electrolytes which may be of interest for future technical application. Theoretical considerations about solid electrolytes are not discussed. It seems that general structural and crystal chemistry considerations may be a useful tool to evaluate new solid electrolytes and to clarify still unsolved problems in this field. Diffuse x-ray scattering can be also a useful tool to evaluate the microscopic diffusion paths in the solid electrolytes.A close cooperation between solid state chemists, electrochemists and physicists is needed. An example of this approach was the discovery of fast ion conduction in NASICON (1).

The ß-alumina family is the object of many industrial projects as far as the pure electrolyte, its fabrication and its reproducible electrochemical properties are concerned. Thus the future research projects on ß-alumina should be more directed towards clarifying the interface problems rather than investigating the electrolyte properties. This material will therefore not be considered in the following discussion except for the proton conducting ß-alumina ceramics. The electrolytes to be considered in this section are classified under the following headings:

(1) polyphase solid electrolytes
(2) non-crystalline electrolytes
(3) crystalline electrolytes
   (a) fluorine conductors
   (b) lithium conductors
   (c) sodium conductors
(4) polymeric solid electrolytes including composite systems
(5) proton conductors
(6) interactions between lithium and aprotic electrolytic solvents

In each class, materials which are under investigation will be discussed and the reasons for their attractiveness will be indicated. Each material will be discussed in terms of the three requirements for practical application. Firstly easy and cheap fabrication is one of the prerequisites (it should be noted that the preparation procedure should be reproducible). A high ionic conductivity with a negligible partial electronic conductivity is a prime condition for a practical solid electrolyte. The chemical and electrochemical stability of the solid electrolyte is necessary for its use as a separator. Thus the electrolyte should not react or decompose with a given cathode or anode material in a given temperature range. Themodynamics allows one to calculate the minimum decomposition voltage for a material. However, it should be noted that kinetic considerations may play an important role and many materials exhibit in fact a higher decomposition voltage than that predicted by thermodynamics. For each class areas will be indicated where questions are still open and further research is required.

II.1 Polyphase Solid Electrolytes

The mixing of an ion-conducting phase with a non-conducting one has been found to enhance the ionic conductivity and these mixtures are termed polyphase solid electrolytes. Such a system is that based on mixture of LiI and $\gamma$-$Al_2O_3$ in which Liang has claimed a maximum conductivity of $2-4 \times 10^{-5} \Omega^{-1}cm^{-1}$ compared to $10^{-7}\ \Omega^{-1}cm^{-1}$ for pure LiI (2). However, other groups have not been able to reproduce this large enhancement and moreover have suggested that the conductivity enhancement relies on the presence of water in the material (3). It has also been suggested that pretreatment of the $\gamma$-$Al_2O_3$ with n-butyl-lithium further enhances the conductivity but in an irreproducible way. Therefore many questions connected with the origin of conductivity enhancement remain open. It is known, however, that a high specific surface area of the particles of the ion-conducting phase is a requirement for enhanced conductivity. This would suggest that the increased conductivity is connected with the interparticle surfaces. Besides this increase in conductivity the LiI-$Al_2O_3$ is stable in contact with Li even at high temperatures and LiI has a thermodynamic decomposition voltage of 2.8 V.

These considerations have led to the development of a number of battery systems. Ambient temperature, low drain batteries with either pure LiI or LiI-$Al_2O_3$ mixtures are in commercial production. The high conductivity ($10^{-1}\Omega^{-1}cm^{-1}$) of the LiI-$Al_2O_3$ system at 300°C has led to the construction of a secondary cell with cathodes of compounds such as $TiS_2$ and $TaS_2$ (2) which intercalate lithium during discharge. More than 40 cycles have been successfully demonstrated with these rechargeable storage batteries.

As mentioned above, many questions as to the influence of the grain size and chemical nature of the non-conducting material on the conductivity and stability of the polyphase system remain open. Therefore, it is suggested that future work in these systems deals with the following topics:

a) variation of the chemical nature of the constituents. For example, exchange of $Al_2O_3$ with $SiO_2$ and an extension to other lithium halides and lithium salts would both seem to be interesting topics.

b) variation of the water content of both conducting and non-conducting phases. It is hoped that this will clarify the role of water which is still in dispute.

c) variation of the physical properties of the components such as the particle surface area and particle size.

II.2 Non-Crystalline Electrolytes

Amorphous 'glass-like' materials have attracted great interest for battery application for a number of reasons. Firstly the preparation of a glass is, in general, easier and cheaper than the fabrication of crystalline ceramics and, in contrast to the latter materials, there is no influence of grain boundaries if a sufficiently high density can be obtained. In addition glasses can be easily shaped into the final form required for a separator (4).

Sodium-conducting borate glasses have been used as separators in Na/S cells as an alternative to ß-alumina ceramics. As the conductivity of this material is low, special high surface area structures are required. In the lithium ion conducting glasses new compounds such as the systems $LiPO_3$-LiX (5), $Li_2Si_2O_5$-$Li_2SO_4$ (6), $B_2O_3$-$Li_2O$-LiX (7) (X = F, Cl, Br, I) and rapidly quenched lithium-niobate and tantalate glasses with no network former (8) have recently been reported. Among these, one of the lithium borate glasses, $B_2O_3$-0.57$Li_2O$-0.85LiCl, has the high ionic conductivity of $10^{-2}$ $\Omega^{-1}cm^{-1}$ at 300°C (8). In addition these glasses are reported to be stable in contact with lithium to a temperature of 300°C and have a melting point between 800 and 900°C. The use of these glasses in lithium-based cells has not yet been reported and the electrochemical stability is unknown, although the decomposition voltage is expected to be high due to the presence of $Li_2O$ and lithium halides. A review of the properties of ionically conducting glasses is given in reference (4).

Further research should be carried out to obtain a better understanding of the short range structural order and the conductivity mechanism. An investigation of the glass forming regions in other systems may result in the discovery of new high conducting materials.

II.3 Crystalline Solid Electrolytes

Most research activities in the last decades have been focussed to the synthesis and properties of crystalline solid electrolytes. The aim of this report is to select material and current topics which may be of interest for future research work. The fluoride, lithium and sodium ion conductors have been chosen. It is obvious that the crystal strucutre plays an important role for the understanding of the microscopic diffusion mechanism in these compounds and therefore the materials are reviewed with respect to their crystal structure.

II.3.a Fluorine Ion conductors

Much research has been devoted to the crystal chemistry, and fluoride ion conductivity of various structures, notably the thysonite ($LaF_3$), and the fluorite ($CaF_2$) structure. These properties have been reviewed recently (9). Contrary to pure alkaline earth fluorides, $LaF_3$ has an unusually high ionic conductivity. Its use in a low-current battery or $O_2$ fuel cell has been described (9). Alkaline earth fluorides readily dissolve large amounts of $YF_3$ and rare earth (III) fluorides. Room temperature conductivity increases of up to $10^8$ then result (10). At 200°C such anion-excess solid solutions can reach conductivity values of about $10^{-2}\ \Omega^{-1}cm^{-1}$. This unusual compositional dependence has recently been described in an Enhanced Ionic Motion model that accounts for fast fluoride conduction in these fluorite-structured solid solutions. More importantly, this behaviour has initiated the development of thin-film batteries with an alkaline earth metal as anode, a solid solution as solid electrolyte, and a suitable metal fluoride as cathode. The present state of the art can be examplified by reviewing data for the cell $Ca/BiF_3$ (11-a). Cells of this type readily develop open-circuit voltages of 3.00 to 3.08 V in the temperature region 25 - 240°C. At 240°C current densities of 0.15 $mA/cm^2$ can be obtained. The major drawback in attaining high current densities in the cells $Ca/BiF_3$ is anode passivation by $CaF_2$. This is generally true for this type of batteries. The studies reported to date provide a clue on how to overcome this problem of anode passivation. If alkaline earth metal-rare earth metal alloys are employed as anode, the discharge product will be a solid solution. As these solid solutions have much larger conductivities than undoped alkaline earth fluorides, performance characteristics of these cells can be expected to be improved substantially. It is, therefore, proposed to study phase diagrams of the systems $MF_2$-$LaF_3$ (discharge product $M_{1-x}La_xF_{2+x}$, M = Ca, Sr, Ba). In addition, it is proposed to study the systems $MF_2$-(Li, Na or K)F (discharge product $M_{1-x}$ $(Li, Na\ or\ K)_xF_{2-x}$) as it is well known that these anion deficient solid solutions are better conductors for fluoride ions than the hosts. In this case the electrical properties of the discharge products also need to be studied in detail.

For the utilization of solid solutions as separators the use of dispersion of the materials in volatile organic solvents seem more promising than co-evaporative techniques (11-b). Anodes, or more generally, metals can easily be covered with the dispersion. After the solvent has evaporated cells can be assembled under mechanical pressure.

II.3.b Lithium Ion Conductors

Lithium-based batteries have high technological and economic importance for a number of reasons. The high electropositive character and low equivalent weight of lithium result in high voltages and energy densities in these systems. Moreover lithium metal is easier to handle than either sodium or potassium although dry conditions are necessary.

For ambient temperature operation, much research has been carried out to develop battery systems based on organic electrolytes and a number of primary systems are commercially available. In recent years a number of insertion or intercalation compounds have been discovered, in which the diffusion coefficient for lithium ions is high, and have been shown to be applicable in secondary lithium-based cells. One system with a $TiS_2$ cathode is available, although the cycle life is limited by the reaction of lithium with the organic electrolyte (cf. section II.6). Thus the development of all-solid state cells based on these materials seems to be attractive. In addition the low melting point of lithium (183°C) would allow the development of a system comparable to that of the Na/S battery if a suitable separator with good conductivity could be found.

A wide range of materials have been found to transport lithium but only a few of these fulfill all of the conditions mentioned in the introduction. None of the crystalline solid lithium ion conductors has reached engineering development. The crystalline lithium ion conducting electrolytes are reviewed in ref. 3.

At ambient temperature the highest conductivities are obtained with the layer structures such as Li-Na ß-alumina and $Li_3N$. Li-Na ß-alumina, which exhibits fast co-ionic conduction, has been found to be unstable against molten lithium thus restricting its application. Lithium nitride, $Li_3N$, is a promising material which has a conductivity of around $10^{-3}\ \Omega^{-1}cm^{-1}$ at 300°K and is stable in contact with lithium. Moreover, although thermodynamics predicts a low decomposition voltage (0.44 V), no decomposition occurs on application of voltages greater than 3 V suggesting a kinetic hindrance to decomposition (12). However the high reactivity of this compound may restrict the choice of cathode.

Because of their extremely low ionic conductivity at ambient temperature, lithium ion conductors with the rocksalt structure such as the lithium halides are interesting for application only in the polyphase systems (cf. section II.1). Similarly the crystalline compounds with antifluorite structures generally exhibit low conductivities and an extraordinary sensitivity to moisture. However one compound in this group which may be interesting for high temperature application is lithium nitride chloride, $Li_{1.8}N_{0.4}Cl_{0.6}$. This material has a conductivity of approximately $5 \times 10^{-3}\ \Omega^{-1}cm^{-1}$ at 600°K. In addition it is stable in contact with lithium and its decomposition voltage is greater than 2.5 V (13).

Another compound which may be interesting for high-temperature application has the formula $Li_{3.6}Si_{0.6}P_{0.4}O_4$. This compound is a member of the structural group of solid Li-ionic conductors which contains $MO_4$-tetrahedra (M = P, Ge, Si). The compound has a conductivity of about $5 \times 10^{-3}\ \Omega^{-1}cm^{-1}$ at 600 K, is stable against lithium, and insensitive to moist atmosphere. However, the extent of electrochemical stability is not yet known, and cells have not been prepared using this compound. Solid solutions based on the $Li_4SiO_4$ structure ($Li_4SiO_4$ with additions of $Li_3PO_4$, $Li_5AlO_4$, or $Li_3VO_4$) (14), and on the lisicon structure ($Li_4GeO_4$ with $Li_2ZnGeO_4$, and/or $Li_3PO_4$) (15) have recently attracted great interest. Future research should be aimed towards determining the composition dependence of the ionic ocnductivity of solid solutions, and the structural variations in the solid solutions.

II.3.c Sodium Ion Conductors

The field of sodium ion conductors has long been dominated by the ß-alumina family. The technical process for the industrial production of closed-end ß-alumina ceramic tubes has been developed in the last few years. The still unsolved problem in ceramic ß-alumina tubes is the statistical occurrence of microcracks which limit the lifetime of the tubes. Future research has to be focussed mainly to this problem. The use of ß-alumina tubes in Na/S cells has been successfully demonstrated. As an alternative material to ß-alumina the NASICON compounds have to be considered (1,16). The compositions $Na_{1+x}Si_xZr_2P_{3-x}O_{12}$ are highly conducting for $1.8 \leq x \leq 2.4$ and reach the conductivity of ß-alumina at the working temperature of the Na/S cell. The advantage of the NASICON compounds is the ease of preparation at a far lower sintering temperature (1200°C) than ß-alumina. The potential usefulness of NASICON for application in Na/S cells has been demonstrated. Structural and electrochemical investigations should be devoted to clarify the phase diagram of NASICON for the composition range $0.4 \leq x \leq 2.8$ with the goal of finding the optimum composition for the use in cells.

With the solid crystalline ionic conductors, a number of general research questions arise. With co-ionic conduction the ion with larger radius expands the diffusion path allowing more rapid motion of the smaller ion. It would seem interesting to determine if this effect could be extended to other systems than ß-alumina. There is also a need for additional understanding of the formation of solid solutions and the occurrence of fast ionic conduction therein.

Dendrite formation in crystalline electrolytes is a limiting factor for the use of these materials in secondary cells. A basic understanding of dendrite formation and its connection with the crystal defect structure is necessary. Similarly the kinetics of transport across the solid electrolyte - solid electrode interface has been the subject of little research.

## II.4 Polymeric Solid Electrolytes Including Composite Systems

Polymeric thin foils with a remarkable ionic conductivity should be a goal of future investigations. This is based on the fact that thin foils can be shaped in any form, that the elasticity compensates for volume changes of the electrodes and that the known production technology of polymers enables a cheap mass production. Polymeric electrolytes have been found which are solid solutions of an alkali metal salt like $LiClO_4$ or NaI in a solvating propylene oxide (17). The physical properties of these compounds range from that of tough thermoplastics to rubbery elastomers. They show cationic mobility with conductivities between $10^{-2}$ - $10^{-3}$ $\Omega^{-1}cm^{-1}$ at about 120°C decreasing rapidly to $10^{-7}$ $\Omega^{-1}cm^{-1}$ at room temperature. These compounds are specially applicable for solid state batteries with intercalation electrodes working at temperatures between 50°C and 100°C. The thermal and electrochemical stability of the compounds is still unknown. Future emphasis should be given to develop new polymeric electrolytes with high room temperature conductivity, to the investigation of the stability of the polymers against the corresponding alkali metals and to the polymer-electrode interface kinetics.

## II.5 Proton Conductors

Over the last 5 years several proton conducting solids have been found which show promise for use in fuel cells, electrolysis cells and gas sensors (18). Solid electrolytes may have favourable properties compared to liquid electrolytes such as phosphoric acid or sulphuric acid, such as the ease of coating and support of electrocatalyst, the stability with respect to water and the possibility of operation at higher temperatures of 200°C - 300°C. Solid oxide ion conductors are at present limited to high temperature operation ($T > 700°C$).

It would be interesting to produce a ceramic type or thin film proton conductor which will have a conductivity in the region of 1 $ohm^{-1}cm^{-1}$ at 100 - 300°C. Most proton conductors known to date are unstable in this region. Some, however, show promise, as for example hydrogenuranylphosphate ($HUOPO_4 \cdot 4H_2O$), $NH_4$-ß- and ß"-alumina and $H_3O$-ß- and ß"-alumina and $KOHO.03H_2O$. The microscopic diffusion mechanism in these compounds is different from the previously mentioned materials. Various mechanisms for the proton diffusion mechanism may be considered. Proton exchange between $H_3O^+$-$H_2O$, $H_2O$-$OH^-$ and $OH^-$-$O^=$ pairs and diffusion of $H_3O^+$ or $NH_4^+$ have been proposed (19). Several examples of the $H_3O^+$-$H_2O$ mechanism are known. Compounds which are stable at higher temperatures are more likely to involve the other mechanisms. The search for new proton conductors should be devoted to materials which are stable in water free atmospheres. The investigation of the proton diffusion mechanism remains necessary for the understanding of these materials.

II.6 Interactions Between Lithium and Aprotic Electrolytic Solvents

Rechargeable, ambient temperature aprotic Li cells have high energy densities and show promise to fill many civilian and military applications. The high reactivity of Li with aprotic solvents is the limiting factor causing low cycling efficiency and rapid passivation on stand in these cells. Despite outstanding progress in the past 7 years, a number of research problems must be solved for the batteries to come into wide use. Among these are:

- Systematic investigations of the reactions between Li and families of organic, electrolytic solvents. The effects of molecular structure and the role of surface films are key parameters.
- Lithium hexafluorarsenate solutions show outstanding behaviour for Li cycling in a number of solvents. This effect is not yet understood and research should be directed to this question, and to what other anions - which are cheaper and/or less toxic - would show similar behaviour.
- Solutions in cyclic ether solvents such as tetrahydrofuran (THF) and 2-methyl-tetrahydrofuran (2Me-THF) show the phenomenon of "recontacting". "Recontacting" means that Li isolated from discharge by an insulating reaction product on its surface can be discharged after a subsequent plate of Li has been applied. Further research is necessary to clarify and optimize the effect.
- Improvements of solution conductivity without loss of ability to cycle Li are always appropriate. Key parameters are solvent structure, solvent composition and electrolyte type and concentration.
- In solvents such as methylacetamide (MA), propylenecarbonate (PC) and tetrahydrofuran (THF), additives which are known to react to modify the Li surface film (precursors), or which merely adsorb ("levelling agents") have shown promising ability to improve Li cycling efficiency. These solvents are too reactive for practical use, and the effects of such additives need to be extended to less reactive media such as 2Me-THF/$LiAsF_6$.

ACKNOWLEDGEMENT

This study group report is the comprehensive summary of the contributions of the following members:

U. v.Alpen, S.B. Brummer, Ph. Colomban, R. Collongues, L. Cot, A. Howe, R.A. Huggins, A. Levasseur, F.W. Poulsen, J.M. Réau and J. Schoonman.

REFERENCES

(1) H.Y.-P. Hong, Mat. Res. Bull. 11, 173 (1976).

J.B. Goodenough, H.Y.-P. Hong and J.A. Kafalas, Mat. Res. Bull. 11, 203 (1976).

(2) C.C. Liang, A.V. Joshi and N.T. Hamilton, J. Appl. Electrochem. 8, 445 (1978).

C.C. Liang, U.S. Patent No. 81083 (1970) and No. 3541124 (1973).

B. Owens and H.J. Hilton, U.S. Patent No. 4007122 (1977).

(3) U. v.Alpen and M.F. Bell, in "Fast Ion Transport in Solids", P. Vashishta and J. Mundy Eds., Elsevier North Holland, N.Y. (1979).

P. Pack, B. Owens and J.B. Wagner Jr., J. Electrochem. Soc., in press.

R.J. Brooks, EC-report on Anglo-Danish battery project.

(4) D. Ravaine and J.L. Sauquet, in "Solid Electrolytes", P. Hagenmüller and W. van Gool Eds., Academic Press, p. 277 (1978).

(5) J.P. Malugani and G. Roberts, Mat. Res. Bull. 14, 1075 (1979).

(6) A. Koné, B. Barreau, J.L. Souquet and M. Ribes, Mat. Res. Bull. 14, 393 (1979).

(7) A. Levasseur, J.C. Brethous, J.M. Réau and P. Hagenmüller, Mat. Res. Bull. 14, 921 (1979) and French Pat. No. 7711376 (1978).

(8) A.M. Glass and K. Nassau, in "Fast Ion Transport in Solids", P. Vashishta and J. Mundy Eds., Elsevier North Holland, N.Y. (1979).

(9) J.M. Réau and J. Portier, see ref. 4.

(10) K.E.D. Wapenaar and J. Schoonman, J. Electrochem. Soc. 126, 667 (1979).

(11a) J. Schoonman, K.E.D. Wapenaar, G. Oversluizen and G.J. Dirksen, J. Electrochem. Soc. 126, 709 (1979).

(11b) J. Schoonman, ibid 123, 1772 (1976).

(12) U. v.Alpen. J. Solid State Chem. 29, 372 (1979).

(13) P. Hartwig, W. Weppner and W. Wichelhaus, see ref. 8.

(14) Y.-W. Hu, I.D. Raistrick and R.A. Huggins, J. Electrochem. Soc. 124, 1240 (1977).

(15) J. Schoonman, J.G. Kamphorst and E.E. Hellstrom, to be published in the proceedings of the NATO Advanced Study Institute "Materials for Andvanced Batteries", Aussois, 9 - 14, 9 (1979).

(16) U. v.Alpen, M.F. Bell, H. Laig-Hörstebrock and G. Bräutigam, see ref. 8.

G. Kafalas and R.J. Cava, ibid.

(17) M.B. Armand, J.M. Chabagno and M. Duclot, see ref. 8.

(18) M.G. Shilton and A.T. Howe, Mat. Res. Bull. 12, 701 (1977).

S.H. Sheffield and A.T. Howe, ibid 14, 929 (1979).

G.S. Farrington and J.L. Briant, ibid 13, 763 (1978).

(19) Ph. Colomban, G. Lucazeau, R. Mercier and A. Novak, J. Chem. Phys. 67, 5244 (1977).

## ELECTRODE MATERIALS

The discussions of the electrode materials study group were aimed at identifying classes of electrode reactions for secondary batteries and recognizing advantages, limitations, and research opportunities for each.

The classification scheme used for electrode reactions is summarized below. An example of each is also shown.

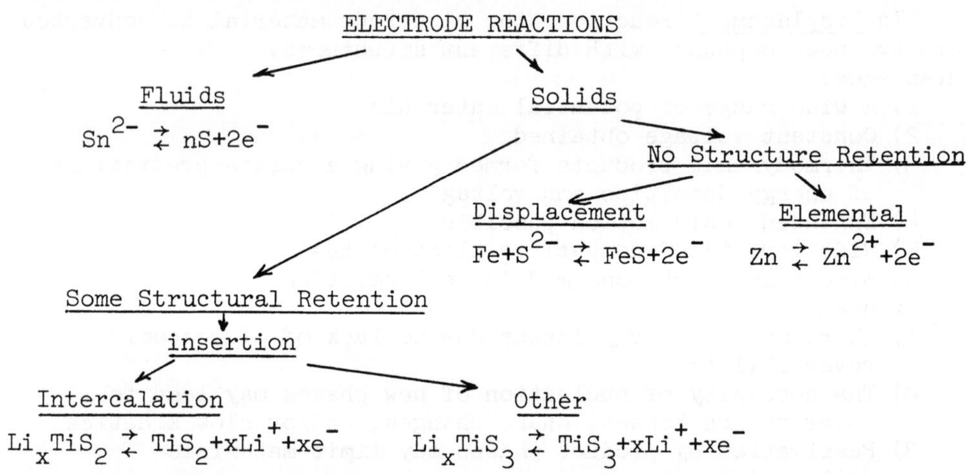


It is recognized and emphasized that electrode reactions may involve more than one mechanism. In this scheme we use the term insertion to refer to the addition of a species to a host structure with retention of the major structural features of the.

host. Intercalation is a special case of insertion in which no strong chemical bonds of the host are broken during the reaction.

By fluid electrode reactions we refer to those in which the electroactive species is a gas, a liquid, or is dissolved in a solvent. The electrode reaction involves electron transfer between the active species and an inert current collector.
Advantages include:
1) High specific energy densities possible
2) Generally fast kinetics
3) No shape change problems
4) Electrodes may be easily regenerated or rejuvenated

Problems may arise from:
1) Self-discharge
2) Active species often highly corrosive
3) Ion selective membranes may be required
4) May form electronic or ionic insulators on current collectors.

Elemental electrode reactions involve the stripping and plating of a metallic element or alloy.
Advantages:
1) The highest specific energy densities
2) Fast kinetics

Disadvantages:
1) Electrolyte compatibility
2) Shape change on cycling

In Displacement reactions the electrode material is converted into two new compounds with different structures.
Advantages:
1) A wide range of potential materials
2) Constant voltage obtained
3) Thermodynamic products formed giving accurate predictions of energy densities and voltage
4) Enhanced ionic motion possible
5) Wide variety of compatible electrolytes
6) Electrolyte not consumed in cell reaction

Disadvantages:
1) There may be energy losses due to lack of microscopic reversibility
2) The necessity of nucleation of new phases may lead to polarization losses, shape changes, and/or slow kinetics
3) Passivation by product phases may limit materials utilization
4) Often require high temperatures.

By electrode reactions which occur with structural retention we include a range of reactions from intercalation to those

involving somewhat larger structural changes.
Advantages:
1) Highly reversible even at low temperature
2) Little or no nucleation losses
3) Usually no ionic or electrically blocking products
4) Materials often have good electronic conductivity
5) Electrolyte composition remains constant if counter-electrode reaction is of similar type or elemental.

Disadvantages:
1) Specific energy densities are moderate
2) Voltage generally not constant as a function of state of charge
3) Electrolyte must be chosen carefully.

## Research Opportunities

### Kinetics

In the general area of kinetics of electrode reactions, several types of studies appear to be important for future research. These areas of study depend on the general class of electrode mechanism outlined earlier.

Fluid electrodes are the best understood kinetically at the present time. Electrocatalysis especially for the case of gases (particularly $O_2$) is a subject of intensive research and will continue.

For solid electrodes more work in the area of understanding porous electrodes is needed (see the paper by S. Atlung). In this regard, stronger interactions with scientists involved with fields such as catalysis may be beneficial. Most solid electrodes are two phase and further theoretical and experimental work is needed to understand the factors influencing nucleation and grain growth.

In the case of insertion electrodes, theoretical and experimental data on self diffusion mechanisms is needed. In particular, the differences and similarities to solid electrolytes and the differences between local and long range diffusion should be addressed.

Perhaps the most important and least understood area of electrode kinetics is the influence of electric field gradients on rate. In particular, several electrodes appear to operate at much higher rates than expected from known self diffusion data (these include Li/Sb, Li/$V_2O_5$, and Li/$NiPS_3$). One possible mechanism of enhanced ionic motion is discussed in the paper by Weppner. Other mechanisms also seem likely.

The development of new experimental techniques to obtain useful kinetic data in multiphase systems should be encouraged.

## Thermodynamic Investigations

The relationship between cathode structure and thermodynamic properties raises several important scientific questions. These include: what makes the intercalation of alkali metals possible in certain solids? When are microdomains or two-phase regions expected? What is the influence of impurities or additives?

A precise knowledge of the thermodynamic properties of electrodes is also essential for the evaluation of the maximum energy density of batteries in both the charged and discharged state. Such information is also useful in the evaluation of the compositional variation of the chemical diffusion coefficient, of phase stabilities and of structural changes with composition. Of particular importance is the compositional variation of the partial molar Gibbs free energies, enthalpies and entropies for solids with both narrow and wide ranges of stoichiometry. The effects of impurities and additives on the partial molar properties is also of interest. Although values of partial molar Gibbs free energies or chemical potentials can be obtained using galvanic-cell techniques, calorimetric investigations are needed for precise determinations of partial molar enthalpies and entropies. Such studies are particularly necessary for multicomponent phases which include not only the elements in the electrode but also those in the electrolyte which chemically interact with the electrode.

Interfacial thermodynamic studies are needed to evaluate the chemical segregation of impurities and additives to grain boundaries and electrode-electrolyte interfaces. Another example of where surface thermodynamic properties may be important is when very small solid particles are used in the electrode.

## New Materials

There are many possibilities for electrodes involving new materials, particularly in the case of insertion electrodes. To date, the insertion electrodes which have been studied most extensively are cathodes of layered or pseudo one-dimensional compounds. In addition, lithium has been the preferred cation. Work on other structures, such as three-dimensional Chevrel structures, or structures containing interconnecting tunnels, may lead to advances. The insertion of other cations, such as sodium, should be studied. Systems which intercalate anions should be studied. In particular, electrodes capable of intercalating $F^-$ would be of interest. Disordered analogs of known crystalline intercalation systems may present further opportunities. The use of insertion systems as anodes should be investigated. A suitable insertion anode would make possible a cell in which Li or Na is alternately inserted in the anode or cathode during charge and discharge (a so-called "rocking-chair" battery).

New liquid electrodes, analogous to $SOCl_2$ or $SO_2$ which might be capable of use in rechargeable systems should be investigated. Electrodes capable of rejuvenation (e.g. Li/Bi, $Zn/Br_2$) may be of interest because of the possibility of extended cycle life.

ACKNOWLEDGMENT

This report was prepared by C. Berthier, C. Cros, J-P. Duchange, M. Fouletier, R. R. Haering, D. W. Murphy, J. Rouxel, B. C. Tofield, W. Weppner, M. S. Whittingham and W. Worrell.

# INTERFACE PROBLEMS

## 1. Introduction

All electrochemical systems involve the transport of charge and matter across interfaces. These same interfaces are also often responsible for limiting the performance and capacity of secondary battery systems. The great variety of interfaces found in batteries has been used as a basis for classifying existing and new batteries (Study Group I : Current Status and New Systems) and different electrode reactions can be arranged according to the type of interfacial process (Study Group IV : Electrode Materials). However the details of particular interfacial reactions are usually specific to that system and generalisations are both difficult to compose and probably not very useful. The useful introductory paper by F.G. Will (Interphase Phenomena in Advanced Batteries) emphasised this conclusion and it was considered more appropriate therefore in this report to prepare in tabular form a list of those interfacial processes that are believed to limit the performance of a selection of existing and advanced battery systems. These interfacial processes are classified under the headings, passivation, corrosion, structure/morphology, miscellaneous, and are accompanied by recommendations for future research. Further remarks on the different battery systems are contained in section 3. In addition to the expertise available at the NATO ADVANCED BATTERY meeting a number of recent publications (1) were consulted to determine the current status of different systems.

## 2. TABLES

| System | Observed Cell operating Limitation | Interfacial Phenomena/Problems. | | | | Recommended Research |
|---|---|---|---|---|---|---|
| | | Passivation | Corrosion | Struc/Morphology | Miscellaneous | |
| Pb - $PbO_2$ | limited cycle life | | ⊖ grid | ⊖ material shedding | | growth mechanism and characterisation of grid corrosion layers; interface between grid and active material, role of different crystal phases of $PbO_2$; correlation between shedding, grid corrosion and crystal phase changes |
| | capacity loss | | | ⊖ densification<br>⊖ particle growth<br>⊖ loss of porosity | | establish mechanism, role of expanders nucleating agents |
| | low Pb utilisation | $PbSO_4$ formation | | | | additives |
| Zn-Ni | short cycle life<br>low efficiency Ni(+) substrate | | | ⊖ dendrite shorting | ⊖ expensive Ni | inhibitors, additives, electrolyte composition and hydrodynamics mechanism of dendrite shorting |
| | capacity loss | | | ⊖ shape change<br>⊖ densification | | loading agents, geometry, establish mechanism. |
| Fe-Ni | low efficiency Ni ⊖ substrate | | | | ⊖ low $H_2$ overvoltage<br>⊖ slow kinetics | additions to ⊖ or electrolyte. alternative substrates/structures |
| | low power density | | | | ⊖ expensive Ni | optimised electrode struc. small particles large surface area. |
| | high self discharge | | ⊖ electrode | | | role of additives to ⊖ electrode or to electrolyte. |

| System | Observed cell operating limitation | Interfacial Phenomena/Problems | | | | Recommended Research |
|---|---|---|---|---|---|---|
| | | Passivation | Corrosion | Struc/Morphology | Miscellaneous | |
| Fe-air | low efficiency | Θ formation of oxide species | | | Θ Low $H_2$ overvoltage.<br>⊕ polarisation. | development of improved air electrodes. |
| | high self discharge | | Θ electrode | | | role of additives to Θ electrode or to electrolyte. |
| | low capacity and power density | | | Θ expansion of active material | | better control of porosity and morphological changes. |
| Zn-air | low efficiency | Θ formation of oxide species | | | ⊕ polarisation | improved air electrode. |
| | low power/charge density and capacity loss. | | | Θ dendrite growth | | role of inhibitors (levellers) |
| $Zn-Cl_2$ $8H_2O$ | life | | ⊕ electrode Ti heat exchanger | | | rate of Ti heat exchanger corrosion. Mechanism of electrode activation |
| | low power/charge density | | | Θ dendrite growth | | electrolyte additives. Dendrite inhibitors. |
| $Zn-Br_2$ | limited cycle life | | Θ electrode | Θ dendrite growth | limited membrane conductivity | props. of complexes, new membranes (low-cost inhibitors |
| | power/charge density | ⊕ $Br_2$ complex films | Θ electrode | | | |
| $Ni-H_2$ | high self discharge electrolyte loss. Containment of high pressure $H_2$ | | | | $O_2-H_2$ recombination<br>thermal management<br>Θ expensive Ni | optimisation of new cell designs inc. development of metallic hydrides |
| | | | | | | |

| System | Observed cell operating limitation | Interfacial Phenomena/Problems: Passivation | Corrosion | Struc/ Morphology | Miscellaneous | Recommended Research. |
|---|---|---|---|---|---|---|
| Li-organic | short cycle life | ⊖ thin films producing loss of area and contact | | ⊖ dendrites at high current density<br>⊕ possible incorporation of electrolyte into insertion electrodes. | | kinetics of formation, morphology and props. of passivating films. modifications of films by additives and synthesis of new electrolytes |
| Li-Al/ FeS | limited cycle life | | ⊕ current collector, particularly for $FeS_2$ cells | | | development of alternative materials for ⊕ current collector, separator, particle retaining felts both technical and economic factors limiting at present. |
| | non-uniform capacity and capacity loss | ⊕ reaction of active mass with electrolyte | reaction of active mass with $Y_2O_3$ particle retaining felt | disintegration of both electrodes | wetting of BN ceramic separator | |
| Na/S | non-uniform performance | | | | Na penetration of electrolyte<br>Na/electrolyte wetting | role of grain boundaries and other microstructural features<br>surface properties of polycrystalline solid electrolytes. |
| | capacity loss | insulating S or $Na_xS$ films | | | | role of additives similar to $C_6$ $N_4$. |
| | cycle life | | current collector at ⊕ electrode | | deleterious effects of impurities and $H_2O$ | influence of impurities on properties of solid electrolytes. |

| System | Observed cell operating limitation | Interfacial Phenomena/Problems | | | | Recommended Research |
|---|---|---|---|---|---|---|
| | | Passivation | Corrosion | Struc/Morphology | Miscellaneous | |
| Solid electrolyte/solid electrode | capacity loss on cycling<br>low charge density | interfacial films due to reactions and presence of $H_2O$ | | loss of contact between electrode and electrolyte, degradation of assembly arising from volume changes | | development of mechanically 'soft' components to maintain integrity of solid/solid interface and to accommodate volume changes; synthesis of new insertion compounds for ⊕ and ⊖ electrodes. |

## 3. Further comments on specific battery systems

### 3.1. Aqueous Systems.

Examination of the tabulated summaries in section 2 suggests that further investigations into a limited number of topics could produce improvements in many aqueous battery systems. These investigations should clearly include the following topics:

(a) Mechanism of additives such as 'expanders' (Pb electrode), and 'levellers' or dendrite inhibitors (Zn electrode) requires clarification. The basic electrochemical processes of adsorption and nucleation in the presence of organic additions are still poorly understood and it is also important to examine the role of impurities that can be present in water or introduced from non-active components such as separators.

(b) Morphological changes can produce disintegra- of the electrode active mass and further investigations are needed to understand the associated reactions of complex highly defective non-stoichiometric compounds. The concepts and techniques of solid state chemistry require to be involved to a much greater degree in this area if progress is to be made.

(c) Modelling of cell performance as a function of operating conditions and particularly of the behaviour of porous electrodes needs further refinement and evaluation. Often the weakest part of the model is the lack of information about the relevant electrode kinetics. Further complications can arise, for example in lead-acid batteries (2):

(a) the positive grid suffers from progressive corrosion which affects the electrical contact and adhesion of the active mass;

(b) the dissolution/precipitation mechanism of the two electrode couples, involving $PbSO_4$, results in particle size changes and changes in active surface area due to the passivating effect of the non-conducting $PbSO_4$;

(c) the molar volume changes between charged and discharged species result in further morphological changes.

(d) the local current density, acid concentration, and temperature influence all the factors mentioned above. A very sophisticated model is thus required to take account of these complicating factors and thus combine relevant data on kinetic and thermodynamic aspects with information about physical parameters such as crystal structure, surface area and porosity. Such a model could then predict the performance and cycle life of a battery as a function of the rate and depth of discharge when carried out continuously or intermittently.

(d) Viable air electrodes still need to be developed. The elusive goal of such an electrode has attractive advantages to battery developers in that air 'is free', presents no storage problems, and does not contribute to the overall weight and volume of a battery. However although the oxidant material is free the measures that have to be adopted to persuade it to react must not be expensive. More detailed examination is required of both existing and novel electro-catalysts. New electron beam instruments (e.g. STEM) now enable micro-diffraction and micro-analysis to be performed on very fine particles ($\sim$ 100 Å diam). These techniques combined with surface characterisation (Auger and ESCA) should ensure that electro-catalysts can be prepared with reproducible properties. It must be emphasised that x-ray diffraction indicates an average structure and by itself is not sufficient to characterise the complex highly defective compounds often used in battery systems. It should be added that development of air electrodes for batteries would also obviously benefit fuel cell technology.

### 3.2. Li-organic electrolyte systems.

A recent review (3) on Li-organic electrolyte systems concluded that 'a long life SECONDARY organic battery has not come into sight upto date and there are several severe problems concerning anode, electrolyte, and cathode which have yet to be solved'. The excellent performance (4) of positive electrode materials such as $TiS_2$ has now focussed attention on the poor cycling efficiency of the lithium electrode arising from the formation of thin passivating films which can isolate the lithium plated during the charge cycle. The morphology of these films is being examined (5,6) by a variety of electron-beam techniques. Other investigations (7) aim to modify the properties of the film by the introduction of surface-active additives, the synthesis of specific solvents, and by the introduction of novel alloys which can incorporate the lithium by a mechanism similar to that prevailing in the positive plate insertion compounds such as $TiS_2$.

### 3.3. Li-Al/Fe System

An operating temperature around 425$^{o}$C and the presence of molten LiCl-KCl electrolyte inevitably produce severe environmental materials problems for the current collectors, separators, lead throughs, etc. The principal difficulties have often been discussed (e.g. 1d,8,9,10) and technological progress has been made as demonstrated by the use of boron nitride cloth separators and molybdenum as the current collector for cells incorporating $FeS_2$ positive electrodes. These and other solutions, however, are not economically viable at present, and there must remain some doubt whether appropriate solutions to the severe materials problems can indeed be obtained.

### 3.4. Na/S system

Interfacial phenomena associated with the Na/S battery have been discussed by Will (1e) at the present meeting and many other general reviews (e.g. 11,12) are available. It would appear that one of the major outstanding problems concerns the lack of uniformity in the cycle life of the beta alumina electrolyte tubes. For supposedly identical cells, the range of cycle lives may be from 50 to 1500 cycles. The early failure after only 50 cycles of a small fraction of the ceramic tubes is at present unexplained. Cracks and the penetration of sodium dendrites have been attributed to the occurrence of local high electric fields and increased interfacial polarisation caused by poor wetting and the incorporation of protonated impurities such as water. The role of grain boundaries emerging at the surface remains obscure and the presence of syntatic $\beta$ and $\beta''$ intergrowth (13) which can also produce inhomogeneities in the sodium ion current density requires further investigation. In addition the different crystal orientations inevitably present at the ceramic electrolyte surface may introduce variations in the current distribution and if this is an important effect then polycrystalline one-and two-dimensional ionic conductors have obvious drawbacks compared to three dimensional materials such as $Na_3 Zr_2 P Si_2O_{12}$. Problems arising from non-uniform surface properties might be reduced by coating the polycrystalline ceramic with a thin layer of a non-crystalline electrolyte such as an ion conducting glass.

Better understanding of grain boundary impedance is also required as complex plane analytical techniques usually utilise a simple equivalent circuit which cannot exactly represent the actual distribution of grain boundary properties expected to be present in a typical ceramic material. The presence of only a small fraction of high resistivity boundaries can have a significant influence upon the current distribution.

### 3.5. Solid Electrolyte/Solid Electrode Systems

The success of Li/$I_2$ (poly-2-vinylpyridine) cells as power sources for cardiac pacemakers has stimulated investigations into solid state battery systems. Although a number of primary cells have been described (14,15) there are very few reports of secondary solid state systems (16). The introduction of insertion electrodes which do not involve the formation of interfacial reaction layers has removed one of the problems associated with the development of solid state cells. Major outstanding difficulties are associated with maintaining the integrity of the solid electrolyte/solid electrode interface at high current densities. The fabrication of mechanically 'soft' components such as LiI/$Al_2O_3$ composite electrolyte and polymeric electrolytes (17) can help to retain good interfacial contact and accommodate volume changes associated with the charge/discharge process. The design of solid state cells needs to be optimised and impoved lithium and sodium ion electrolytes are required for ambient temperature operation. Further investigations are also required to determine the factors influencing ion transport across solid electrolyte/ solid insertion electrode interfaces (18).

## 4. Acknowledgements

This summary was prepared in consultation with S. Atlung, G. Eichinger, M. Garreau, M. Kleitz, M. Lazzari, C.C. Liang, B.C.H. Steele, D.S. Tannhauser and F.G. Will.

## 5. References

1. (a) 'Electrode Materials and Processes for Energy Conversion and Storage', ed. J.D.E. McIntyre, S. Srinivason and F.G. Will (Electrochemical Soc. 1977).
(b) 'Battery Design and Optimisation', ed. S. Gross (Electrochemical Soc. 1979).
(c) D.A.J. Rand, J. Power Sources 4, 101, 1979.
(d) F. Will 'Interface Phenomena in Advanced Batteries' Contribution in present Proceedings of Aussois meeting.

2. W.H. Tiedemann and J. Newman. p.23 of ref. 1 (b).

3. J.O. Besenhard and G. Eichinger, J. Electroanal, Chem. 68, 1, 1976, ibid. 72, 1, 1976.

4. M.S. Whittingham, Prog. in Solid State Chem., 12, 1, 1978.

5. A.N. Dey, Thin Solid Films 43, 131, 1977.

6. M. Garreau and J. Thevenim, Microscopic et Spectro scopic Electron, 3, 27, 1978
J.P. Contour, A. Saless, M. Froment, M. Garreau J. Thevenim and D. Warin, Mocroscopic et Spectroscopic Electron, 4, 483, 1979.

7. S.B. Brummer, V.R. Koch and R.D. Rauk, 'The Recharging of the Lithium Electrode in Organic Electrolytes', Contribution in present Proceedings of Aussois meeting.

8. D.R. Vissers, 'Lithium-Aluminum/Iron Sulphide Batteries', Contribution in present proceedings of Aussois meeting.

9. J.E. Battles, J.A. Smaga and K.M. Myles, Met. Trans. 9A, 183, 1978.

10. High Performance Batteries for Electric Vehicle Propulsion and Stationary Energy Storage, Argonne National Lab. Progress Reports, e.g. ANL 79-39.

11. I. Wynn Jones, 'Sodium-sulphur Batteries', Contribution in present Proceedings of Aussois meeting.

12. G.J. May, J. Power sources, 3, 13, 1978.

13. J.O. Bovin, Acta Cryst, A35, 572-580, 1979
J.O. Bovin, Naturwissenschaften 66, 576, 1979.

14. B.B. Owens and P.M. Skarstad, p. 61 in 'Fast Ion Transport in Solids', ed. P. Vashishta, J.N. Mundy, and G.K. Shenoy (North Holland, 1979).

15. C.C. Liang, 'Solid State Batteries' in Applied Solid State Science, Vol. 4, p.25, ed. R. Wolfe (Academic Press 1974).

16. C.C. Liang, A.V. Joshi and N.E. Hamilton, J. Appl. Electrochem. 8, 445, 1978.

17. M.B. Armand, J.M. Chabagno and M.J. Duclot, p.131, in 'Fast Ion Transport in Solids' ed. P. Vashishta, J.N. Mundy and G.K. Shenoy. (Academic Press, 1979).

18. K.Y. Cheung, B.C.H. Steele and G.J. Dudley, p. 141, ibid.

PARTICIPANTS

Dr. W. A. Adams
Defense Research Establishment Ottawa
Department of National Defense
Ottawa, Ontario K1A 0K2
Canada

Dr. U.v. Alpen
Max-Planck-Institut für Festkörperforschung
Heisenbergst. 1
D-7000 Stuttgart 80
W. Germany

Dr. M. Armand
Laboratoire D'Energetique Electrochimique
ENSEEG Domaine Université
38401 St. Martin d'Heres
France

Dr. S. Atlung
Fysisk-kemish Institute
Building 206
Technical University of Denmark
Lyngby, DK 2800
Denmark

Dr. A. Bélanger
IREQ
1800 Montée Ste Julie
Varennes, P.Q. J0L 2P0
Canada

Dr. C. Berthier
Laboratoire de Spectrometrie Physique
Université Scientifique et Medicale de Grenoble
Domaine Universitaire
38401 St. Martin d'Heres
France

Dr. H. Böhm
AEG-Telefunken
Institut für Physikalische Chemie
Goldsteinstr. 235
D-6000 Frankfurt 72
W. Germany

Dr. J. Broadhead
Bell Labs.
600 Mountain Avenue
Murray Hill, New Jersey 07974
U.S.A.

Dr. S. B. Brummer
EIC Corp.
55 Chapel Street
Newton, Mass. 02158
U.S.A.

Dr. E. J. Casey
Defence Research Establishment Ottawa
Department of National Defence
Ottawa, Ontario K1A 0K2
Canada

Dr. S. Caulder
Naval Research Laboratory
Code 6130
Washington, D.C. 20375
U.S.A.

Professor Collongues
Laboratoire de Chimie Appliquée de l'Etat Solide
Ecole Nationale Supérieure de Chimie de Paris, II
Rue Pierre et Marie Curie
75231 Paris Cédex 05
France

Dr. P. Colomban
Laboratoire de Chimie Appliquée de l'Etat Solide
Ecole Nationale Supérieure de Chimie de Paris, II
Rue Pierre et Marie Curie
75231 Paris Cedex 05
France

Professor L. Cot
Ecole Nationale de Chimie - Chimie Minerale
8, Rue de l'Ecole Normale
34075 Montpellier
France

Dr. C. Cros
Laboratoire de Chimie du Solide du CNRS
Université de Bordeaux I
351 Cours de la Liberation
33405 Talence Cédex
France

Dr. J.-P. Duchange
S.A.F.T.
Rue Georges Leclanché
86009 Poitiers
France

Dr. G. Eichinger
Anorganisch-Chemisches Institut
der Technischen Universität München
Lichtenbergstr. 4
D-8046 Garching
W. Germany

Dr. M. Fouletier
Laboratoire D'Energetique Electrochimique
ENSEEG Domaine Université
38401 St. Martin d'Heres
France

Prof. Dr. M. Garreau
33, Ave. de Lattre de Tassigny
92340 Bourg la Reine
France

Professor J. B. Goodenough
Inorganic Chemistry Dept.
University of Oxford
South Parks Road
Oxford, OX1 3QR
U.K.

Professor R. R. Haering
Dept. of Physics
University of British Columbia
2075 Westbrook Mall
Vancouver, B.C. V6T 1W5
Canada

Dr. A. Howe
Chemistry Dept.
Leeds University
Leeds
U.K.

Professor R. Huggins
Dept. of Material Science
Stanford University
Stanford, California 94305
U.S.A.

Dr. I. W. Jones
Chloride Silent Power
Davy Road
Astmoor, Runcorn
Ceshire, WA7 1PZ
U.K.

Dr. M. Kleitz
Laboratoire D'Energetique Electrochimique
ENSEEG Domaine Université
38401 St. Martin d'Heres
France

Dr. M. Lazzari
Istituto Di Chimica Fisica
Electrochimica E Metallurgia
Del Politecnico Di Milano
Milan, Italy

Dr. A. Levasseur
Laboratoire de Chimie du Solide du CNRS
Université de Bordeaux I
351 Cours de la Liberation
33405 Talence Cédex
France

Dr. C. C. Liang
Wilson Greatbatch Ltd.
10,000 Wehrle Drive
Clarence, New York 14031
U.S.A.

Professor G. Mamantov
Dept. of Chemistry
University of Tennessee
Cumberland Avenue, S.W.
Knoxville, Tennessee 37916
U.S.A.

Dr. A. LeMehaute
Groupe d'Electrochimie
Laboratoires de Marcoussis
Route de Nozay
91460 Marcoussis
France

Dr. D. W. Murphy
Bell Labs.
600 Mountain Avenue
Murray Hill, New Jersey 07974
U.S.A.

Dr. F. W. Poulsen
Metallurgy Department
Riso National Laboratory
P.O. Box 49
DK-4000 Roskilde
Denmark

Dr. J-M. Reau
Laboratoire de Chimie des Solides du CNRS
Université de Bordeaux I
351 Cours de la Liberation
33405 Talence Cedex
France

Professor J. Rouxel
Laboratoire de Chimie des Solides
Université de Nantes
2, Rue de la Houssiniére
44072 Nantes Cédex
France

Dr. J. Schoonman
Dept. of Solid State Chemistry
Physics Laboratory
State University
Sorbonnelaan 4
Utrecht
The Netherlands

Dr. B. C. H. Steele
Dept. of Metallurgy and Materials Science
Royal School of Mines
Prince Consort Road
London SW72BP
U.K.

Dr. S. Sudar
Rockwell International, Energy Systems Group
8900 De Soto Avenue
Canoga Park, California 91304
U.S.A.

Dr. D. S. Tannhauser
Physics Dept.
Technion, Haifa
Israel

Dr. B. C. Tofield
Building 552
Materials Development Division
A.E.R.E.
Harwell
Oxford, OX11, ORA
U.K.

Dr. D. R. Vissers
Chemical Engineering Division
Argonne National Laboratory
9700 South Cass Avenue
Argonne, Illinois 60439
U.S.A.

Dr. E. Voss
VARTA Batteries AG
Gundelhardtstr. 72
D-6233 Kelkheim (Taunus)
W. Germany

Dr. G. Weddigen
Zentrales Forschungslabor
Brown Boveri AG
Eppelheimer Str. 82
Postfach 10 13 32
D-6900 Heidelberg 1
W. Germany

Dr. W. Weppner
Max-Planck-Institut für Festkörperforschung
Heisenbergstr. 1
D-7000 Stuttgart 80
W. Germany

Dr. M. S. Whittingham
Exxon Research and Engineering Co.
Corporate Research Laboratories
P.O. Box 45
Linden, New Jersey 07036
U.S.A.

Dr. F. G. Will
Gen. Electric Res. and Development Center
P.O. Box 8
Schenectady, New York 12301
U.S.A.

Professor W. Worrel
Dept. of Metallurgy
University of Pennsylvania
Philadelphia, Pennsylvania 19104
U.S.A.

INDEX

Absorbed species on reactive surfaces, 39
Active material/melt interface, 178
Adams, W. A., 211
Additives, surface-active, 126-130
Adsorbed species, detection of, 41
Alkali metals, reactions with amorphous sulfides, 293
Alloying substrates, 130
Aluminum nitrate, 44
Aluminum metal and negative electrode development, 328
Amorphous cathodes for lithium batteries, 291-299
Anion conduction, 233
Anodic pulses activation phenomenon, 183
Aprotic electrolytic solvents, 340
Aqueous systems
  electrochemical storage, 29-46
  suggested investigation, 353-355
Armand, M. B., 145
Atlung, S., 275
Automobile, electric
  ambient temperature battery for, 23
  batteries for, 11
  battery performance requirements, 22
  driving profiles of, 19
Automobile, electric (continued)
  energy consumption by, 17
  energy efficiency comparison, 14-16
  energy requirements of, 20
  Li/MS batteries for, 47
  urban range of, 21

Batteries
  exploratory systems, 312-314
  lithium, amorphous cathodes for, 291-299
Battery
  design with intercalation cathodes, 275-289
  development, 309-314
  electric vehicle, 47
  energy storage, 47
  high temperature, 8-11
  lead-acid, 199-203, 312
  lithium-aluminum/iron sulphide, 47-90
  lithium/metal sulfide, performance, 48
  Ni-Cd, 39, 44
  performance of baseline systems, 310, 311
  rechargeable, aqueous, electromechanical, 29-46
  research recommendations, 312-314
  seawater, 30
  sodium sulfur, 164-170
  state of the art, 309-314
  testing, 50
  zinc/nickel oxide, 23

Battery Energy Storage Test (BEST), 50
Battery systems, requirements of, 3-27
Belanger, A., 211
Bell, M. F., 263
Beta alumina as solid electrolyte, 97
Bismuth oxide cathode, 267
BN, 44
Bonino, F., 301
Brummer, S. B., 123

Cairns, Elton, J., 3
Calasanzio, D., 223
Calcium/metal sulfide cells, 328
Carbon felt/melt interface, 172
Carbon fibre
  felt, vitreous 44
  paper, non-conducting, 44
Casey, E. J., 29
Cathode
  porous, 278-280
  selenium IV, 223-227
Cathodes, amorphous, for lithium batteries, 291-299
Cathodes, intercalation, and battery design, 275, 289
Cation conduction, 233
Cationic conduction, noncrystalline, 103
Caulder, S. M., 199
Charge carriers, 35
Childs, P. E., 235
Chlorine electrode interface, 189
Chlorinehydrate interface, 187-189
Chloroaluminate melts, 117
Chloroaluminates, molten, selenium cathode in, 223-227
Cost, high temperature systems, 322
Coulometric titration curve, 271
Crystalline layered compounds, 291-299
Crystalline solid electrolytes, 335
Current collector/active material interface, 177
Current collector/melt interface, 176
Cyclic ethers and Li reaction, 135

Dichalcogenides, 291
Differential scanning calorimetry, (DSC), 42
Divanadium pentasulfide, 293
Dopants, effect on ionic conductivity, 253-256
Dubois, A. R., 211

Eichinger, G., 241
Electric utilities
  energy storage, 7
  load profile, 5
Electrocatalysts, 35
Electrochemical spectroscopy, 38-42
Electrode
  FeS, 58-62
  $FeS_2$, 54-58
  kinetics of, Li-Al, 52
  Li, progressive film formation on, 185
  Li-Si, 53
  Li, Tafel behavior of, 182
  lithium recharging in organic electrolytes, 123-141
  materials, 343-347
  negative, material, 51
  positive, of transition metal sulfides, 54
  reactions, classification of, 343-345
  thermodynamics of, 346
Electrodes
  air, 40
  enhanced ionic motion in, 269-274
  intercalation, 145-161
  micro-reference, 205-209
Electrointercalation and diffusion of silver in tantalum disulphide, 301-305
Electrolysis/fuel cells, exploratory systems, 313
Electrolyte
  liquid, 313

Electrolyte (continued)
lithium electrode recharging in organic, 123-141
purification, 131
research, 314
Electrolytes
free radicals in, 40
low melting, 118
low moisture, 37
molten LiCl-KCl eutectic, 51
molten salt, 111-119
propylene carbonate based, 180-186
transition region, 36-38
Electrolytes, solid, 91-105, 331-342
beta alumina use in, 97
cationic conduction, non-crystalline, 103
characteristics of, 92-97
fast ionic conduction, 93
in hydride-air batteries and hydrogen-oxygen fuel cells, 235-240
ion behavior in, 229-233
in $Li_4GeO_4$-$Li_2ZnGeO_4$-$Li_3PO_4$, 247-252
lithium iodide-based materials in, 99
lithium orthosilicate in, 104
NASICON in, 100
orthophosphate in, 104
proton conductors in, 100-102
research suggestions, 314
sodium borate glass, use in, 98
stability requirements, 93-97
Electrolytic solvents, aprotic, 340
Electron spin resonance spectrum, 40
Energy density cell development, 211-221
Energy economy, United States, 3
Energy storage
battery, stationary, 47, 83
capabilities, 5
off-peak, 9, 13
systems, aqueous electrochemical, 29-46
Energy storage (continued)
systems for wind and solar power, 29-46
Epelboin, I., 257
Eutectic LiCl-KCl, 211

Fast Fourier transform infrared scan, 39
FeS charge/discharge voltammogram, 179
Film formation on Li electrodes, 185
Films, surface insulating, 253-256
Filters, water, 33
Fluorine ion conductors, 336
Free radicals in electrolytes, 40
Froment, M., 257

Garreau, M., 257
Gauthier, M., 211

Halide eutectic lithium melts, 211
Hellstrom, E. E., 247
High temperature systems, 315-330
Hooper, A., 253
Host lattice strain on intercalation systems, 283-289
Howe, A. T., 235
Huggins, Robert A., 91
Hydride-air batteries, proton conductor in, 235-240
Hydroelectric storage, 7
Hydrogen-oxygen fuel cells, proton conductor in, 235-240
Hydrogen uranyl phosphate tetrahydrate, 235-240

Impedance, a.c.technique in $Li$-$Bi_2O_3$ battery system, 263-268
Insertion compounds research, 314
Intercalation cathodes, battery design dynamics, 275-289
Intercalation electrodes, 145-161
high temperature, 219
kinetics of, 159-161
thermodynamics, 150-159
Intercalation reactions, 147-150

Intercalation systems, strain effects on, 283-289
Instrumentation for high temperature research and development, 329
Interface
  active material/melt, 178
  carbon felt/melt, 172
  chlorine electrode in $Zn-Cl_2 \cdot 8H_2O$ system, 189
  chlorinehydrate in $Zn-Cl_2 \cdot 8H_2O$ system, 187-189
  current collector/active material, 177
  current collector/melt, 176
  lithium in propylene carbonate-based electrolytes, 182-185
  in organic electrolytes, 185
  phenomena, 163-194
  planar-carbon/melt, 171
  separator/melt, 177
  sodium/beta alumina, 164-170
  sulfur electrode, 170
  zinc electrode/electrolyte in zinc bromine system, 191-194
Interfaces, 349-357
Interfacial phenomena problems, 350-343
Ion exchange membranes, 312
Ionic conductivity, effect of dopants on, 253-256
Ionic motion in solid electrode materials, 269-274
Ions
  behavior of in solid electrolytes, 229-233
  polarizability, 230
IREQ-2 cell charge and discharge curves, 216, 218

Jacobsen, T., 275
Jacobson, A. J., 291

Kamphorst, J. D., 247
Koch, V. R., 123

Lachman, W. L., 45
Lanthanum-nickel, 44
Laser-Raman technique, 41
Lazzari, M., 301
Lead-acid battery, 199-203
  competitive position of, 200
  energy density, 200-202
  research, 202
  research recommendations, 312
Lead calcium alloys, 44
Lead chloride, 30, 44
Levelling agents, 126
Lisicon, 247-252
Lithium
  anode, 267
  anode voltage recovery, 184
  and aprotic electrolytic solvents, 340
  batteries, amorphous cathodes for, 291-299
  cycling, 126-133
  electrodeposition of, 37
  intercalation systems, 283
  iodide-based materials, 99
  ion conductors, 336
  orthosilicate, 104
  plated, isolation of, 125
  reaction with cyclic ethers, 135
  rechargeability of, 37
  substrate, 258
Lithium electrode, recharging of, 123-141
Lithium electrodes in propylene carbonate, 257-261
Lithium nitride crystals, analysis of surface insulating films on, 253-256
Lithium-Aluminum electrode phase diagram, 52
Lithium-aluminum/iron sulphide battery
  bicell design, 71-76
  commercialization of, 84-86
  compatibility tests, materials, 64-67
  development and testing, 71-87
  energy storage, 83
  feedthrough components, 71
  Mark IA, 81-83
  Mark II, 84-86
  materials and components of, 63-71

Lithium-aluminum/iron sulphide battery (continued)
multiplate design, 75-80
oxidation potentials in, 67
separators, 68-71
suggested investigation, 355
Lithium-aluminum substrate, 259
LiAl/LiCl-KCl/FeS cells, 211-221
Li-$Bi_2O_3$ battery system, 263-268
LiCl-KCl eutectic, 114-116
$Li_4GeO_4$-$Li_2ZnGeO_4$-$Li_3PO_4$ system, solid electrolytes in, 247-252
Li/$Li_xMO_2$ cells, 283
Li/$Li_xMX_2$ cells, 283
Li/MS based systems, 315-330
Lithium/metal sulfide battery, 48
chemistry of, 51
electrochemical reaction, 50
micro-reference electrodes for, 205-209
program goals, 49
Lithium-organic electrolyte system, 180-186
suggested investigation, 355
$Li_4Si$/LiCl-KCl/$FeS_2$ battery, 8-12
Li/$SOCL_2$ cells, passivating film in, 241-246
$Li_{14}Zn(GeO_4)_4$, 247-252

Magnesium metal and negative electrode development, 328
Mamantov, G., 111, 223
Marassi, R., 223
Materials, exotic, 43
Methylacetate, 186
2-Methyltetrahydrofuran/$LiAsF_6$, 133-140
Micrograph, Zn deposit, 192-194
Micro-reference electrodes, 205-209
Molten salt electrochemical studies, 211-221
Molten salt electrolytes
in high temperature systems, 327
Molten salts
chloroaluminate, 117
Molten salts (continued)
physical properties of, 112
polysulfide, 116
solvent LiCl-KCl eutectic, 114
Molybdenum disulfide, 293, 294
Molybdenum trisulfide, 292, 296, 297
Morin, F., 211
$MoS_2$, 291, 293, 294
$MoS_3$, 292, 296, 297

NAFION, 312
NaI-doped polyethylene oxide, 44
Na/S based systems, 315-3350
suggested investigation, 355
NASICON, 100
Nazirpsio, 44
Non-crystalline electrolytes, 335

Organic electrolytes, lithium electrode recharging, 123-141
Orthophosphate, 104
Oxide-based glass and ceramic, 44

Passivating film in Li/$SOCL_2$, 241-246
Photoassisted electrolysis, 34
Photoelectrochemical conversion, 35
Planar-carbon/melt interface, 171
Polarizability scale, 229-233
Polyethersulfone, 44
Polymeric solid electrolytes, 339
Polyphase solid electrolytes, 334
Polypropylene fibre mattes/sheets, 44
Polyphenylene sulfide/oxide, 44
Polysulfide melts, 116
Porous cathodes, 278-280
Positive electrode development, 328
Precursors, 127-130
Propylene carbonate-based electrolytes, lithium interfaces in, 182-185
Propylene carbonate, lithium electrodes in, 257-261
Proton conductivity, 100-102

Proton conductor
  in hydride-air battery and hydrogen oxygen cell, 235-240
  research, 339
  as transition region electrolyte, 37
Pulse-electrochemical technique, 35
Pulse-photo techniques, 35
Pyrodistillation, 32

Radiation-grafted copolymers, 43
Rauh, R. D., 123
Reactive surfaces, absorbed species on, 39
Redey, Laszlo, 205
Redox systems, 314
Reference electrodes, 205-209
Research
  centers, high temperature systems, 317
  electrode reactions, 345-347
  electrolyte, 314
  high temperature systems, 327-329
  insertion compounds, 314
  lead acid battery, 202
  liquid electrodes, 347
  solid-solid interfacial, 314
  suggested, 353-357
Ruthenium sulfate, 34

Safety, high temperature systems, 315-330
Scavenger, 130
Schoonman, J., 247
Seawater battery, 30
Selenium (IV) Cathode in molten chloroaluminates, 223-227
Separator/melt interface, 177
Silicate-based glasses and ceramics, 44
SiC, 44
Silver electrointercalation and diffusion of in tantalum disulphide, 301-305
Silver salt, 30
Simon, A. C., 199
Sodium/beta alumina interface, 164-170
Sodium borate-based glass, 98
Sodium ion conductors, 338
Sodium rechargeability, 37
Sodium-sulfur system, 164-170
Sofield, C. J., 253
Solar power, energy storage for, 23-26
Solid electrolytes, 91-105, 331-342
  advances in, 332
  beta alumina use in, 97
  capability prediction, 332
  cationic conduction, non-crystalline, 103
  characteristics of, 92-97
  crystalline, 335
  fast ionic conduction, 93
  fluorine ion, 336
  in hydride-air batteries and hydrogen fuel cells, 235-240
  ion behavior in, 229-233
  in $Li_4GeO_4$-$Li_2ZnGeO_4$-$Li_3PO$, 247-252
  lithium iodide-based materials in, 99
  lithium ion, 336
  lithium orthosilicate in, 104
  materials problems, 333
  NASICON in, 100
  non-crystalline, 335
  orthophosphate in, 104
  polymeric solid, 339
  polyphase, 334
  proton, 339
  proton conductors in, 100-102
  research suggestions, 314
  sodium borate glass, use in, 98
  sodium ion, 97, 100, 338
  stability requirements, 93-97
  state of the art, 331
Solid electrolyte/solid electrode systems, suggested investigation, 357
Solid-ion conductors, 44

Solid-solid interfacial research, 314
Solvent modification, 133-140
Space requirements, battery stations, 7
Spectroscopy, electrochemical, 38-42
Storage capabilities, 5
  hydroelectric, 7
$S_4N_4$, 45
Sulfides, soluble, 34
Sulfur electrode interface, 170
Sulfuric acid, anhydrous, 37

Tafel behavior of Li electrode, 182
Tantalum disulphide, electro-intercalation of silver in, 301-305
Tetrahydrofuran, 185
Thevenin, J., 257
$TiS_2$, 211
$TiSe_2$, 211, 220, 221, 291
Titanium alloys, new, 44
Titanium disulfide, 217
Tofield, B. C., 253
Transition region electrolytes, 36-38
Turbine, gas, 7
Turner, J. F., 253

United States energy economy, 3

$VS_2$, 218, 220
$VS_x$, 211
$V_2S_5$, behavior in lithium cells, 298
Vanadium pentasulfide in lithium cells, 297, 298
Vehicles, electric, see Automobile, electric
Vincent, C. A., 301
Vissers, Donald R., 47, 205
von Alpen, U., 263
Voltage recovery, Li anode, 184
Voltammetry on FeS and $FeS_2$, 217-219
Voltammogram
  FeS charge/discharge, 179

Warin, D., 257
Water
  filters, 33
  oils in, 33
  photoassisted electrolysis of, 34
  purity of, 31-33
Weppner, W., 269
West, K., 275
Whittingham, M. S., 291
Wind power, energy storage for, 23-26

Zinc bromine system, 190-194
  zinc electrode/electrolyte interface in, 191-194
Zinc-chloride hydrate system, 186-190
  chlorine electrode interface in, 189
  chlorinehydrate interfaces in, 187-189
Zinc electrode/electrolyte interface in zinc bromine system, 191-194
Zinc/nickel oxide battery, 23

MIX
Papier aus verantwortungsvollen Quellen
Paper from responsible sources
FSC® C105338

If you have any concerns about our products,
you can contact us on
**ProductSafety@springernature.com**

In case Publisher is established outside the EU,
the EU authorized representative is:
**Springer Nature Customer Service Center GmbH**
**Europaplatz 3, 69115 Heidelberg, Germany**

Printed by Libri Plureos GmbH
in Hamburg, Germany